MICROELECTRONICS
TO NANOELECTRONICS

Materials, Devices
& Manufacturability

MICROELECTRONICS
TO NANOELECTRONICS
Materials, Devices
& Manufacturability

Edited by
ANUPAMA B. KAUL

CRC Press
Taylor & Francis Group
Boca Raton London New York

CRC Press is an imprint of the
Taylor & Francis Group, an **informa** business

CRC Press
Taylor & Francis Group
6000 Broken Sound Parkway NW, Suite 300
Boca Raton, FL 33487-2742

First issued in paperback 2017

Version Date: 20120518

ISBN 13: 978-1-138-07237-4 (pbk)
ISBN 13: 978-1-4665-0954-2 (hbk)

Library of Congress Cataloging-in-Publication Data

Microelectronics to nanoelectronics : materials, devices & manufacturability / editor,
Anupama B. Kaul.
 p. cm.
 Summary: "This book gives a comprehensive cross-section of promising MEMS and
nanotechnologies that are important to the future growth of the electronics industry.
International experts in academia and leading industrial institutions present the most recent
scientific developments. They highlight new technologies that have successfully transitioned
from the laboratory to the marketplace as well as technologies that have near-term market
applications in electronics, materials, and optics. The book also provides up-to-date references
on the latest advances in this evolving field"-- Provided by publisher.
 Includes bibliographical references and index.
 ISBN 978-1-4665-0954-2 (hardback)
 1. Nanotechnology. 2. Microelectronics. 3. Nanoelectronics. I. Kaul, Anupama B.

T174.7.M498 2012
621.3815--dc23   2012016100

Visit the Taylor & Francis Web site at
http://www.taylorandfrancis.com

and the CRC Press Web site at
http://www.crcpress.com

In Loving Memory of

Mrs. S. P. Kaul

Justice J. N. Bhat

I dedicate this book to

Ashish, Ishani, and Arnav

Contents

About the Cover

Images from top to bottom: A transmission electron microscopy (TEM) image of a single viral-templated nanocrystalline gold nanowire. A genetically modified M13 virus with an affinity for gold was used to bind 5 nm gold nanoparticles. Electroless gold deposition was subsequently used to control the size and connectivity of the resulting nanowires. (Image courtesy of Professor Elaine Haberer, University of California–Riverside.)

TEM micrograph of a multi-walled carbon nanotube in a Y-junction form, which can be used as a prototypical nanoelectronic element constituting a switch or a transistor, depending on the transmission characteristics of the constituent branches. The Fe-Ti catalyst particles formed in the CVD growth, along with the presence of topological defects at the junction region, can influence the electrical transport characteristics. (Image courtesy of Professor Prabhakar R. Bandaru, University of California–San Diego, La Jolla.)

An artistic rendition illustrating an array of suspended nanowires assembled via a hybrid bottom-up/top-down approach: chemically synthesized nanowires integrated with lithographically patterned electrodes, which are promising for resonant sensing and high-speed nanomechanical computing. (Image courtesy of Professor Philip Feng, Case Western Reserve University, Cleveland, Ohio.)

A scanning electron microscopy (SEM) image of monolayer thick graphene sheet connected with electrical leads. Such devices show the ultimate in chemical sensitivity where adsorption of individual gas molecules can be detected. Top inset shows the hexagonal motif of carbon atoms in graphene which serves as the building block for other carbon-based nanomaterials such as CNTs and bucky balls. (SEM image courtesy of Dr. Konstantin Novoselov, University of Manchester, United Kingdom. Adapted from F. Schedin, et al. *Nature Materials* 6, 652, 2007. Copyright © Nature Publishing Group.)

A TEM micrograph of a singly wound coiled carbon nanotube (CNT) synthesized with thermal chemical vapor deposition (CVD) through the use of indium- and tin-based catalysts. (Image courtesy of Professor Prabhakar R. Bandaru, University of California–San Diego, La Jolla.)

SEM image of a single, vertically oriented carbon nanofiber (CNF) centered within high-aspect-ratio electrodes. Such structures, formed with low-cost, wafer-scale approaches using a hybrid combination of top-down and bottom-up nanofabrication techniques, are a stepping stone for the realization of controlled architectures for three-dimensional (3D) electronics. (Image courtesy of Dr. Anupama B. Kaul, Jet Propulsion Laboratory, California Institute of Technology, Pasadena.)

Background is an example of packaged and integrated components that have been created using micro- and nanofabrication techniques and form the backbone of the semiconductor industry for microprocessors and other system-level applications.

Foreword

It is a pleasure to write the Foreword for this book because of its timeliness and its excellent breadth and depth of coverage that capture most key facets of the field and also because its editor, Dr. Anupama Kaul, is an esteemed collaborator of mine in micro-electro-mechanical (MEMS) and nanotechnologies research. Indeed, her expertise and excellent reputation in the field derive from her extensive work on the development of carbon nanotube (CNT)- and carbon nanofiber (CNF)-based devices conducted at the Jet Propulsion Laboratory, California Institute of Technology. In particular, Dr. Kaul's contributions encompass nanoelectronic devices based on carbon-based nanomaterials for three application areas: (1) nano-electro-mechanical (NEM) switches and resonators; (2) physical sensors; and (3) optical absorbers aimed at applications in extreme environment electronics for planetary missions, miniaturized sensors as interfaced with vacuum micro-cavities for high-frequency vacuum electronics, and broadband optical absorbers operational from the UV to IR ranges. It is thus not surprising, given Dr. Kaul's excellent technical background and experience, that she is uniquely qualified to edit such an ambitious piece of work, gathering contributions from a stellar roster of top researchers in the field.

In this book, the reader will find that the science and engineering pertaining to virtually all areas of application where MEMS and nanotechnologies hold great potential to provide *enabling* advantages are addressed. This diversity of exposition intertwines the various relevant disciplines, e.g., from fabrication techniques to devices, circuits, and systems, and a plethora of applications from electronics to computing to science to sensing to energy. Thus, the book has something for everyone!

To place the book into perspective, we should recall that many attribute progress in the fields of MEMS and nanotechnologies to physicist Richard Feynman's 1959 contribution titled, *There is Plenty of Room at the Bottom*. Accordingly, the theme of miniaturization permeates these technologies. In this context, the book begins by addressing Moore's law and its impact on transistor scaling and reliability; then it proceeds to address MEMS technology and it spectacular successes typified by thermal ink-jet printing and timing resonators. Several authors deal with MEMS, but at a lower-size scale, and discuss nanoscale electromechanical devices enabled by nanowire structures before closing the MEMS topic with an in-depth exposition of silicon etching and etch techniques for NEMs and MEMs and their exploitation in materials and devices for ultra energy-efficient digital integrated circuits.

Subsequently, the interdisciplinarity and versatility of the field are revealed in chapters on viral-templated materials and devices—a technique exploited for uncovering new knowledge through explorations in biology.

At this point the book turns to CNTs, the quintessential nanotechnology materials. In particular, the principles and methods for integration of carbon nanotubes in miniaturized systems, heterogeneous integration of CNTs on complementary metal oxide semiconductor circuitry and sensing applications, and CNT Y-junctions pertaining to electronic systems implementation are addressed. As is well known, the miniaturization of electronic devices, circuits, and systems is accompanied by issues surrounding heat dissipation. This all-important aspect of exploiting nanotechnologies is addressed by exploring nanoscale effects in multiphase flows and heat transfer. The book concludes with discussions of nanoengineered materials for applications in electronics, biology, and energy harnessing.

As can be appreciated from the above summary, the book does indeed have something for everyone. While the theme of "making everything small," as envisioned by Feynman will undoubtedly continue to present a rich set of challenges, this book marks a milestone by capturing the current status of research and development. Therefore, it is of extreme value to both established researchers and newcomers who aim to discover and exploit the vast space available at the bottom!

Dr. Héctor J. De Los Santos
Karlsruhe Institute of Technology
Karlsruhe, Germany

Preface

Our desire for gadgets smaller, faster, and cheaper continues to fuel the miniaturization revolution that is now well into the nanoscale regime (defined as functional length scales smaller than 100 nm). In this regime, broadly defined as nanotechnology, novel properties often emerge that in many instances can be exploited to enhance the performance of devices, components, and systems for applications targeted toward a broad array of sectors spanning sustainable energy, climate change, homeland security, and healthcare. However, nowhere are the phenomenal advances in and benefits of miniaturization more apparent than in the electronics industry. Historically, the first digital computer, the Electronic Numerical Integrator and Computer (ENIAC) built in 1946, occupied an entire room, weighed 30 tons, and consumed 200 kilowatts of power. In contrast, a modern-day computer is handheld and operates on a 3- to 4-volt battery.

Needless to say, this 60-year journey in the miniaturization revolution has been filled with radical innovations in new materials, structures, devices, and integration schemes at every juncture that have helped shape the electronics industry into the trillion dollar empire it is today. One example of such innovation was replacing the vacuum tubes in the ENIAC with solid-state transistors, for which William Shockley, John Bardeen, and Walter Brattain received the Nobel Prize in Physics in 1956. While the smaller size of the transistor was a great benefit to circuit designers at the time, circuit complexity was limited by the unreliable and manually soldered connections, and the long wiring lengths also limited operational speeds. Yet another radical innovation came about in 1958, when Jack Kilby of Texas Instruments proposed making all the components from a single block of material instead of manufacturing discrete components; this made the circuits smaller and faster. Kilby's monolithic integration scheme and invention of the integrated circuit (IC) won him the Nobel Prize in Physics in 2000, over 40 years later.

Our voyage into the world of things smaller and faster does not seem to end here, however. While incremental advances have enabled a doubling of the density of ICs on a silicon (Si) chip every 2 years according to Moore's law, by 2018 however, we will find ourselves at a technological crossroads again, much like our predecessors. Critical dimensions in the Si transistor will approach the size of atoms and quantum mechanical effects will predominate, eventually prohibiting device operation altogether. Just as the forefathers of the modern-day computer did, scientists know that novel and drastically different solutions are necessary to overcome fundamental limitations and thus sustain electronics commerce and propel the larger global economy forward in the coming decades.

Such solutions and disruptive technologies will more than likely emerge from the realms of nanotechnology. A testament to nanotechnology's promise is perhaps best exemplified by the 2010 Nobel Prize in Physics awarded to Andre Geim and Konstantin Novoselov for their ground-breaking work on the nanomaterial graphene—a monolayer thick sheet of carbon atoms exhibiting remarkable electronic, mechanical, and optical properties; such properties are not only attractive for computation, but also exhibit far-reaching implications in other areas such as sensors, structural materials, and biology. The broad importance of nanotechnology to our overall economy is also mirrored by the global commitment to invest in nanotechnology research; for example, in the United States alone, federal support for the National Nanotechnology Initiative (NNI) formed in 1999 is expected to remain at $1.8 billion in 2013.

The purpose of this book is to provide a synopsis of the exciting work that is currently in progress in the area of micro- and nanotechnologies targeted primarily for electronics applications and pursued in academia, government laboratories, and industry. A broad spectrum of micro- and nanotechnologies is showcased here, from early laboratory investigations to more mature technologies that have already reached the commercial markets. Since nanotechnology is inherently multidisciplinary, this book is intended to serve as a reference for students in upper division undergraduate or graduate courses in electrical engineering, materials science, physics, chemistry, and bioengineering and for working professionals.

We start in Chapter 1 with a description of current state-of-the-art Si transistor technology in the context of Moore's law and the impact of continued scaling on transistor performance. With billions of transistors now in use in modern microprocessors, reliability issues in complementary metal–oxide semiconductor (CMOS) transistors are highlighted, and the near-term approaches used to mitigate these issues are also discussed. Highly scaled transistors are also more vulnerable to failure in harsh environments such as radiation, as discussed in Chapter 2. Nowhere is the need for radiation-hard electronics greater than in space-based electronic systems that must withstand large doses of integrated fluxes of ionizing and non-ionizing radiation over long mission durations; such issues are particularly apparent and exacerbated in scaled transistors, as the discussion in Chapter 2 highlights.

The Si microelectronics industry also gave birth to the field of micro-electro-mechanical systems (MEMS) when Kurt Petersen at International Business Machines (IBM) demonstrated the first micro-machined pressure sensor in 1970 using batch fabrication. Since then, many novel MEMS-based devices and components have been developed. They are no longer confined to the realms of the laboratory and have been commercialized and continue to generate revenue. Chapter 3 presents an overview of surface micro-machined ink-jet print heads as quintessential examples of one such successful MEMS technology. Ink-jet MEMS represents one of the most profitable MEMS technologies and is now enabling applications in other areas such as microfluidics and biological and chemical sensing. Further, in Chapter 4 another example of MEMS components is provided, specifically mechanical resonators that surpass the performance of electronic resonators based on solid-state transistors. Such high-frequency mechanical resonators have applications serving as frequency standards and filters in communication systems. With dimensions of mechanical structures in MEMS now reaching the nanoscale regime, Chapter 5 reviews the design, fabrication, and characterization of nano-electro-mechanical (NEM) resonators that can serve as sensitive mass detectors that are particularly appealing for biological sensing applications.

Chapters 6 and 7 provide examples of the latest developments in nanofabrication derived from the top-down and bottom-up approaches, respectively. The top-down approach has traditionally sustained the microelectronics industry. Features are formed using combinations of deposition, lithography, and etching, but new techniques and tools are necessary to realize feature sizes in the nanoscale regime. Top-down pattern transfer etching techniques using state-of-the-art inductively coupled plasmas (ICPs) are discussed in Chapter 6. The chapter explains how highly anisotropic, high-aspect-ratio Si nanostructures are formed using wafer-scale approaches for MEMS and NEMS applications. Conversely, in the bottom-up approach, nanoscale building blocks can be hierarchically assembled to form complex structures, akin to the way nature uses proteins to build complex biological systems, for example. Chapter 7 presents an example of one such bottom-up approach. Select groups of viruses, each with unique structural traits, are employed to serve as templates for the bottom-up assembly of structures such as conductive nanowires for nanoelectronics, films for photovoltaics, and nanocomposites for sensors.

Chapters 8, 9, and 10 cover techniques and approaches used for integrating low-dimensionality, bottom-up synthesized materials such as CNTs (discovered in the early 1990s) with more mature technologies. Chapter 8 provides a comprehensive review of CNTs and the methods used to integrate them within the well-established semiconductor processing platform. Emphasis is placed on large-scale integration and manufacturability. Then, in Chapter 9, approaches used to integrate single-walled carbon nanotubes (SWCNTs) with CMOS circuitry are presented. The use of such hybrid SWCNT–CMOS components is discussed in the context of thermal and chemical sensing mechanisms; individual SWCNTs are decorated with deoxyribonucleic acid (DNA) to enhance sensitivity. In Chapter 10, the design, modeling, and analysis of NEMS-based ICs are discussed. These ICs appear to be more energy efficient than their CMOS-based counterparts and they rely on nanomechanical switches. Computation based on mechanical components dates as far back as the 1800s when Charles Babbage proposed the difference engine. It is interesting to note that we are now revisiting ideas formulated almost two centuries ago in that mechanical structures, albeit in the nanoscale regime, are serving as building blocks for electronic computation.

While CNTs are linear and uniform, nonlinear CNT topologies such as SWCNT-based Y-junctions can be utilized to form novel architectures for electronic computation, which is the focus of Chapter 11. Such branched nanoelectronic architectures can constitute a switch or transistor, depending on the transmission characteristics of the constituent branches, and function vastly different than solid-state transistors.

The increased integration densities and higher clock speeds in Si ICs also created challenges for managing the large amounts of power or heat generated during microprocessor operation. In Chapter 12, thermal management schemes, specifically with phase change heat transfer techniques utilizing nanoscale materials and surfaces, are examined as viable cooling technologies. Finally, Chapter 13 surveys the broader applications of nanoengineered materials in the areas of electronics, biology, and energy harnessing.

Acknowledgments

I would like to express my sincere gratitude to all the distinguished contributors in this book for their comprehensive coverage of topics of current research in the exciting area of micro- and nanotechnologies that fill a wide variety of applications in electronics. I would also like to particularly thank Dr. Konstantin Novoselov (2010 Nobel Laureate, University of Manchester, UK), Professor Philip Feng (Caltech and Case Western Reserve), Professor Prabhakar Bandaru (University of California–San Diego), Professor Elaine Haberer (University of California–Riverside), and Professor Mehmet Dokmeci (Northeastern University) for providing the images that appear in the collage on the front cover.

In this regard, I would also like to acknowledge the support of my colleagues at the Jet Propulsion Laboratory (JPL), particularly Krikor G. Megerian and Robert Kowalczyk, for their technical assistance that enabled the images shown in the front cover to be made.

In addition, I would also like to thank Ashley Gasque, associate editor at CRC Press for her guidance and assistance in the publication of this book and Joselyn Banks-Kyle, project coordinator and Iris Fahrer, project editor at CRC Press for their editorial assistance in arranging the materials for this book.

I also gratefully acknowledge my family for their support through the years. I thank them from the bottom of my heart for their love, patience, and joyful spirit that sustain me and for which I am truly grateful.

Editor

Dr. Anupama B. Kaul obtained her BS degrees with honors in physics, as well as engineering physics, from Oregon State University. Her MS and PhD degrees were in materials science and engineering, with minors in electrical engineering and physics, from the University of California–Berkeley. Presently, she is a program director at the National Science Foundation (NSF) in the ECCS Division within the Engineering Directorate, where she is serving under the Intergovernmental Personnel Act (IPA) from the Jet Propulsion Laboratory (JPL), California Institute of Technology (Caltech). Prior to joining JPL, Dr. Kaul held industrial research positions at Motorola Labs and the R&D Division of Hewlett-Packard Company, where she worked on RF micro-electro-mechanical-systems (MEMS) components for wireless applications, and surface micro-machined ink-jet print heads, respectively. Her research interests are in characterizing the properties of functional micro- and nanoscale materials, developing bottom-up assembly and top-down nanofabrication techniques, and integrating such materials into novel devices and components for applications in electronics, sensing and energy harnessing.

While at JPL, Dr. Kaul has supported programmatic-review activities for technology development in support of various NASA proposal calls in planetary science. She has written more than 70 journal, conference and NASA technology brief publications, and received 6 NASA Patent Awards, a NASA Team Accomplishment Award, and holds 9 issued and pending patents. Dr. Kaul is a senior member of the Institute of Electrical and Electronics Engineers (IEEE) and has also contributed to several invited book chapters and has served as a guest editor for various journals.

Presently, Dr. Kaul serves as the American editor for *Nanoscience and Nanotechnology Letters*, associate editor of *Reviews in Advanced Sciences and Engineering* and is on the editorial boards of the *Journal of Nanoengineering and Nanomanufacturing* and the *Open Process Chemistry Journal*. Dr. Kaul has organized symposia for the Materials Research Society (MRS) on nanomaterials and devices, and serves as one of the track chairs in nanoelectronics for the IEEE NANO 2012. In addition, Dr. Kaul has also organized and chaired sessions for other conferences sponsored by the Nano Science and Technology Institute (NSTI) and SPIE. She is listed in the *Who's Who in America* and the *Who's Who in Science and Engineering*. Dr. Kaul has served as a reviewer on proposals for the NSF, NASA, and JPL, in addition to being a technical reviewer for leading journals within the Nature Publishing Group, the American Chemical Society, IEEE, and the MRS. Dr. Kaul was selected to participate in the 2012 US Frontiers of Engineering Symposium by the National Academy of Engineering. She is the Nanoelectronics Track

Chair for the IEEE NANO 2012 sponsored by the IEEE and has given invited and keynote talks at various international conferences sponsored by SPIE, MRS and NSTI.

Contributors

Prabhakar R. Bandaru
Department of Mechanical and
  Aerospace Engineering
University of California, San Diego
La Jolla, California

Debjyoti Banerjee
Department of Mechanical
  Engineering
Texas A&M University
College Station, Texas

Kaustav Banerjee
Department of Electrical and
  Computer Engineering
University of California,
  Santa Barbara
Santa Barbara, California

Paul Benning
Hewlett-Packard Company
Corvallis, Oregon

Rob N. Candler
Department of Electrical
  Engineering
University of California,
  Los Angeles
Los Angeles, California

Chia-Ling Chen
Department of Mechanical and
  Aerospace Engineering
University of California,
  Los Angeles
Los Angeles, California

Michelle Chen
Department of Physics and
  Engineering
Point Loma Nazarene University
San Diego, California

Daniel S. Choi
Department of Chemical and
  Materials Engineering
University of Idaho
Moscow, Idaho

Hamed F. Dadgour
Department of Electrical and
  Computer Engineering
University of California,
  Santa Barbara
Santa Barbara, California

Michael F.L. de Volder
Imec and KULeuven
Heverlee, Belgium

Mehmet R. Dokmeci
Harvard–MIT Health Sciences
  Technology
Harvard University
Cambridge, Massachusetts

Philip X.-L. Feng
Department of Electrical
  Engineering and Computer
  Science
Case Western Reserve University
Cleveland, Ohio

Elaine D. Haberer
Department of Electrical
 Engineering and Materials
 Science and Engineering
 Program
University of California, Riverside
Riverside, California

A. John Hart
Department of Mechanical
 Engineering
University of Michigan
Ann Arbor, Michigan

M. David Henry
Department of Applied Physics
California Institute of Technology
Pasadena, California

Matthew A. Hopcroft
Hewlett-Packard Laboratories
Palo Alto, California

Allan H. Johnston
Jet Propulsion Laboratory
California Institute of Technology
Pasadena, California

Bongsang Kim
Sandia National Laboratories
Albuquerque, New Mexico

Amit Marathe
Microsoft Corporation
Silicon Valley Campus
Mountainview, California

Eric R. Meshot
Department of Mechanical
 Engineering
University of Michigan
Ann Arbor, Michigan

Tanya Nigam
Global Foundries
Sunnyvale, California

Naresh Pachauri
Department of Chemical and
 Materials Engineering
University of Idaho
Moscow, Idaho

Sei Jin Park
Department of Mechanical
 Engineering
University of Michigan
Ann Arbor, Michigan

Susan Richards
Hewlett-Packard Company
Corvallis, Oregon

Leif Z. Scheick
Jet Propulsion Laboratory
California Institute of Technology
Pasadena, California

Axel Scherer
Department of Applied Physics
California Institute of Technology
Pasadena, California

Donghyun Shin
University of Texas at Arlington
Arlington, Texas

Navdeep Singh
Department of Mechanical
 Engineering
Texas A&M University
College Station, Texas

Sameer Sonkusale
Department of Electrical and
 Computer Engineering
Tufts University
Medford, Massachusetts

James Stasiak
Hewlett-Packard Company
Corvallis, Oregon

Sameh H. Tawfick
Department of Mechanical
    Engineering
University of Michigan
Ann Arbor, Michigan

Kok-Yong Yiang
Global Foundries
Sunnyvale, California

Zhikan Zhang
Department of Chemical and
    Materials Engineering
University of Idaho
Moscow, Idaho

1

Moore's Law: Technology Scaling and Reliability Challenges

Tanya Nigam, Kok-Yong Yiang, and Amit Marathe

CONTENTS

ABSTRACT Semiconductor technology involves continued scaling of semiconductor processes to the deep sub-micron and nanometer levels according to Moore's law, as well as the addition of new and complex materials and process modules. Aggressive scaling entails numerous challenges involving power dissipation, variability, reliability, yield, and manufacturing. Often, the scaling of a technology is performed without proportionately reducing the power supply voltage to enable higher performance. Such an approach presents great challenges to device engineers, reliability engineers, and process integration engineers. As a result, trade-offs are usually required among reliability, design, and process development. The chapter highlights these numerous challenges in technology scaling for transistors and interconnects with special emphasis on the key reliability issues and the different wear-out mechanisms. The reliability of each new process module and how it interacts with other modules will be critical to the final reliability of the entire process. As the technology becomes more complex and aggressively scaled, reliability becomes critical as the technology is pushed to the limits to squeeze out every ounce of performance. Since generic technology reliability specifications can often be limiting and overly punitive, appropriate product reliability models are required to fully enable designs without compromising reliability.

1.1 Introduction: Technology Scaling Challenges

Silicon-based semiconductor device dimensions have been scaled continuously over the last 35 years. Current CMOS-based technologies have device dimension in the sub-100-nm range with gate dielectric thicknesses in the 1- to 2-nm range. Such advances largely followed the industry's governing tenet, i.e., Moore's law, which states that the number of transistors on a chip doubles every 2 years, as shown in Figure 1.1. To meet these needs, device dimension have shrunk 0.7 times per generation to improve performance by doubling frequency and reducing gate delay (Borkar, 1999; Paul et al., 2006, Taur et al., 1997).

Table 1.1 shows the scaling of typical device dimensions for different CMOS generations (Ghai et al., 2003). For the 70-nm technology node, the typical gate length is only 35 nm while the electrical gate oxide thickness (t_{OX}) is 1.6 nm and the source drain extension (SDE) depth is 17 nm. Modern microprocessors are manufactured with billions of transistors. Keeping power dissipation, variability, and reliability under control is therefore critical.

1.1.1 Power Dissipation

The active power dissipated in a CMOS chip is given by

$$Power = C \times V_{DD}^2 \times f \tag{1.1}$$

where, C is the capacitance, V_{DD} is the supply voltage, and f is the frequency of the circuit.

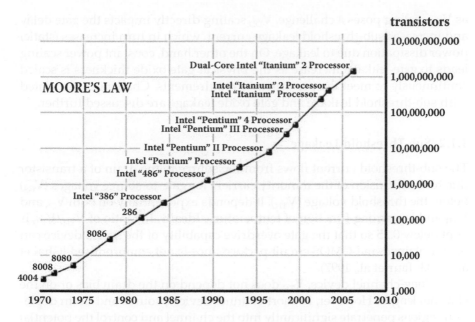

FIGURE 1.1
Plot of CPU transistor counts against dates of introduction. The line corresponds to exponential growth, with transistor count doubling every 2 years. (From http://download.intel.com/technology/silicon/Neikei_Presentation_2009_Tahir_Ghani.pdf)

There are two approaches to technology scaling: (1) constant power scaling where V_{DD} is not scaled and a reduction in capacitance is negated by an increase in f, and (2) reducing power where dissipation by scaling V_{DD} by 0.7 times, leading to a 50% reduction in active power for the scaled technology. The semiconductor industry has followed constant power scaling

TABLE 1.1

Scaling Projection of Transistor Parameters for Different Technology Generation Levels

	Generation Level (nm)				
Parameter	180	130	100	70	Scaling Factor
L_{GATE} (nm)	100	70	50	35	0.7×
V_{DD} (V)	1.5	1.2	1.0	0.8	0.8×
t_{ox} (e) (nm), t_{ox} (phys) (nm)	3.1, 2.1	2.5, 1.5	2.0, 1.0	1.6, 0.6	0.8×
SDE depth (nm)	50	35	24	17	0.7×
SDE under diff (nm)	23	16	11	8	0.7×
L_{MET} (nm)	55	40	27	20	0.7×
Channel doping (× 10^{18} cm^{-3})	1	1.6	2.6	4	$1/(0.8)^2 = 1.6×$
I_{DSAT} (relative)	1	1	1	1	1×
I_{OFF} (nA/μm), 25°C	20	40	80	160	2×

Source: Ghai, T. et al. 2000. *Proceedings of Symposium on VLSI Circuits.* IEEE, Honolulu, pp. 174–175. With permission.

as V_{DD} scaling poses a challenge. V_{DD} scaling directly impacts the gate delay and increases sub-threshold leakage current, which in turn increases (static) power dissipation due to leakage. On the other hand, constant power scaling leads to gate leakage increase as the physical gate oxide thickness is scaled continuously to meet the performance requirements. Challenges associated with sub-threshold leakage and gate oxide leakage are discussed further.

1.1.2 Sub-Threshold Leakage

The sub-threshold current flows from the source to the drain of a transistor due to the diffusion of the minority carriers for gate-to-source voltages (V_{GS}) below the threshold voltage (V_{TH}). It depends exponentially on both V_{GS} and V_{TH} and is a strong function of temperature. Ideally, the ratio of V_{TH}/V_{DD} is kept below 0.25 so that the gate overdrive capability of the scaled device can be maintained and CMOS circuit performance is not compromised (Ghai et al., 2003; Taur et al., 1997).

In a long-channel device, V_{TH} does not depend on the drain bias or on the channel length. However, in short-channel devices, source and drain depletion regions penetrate significantly into the channel and control the potential and the field inside the channel. This is known as the short channel effect (SCE). As a result of SCEs, V_{TH} reduces via (1) a reduction in channel length (V_{TH} roll-off), and (2) an increase in drain bias (drain induced barrier lowering or DIBL; Taur and Ning, 1998). This results in increased sub-threshold currents in short-channel devices. In order to keep SCEs under control, both the gate oxide thickness and the depletion width of the transistor must be reduced. The latter requires tailoring of the channel doping profile by implanting retrograde wells while the former directly leads to reliability challenges associated with gate oxide thickness scaling.

1.1.3 Gate Leakage

Gate leakage increases exponentially with a decrease in the gate oxide thickness and an increase in the potential drop across oxide. It exhibits a weak temperature dependence. Gate current is primarily due to the tunneling of electrons (or holes) from the silicon bulk through the gate oxide potential barrier into the gate (or vice versa). Figure 1.2 shows how reducing the gate oxide thickness leads to an increase in tunneling current (Massoud et al., 1996). Gate leakage is critical during the off-state of the devices and results in the standby power dissipation of the chip.

The maximum tolerable gate leakage current for a 10 mm^2 chip is 1- to 10-A/cm^2 at V_{DD} = 1 V. As shown in Figure 1.2, SiON-based sub-2-nm oxides exceed this threshold. Based on Table 1.1, a gate oxide thickness of 1 nm and below is required for sub-50-nm technologies. Therefore, the use of conventional SiON-based gate stacks becomes a major challenge.

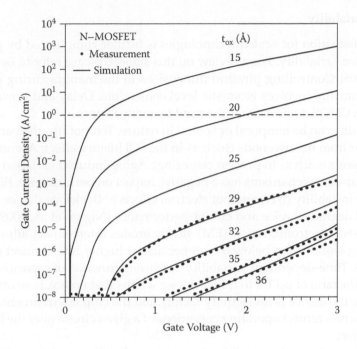

FIGURE 1.2
Measured and simulated I_G–V_G characteristics under inversion conditions for different oxide thicknesses. Dotted line indicates 1-A/cm² limit for leakage current discussed in text. (From Lo et al., 1997. With permission.)

One possible solution is to use dielectric films with higher dielectric constants (known as high-K materials, e.g., HfO_2 or Al_2O_3 etc.) such that the physical gate stack thickness can be increased, leading to a lower gate leakage current despite a lower equivalent gate oxide thickness as measured in inversion. The SiO_2 equivalent oxide thickness is:

$$EOT = t_{HK} \times \varepsilon_{SiO_2} / \varepsilon_{HK} \qquad (1.2)$$

where ε_{SiO2} and ε_{HK} are the dielectric constants of SiO_2 and the high-K material, respectively.

Integrating high-K dielectric into CMOS technology is a major challenge and extensive work has been done to screen and select high-K dielectrics (Wilk et al., 2001; Huff and Gilmer, 2004; Gusev, 2006). Over the last 10 years, hafnium oxide with a dielectric constant of 25 (HfO_2, HfSiON, HfON, and HfSiO) emerged as a viable candidate to replace SiON. In addition to replacing the gate dielectric, it is also essential to replace the poly-Si gate with a metal gate to reduce the effective decrease in inversion capacitance due to depletion of the poly-Si at the high fields typically present in modern devices.

1.1.4 Variability

Power dissipation for scaled technologies is further complicated by process and device variability. For a review on this subject, please refer to Bernstein et al. (2006). Controlling physical dimensions in the manufacturing process now commonly involves atomistic-level constraints. Delay and power variability in CMOS devices are influenced by many factors.

Variability can be temporal or spatial in nature. Temporally, the variability can occur from nanoseconds (such as in the SOI history effect; Asenov et al., 2003) to years (such as in process centering). Aging-induced variation arising from wear-out mechanisms has a negative impact on performance. Bias temperature instability (BTI) and hot electron effects both elevate device thresholds and degrade device and circuit performance (Nigam et al., 2009; Frank et al., 2006). Electromigration (EM) slowly erodes interconnect admittance, becoming more severe below 65 nm because of higher interconnect current densities. Time-dependent variability is a strong function of capacitive loading and the ratio of p-FET to n-FET device widths (beta ratio), how often and how long the device is on (activity factor), and the chip environmental (voltage and temperature) operating conditions of a given circuit over the lifetime of a product.

Spatial variation refers to lateral and vertical differences from intended device dimensions and film thicknesses (Stine et al., 1997). Spatial variation modes exist between devices, between circuits, between chips, and across wafers, lots, and lifetimes of various fabrication systems. Such a variation is often referred to as extrinsic variability. Additionally, intrinsic device variations also occur due to atomic-level differences between devices that exist even though the devices may have identical layout geometries and environments. These stochastic differences appear in the dopant profiles, film thickness variations, and line-edge roughness. A typical example of such variations is V_{TH} variation due to the atomistic nature of the dopants in MOSFETs. The implant and annealing processes result in the placement of a random number of dopants in the channel (described by a Poisson distribution).

Variations are classified as (1) those that involve the chip mean, (2) those that vary within the chip but have local or chip-to-chip correlation, and (3) those that vary randomly from device to device. Chip mean variations can be caused by spatial variation (L_G, V_{TH}, and t_{OX}) and temporal variation (operating temperature, activity factor, and device degradations such as BTI and HCI). Within-chip variation comes from pattern-density or layout-induced transconductance on die hot spots and hot spot-induced BTI. Finally, device-to-device variation comes from atomistic variations in dopant distributions, line edge roughness, and BTI- or (hot carrier injection) HCI-induced V_{TH} distributions.

1.2 Transistor Reliability

Reliability is the probability that a product will perform a required function under stated conditions for a stated period of time. A typical example of reliability is gate oxide integrity. An oxide is defined as reliable if it maintains its insulating properties for 10 or 25 years at a specified bias, temperature, chip area, and failure fraction.

Reliability studies typically require accelerated testing conditions such that the physical mechanism responsible for breakdown can be studied in a time frame much shorter than the targeted lifetime. Based on stress at accelerated conditions, projections are made toward the use conditions. Intrinsic reliability studies revolve around generation of material defects that lead to product failure. Since defect generation is random in nature, the statistical nature of defect generation and its impact on reliability must be understood. Strong et al. (2009) performed an extensive review of device reliability.

1.2.1 Time-Dependent Dielectric Breakdown

Time-dependent dielectric breakdown (TDDB) occurs during the off-state when the voltage across the gate dielectric is high. TDDB failure was traditionally catastrophic and caused the gate dielectric to lose its insulating properties after the breakdown event, leading to a functional failure of the chip. As technology is scaled downward, TDDB is no longer automatically considered catastrophic since the dielectric does not fully lose its insulating properties for sufficiently thin gate oxides.

Typically, a dielectric breakdown leads to the formation of a leakage path through the oxide. Figure 1.3 illustrates the defect generation and eventual breakdown of a dielectric. As a high voltage stress is applied, defects are randomly generated in the bulk of the oxide (Stage 1). During TDDB, bulk defects are created due to the breaking of Si-O bonds and the defects are permanent under typical operating conditions.

Once a critical density of defects is reached, a localized leakage path is formed (Stage 2, often called soft breakdown or SBD). According to Weir et al. (1997), the formation of such a leakage path does not lead to a complete loss of insulating properties. If the stress continues, gate leakage further increases and finally the dielectric breaks down, resulting in ohmic conduction through the gate stack (Stage 3, hard breakdown or HBD).

Since the defect generation and breakdown path creation are random in nature, TDDB requires statistical description by means of a distribution. Dielectric breakdown is described very well by Weibull statistics. Gate oxide breakdown represents the weak link phenomena as the different areas of the gate oxide are competing with each other for the formation of breakdown path. The area with highest density of defects leads to breakdown. Weibull statistics is described by

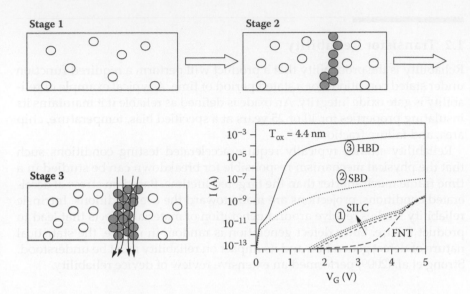

FIGURE 1.3
Defect generation in a vertical cross-section of gate oxide. Stage 1: defects are generated in the oxide. Stage 2: a local conducting path is formed (soft breakdown or SBD). As the stress continues, a final hard breakdown with complete loss of insulating properties occurs. The current voltages for the three stages of oxide wear-out are also shown.

$$F(t_{BD}) = 1 - \exp(-(\frac{t_{BD}}{\tau})^{\beta}) \tag{1.3}$$

where $F(t_{BD})$ is the cumulative failure probability and t_{BD} is the time to breakdown. The characteristic time to breakdown τ is the time for 63rd percentile and β is the Weibull shape factor. An important property of Weibull slope is that it describes three different failure rates that occur during product life. The extrinsic failure is described by Weibull shape $\beta < 1$ which corresponds to decreasing failure rate, random fails for $\beta = 1$ with constant failure rate, and intrinsic fails for wearout with $\beta > 1$. Typical extrinsic and intrinsic Weibull distributions are shown in Figure 1.4.

The observed intrinsic distribution of oxide breakdown can be modeled using percolation theory (Degraeve et al., 1998; Stathis et al., 1999). Under the percolation approach, the defect generation continues until a critical defect density is reached. To model the probability for reaching a critical defect density, defects are randomly generated using Monte Carlo (MC) simulation until a local connecting path is formed as shown in Figure 1.5a. Using this approach, the reduction in Weibull slope for thinner gate oxides was predicted for the first time in agreement with the experimental findings (see Figure 1.5b).

A three-dimensional (3D) analytical model has also been proposed to model the defect generation in gate oxide (Sune 2001; Krishnan and Nicollian, 2007.

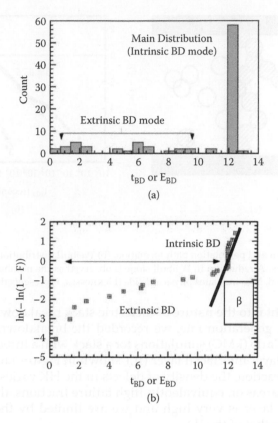

FIGURE 1.4
Probability distribution (a) and Weibull distribution (b) of gate oxide breakdown data. The sharp part of the distribution corresponds to the intrinsic breakdown while shallow distribution is extrinsic. The Weibull shape factor is given by β and τ corresponds to $\ln(-\ln(1 - F)) = 0$.

The analytical model is similar to the MC approach and also predicts the decrease in Weibull slope as a function of thickness. The analytical expression for Weibull shape factor that decreases as a function of oxide thickness is given by:

$$\beta = \alpha \frac{t_{OX}}{a_0} \quad (1.4)$$

where α is the defect generation rate, t_{OX} is the gate oxide thickness, and a_0 is the defect size.

The percolation approach using MC simulations that was applied successfully to single SiO_2 layers can also be extended to any dual or multilayer system as in high-K gate stacks (Nigam et al., 2009). For this work, the kinetic MC technique was used so that the defect generation rate in the two layers could be varied independently. Results of the simulation are shown in Figure 1.6.

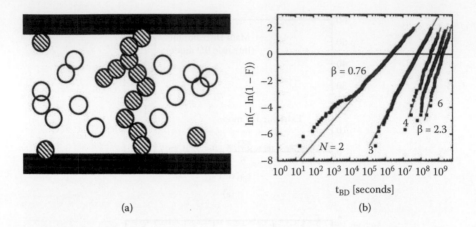

(a)   (b)

FIGURE 1.5

(a) Defect generation and percolation path formation. (b) Weibull distribution as a function of gate oxide thickness. A reduction in Weibull slope is observed as the number of layers in the gate oxide is reduced. N^*a_0 indicates physical oxide thickness; a_0 represents defect size.

To gain insight into the nature of dielectric stack breakdown with a non-uniform defect generation rate, we recorded the breakdown path during kinetic Monte Carlo (kMC) simulations for a stack with a three-layer high K (HK) and two-layer interfacial layer (IL), shown in Figure 1.6a. Depending on the failure fraction, the density of defects in the HK varies significantly. For very small areas or, equivalently, high failure fractions, the defect density in the HK layer is very high and we are limited by the IL (and the Weibull slope is that of the IL).

For low failure fractions or large areas, where lucky events dominate, the breakdown path resembles the case of a uniform oxide with few defects in the HK layer, in agreement with Krishnan and Nicollian (2007). In this case, the Weibull slope is determined by the complete stack thickness. Figures 1.6b through e show the defect distribution around the breakdown path. Figure 1.6f shows the 300× higher defect density at breakdown in the HK layer as compared to the IL for the kMC simulation of Figure 1.6a.

The above approach of using a percolation model for multiple layers may explain the low Weibull slopes observed for HK dielectrics (Bersuker et al., 2008; Degraeve et al., 1999) and the assessment that the IL determines TDDB for these stacks. This is true only for small area devices.

The possibility of a large number of defect paths in the HK layer prior to stack breakdown was proposed by Okada et al. (2007), but its impact on the statistics of the HK stack breakdown was not addressed. The dual layer model of Nigam et al. (2009) showed that if defect generation rate in the HK is higher than that in the IL, the TDDB distribution becomes bimodal. For large areas or small failure fractions, the Weibull slope increases to a value consistent with the complete stack thickness and measured data. This significantly enhances TDDB predictions for typical products that use high-K gate

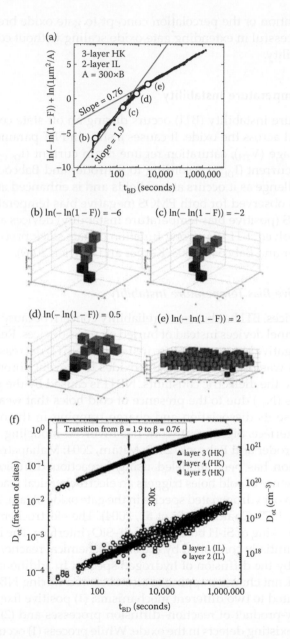

FIGURE 1.6
(a) Simulated TDDB distribution for a dielectric stack with a two-layer IL and three-layer HK. The Weibull slope for large failure fractions is that of a two-layer dielectric; the slope for small failure fractions is that of a five-layer dielectric. (b)–(e) Breakdown path shapes typical for various failure fractions (f). (f) D_{ot} in the HK layer and IL as a function of t_{bd}.

stacks. Application of the percolation concept to gate oxide breakdown has been very successful in extending gate oxide scaling without compromising product reliability.

1.2.2 Bias Temperature Instability

Bias temperature instability (BTI) occurs during an off-state condition with a uniform field across the oxide. It causes a shift in FET parameters such as threshold voltage (V_{TH}), saturation regime drain current (I_{DSAT}), and linear regime drain current (I_{DLIN}) according to Schroder and Babcock (2003). BTI is a major challenge as it occurs at low fields and is enhanced at higher temperatures. It is observed for both PMOS (negative bias temperature instability) and NMOS (positive bias temperature instability) devices as technology scales have evolved. BTI is a strong function of gate stack processing conditions (Schroder and Babcock, 2003; Kerber and Cartier, 2009).

1.2.2.1 Negative Bias Temperature Instability

For PMOS devices, BTI has become a reliability concern because of the switch to surface channel devices instead of buried channel devices. Reliability margins due to negative bias temperature instability (NBTI) decreased further as more nitrogen was incorporated in the oxides to prevent boron penetration and to increase the dielectric constants. NBTI is caused by the generation of interface states (N_{IT}) due to the presence of cold holes that weaken the Si-H bond and cause its dissociation and charge trapping in the oxide close to the Si–SiO$_2$ interface. Significant work has involved modeling and explaining NBTI (Schroder and Babcock, 2003; Mitani, 2004; Mahapatra et al., 2003).

N_{IT} generation has been modeled using a reaction–diffusion process in which the presence of cold holes triggers an electrochemical reaction coupled to the diffusion of hydrogenated species in the gate oxide (Alam, 2003; Jeppson and Svensson, 1977; Chakravarthi et al., 2004). The electrochemical reaction leads to the breaking of Si-H bonds at the Si–SiO$_2$ interface. The interface generation rate is initially controlled by the electrochemical reaction process and subsequently by the diffusion of hydrogen species. In additional to N_{IT} generation, significant charge trapping is also observed during NBTI stress and may be attributed to two different mechanisms: (1) positive fixed-charge formation as a by-product of reaction–diffusion processes and (2) charge trapping into pre-existing defects in the oxide. While process (1) occurs sequential to N_{IT} generation, process (2) occurs parallel to N_{IT} generation.

NBTI relaxes after the stress bias is removed (Chen et al., 2002). The cause of recovery is still controversial. It can be attributed to both re-passivation of interface states by the available hydrogenated species in the oxide and/or de-trapping of the trapped charge in the oxide. NBTI relaxation makes measurement of the true NBTI component very challenging because delays

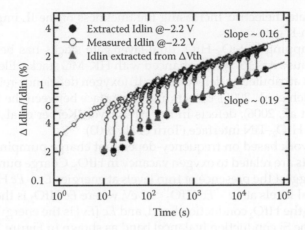

FIGURE 1.7
Continuous measurement of $\%I_{dlin}$ during stress with a $V_{DD} = -0.1$ V (open circle). Closed circle shows $\%I_{dlin}$ obtained from an IV measurement. Also shown is the extracted $\%I_{dlin}$ from a V_{TH} measurement (triangles). Based on the reaction–diffusion model, the slope of 0.16 corresponds to the diffusion of molecular hydrogen and not atomic hydrogen, consistent with the obtained activation energy.

in the range of a few milliseconds already cause significant relaxation. This directly impacts the ability to measure the true NBTI component.

Typically the impact of accelerated voltages on I_{DLIN} is measured after a certain delay. In order to minimize the amount of recovery, an "on-the-fly" technique in which I_{DLIN} is measured during stress (open circles) was used, as shown in Figure 1.7. The figure also shows the amount of I_{DLIN} degradation measured after a delay of 1 second (closed circles). Delay reduces the magnitude of damage and increases the time dependence of the NBTI-induced degradation. The higher slope obtained due to measurement delay can be explained by the stress time dependence of relaxation.

Note the delay between two stress cycles is fixed while the stress time increases exponentially; hence the ratio of t_{Relax}/t_{Stress} decreases as the stress progresses. Since the fractional recovery shows a universal curve as a function of t_{Relax}/t_{Stress} (Denis et al., 2006; Grasser et al., 2007), a lower t_{Relax}/t_{Stress} corresponds to less fractional recovery at longer stress times and a smaller impact of relaxation. This has significant implications for predicting product level reliability as discussed in Section 1.2.2.3.

1.2.2.2 Positive Bias Temperature Instability

For NMOS devices, the switch to higher dielectric constant materials leads to charge trapping in high-K materials. A high-K gate stack is a multilayer structure consisting of an interfacial SiO_2 layer (IL), a deposited high-K layer, a metal gate, and a poly-Si layer. The presence of IL is critical for the success of a high-K gate dielectric. Increasing the thickness of the IL improves reliability of high-K gate stacks.

Electron trapping in SiO_2–HfO_2 dual-layer gate stacks has been studied intensely because it is strongly enhanced with HK–MG stacks. Electron trapping has been attributed to defects in the IL (oxygen deficiency-related defect precursors; Heh et al., 2006), defects at the interface between the two dielectrics (Casse et al., 2006), defects in the HfO_2 layer (Kerber et al., 2003), and defects at the HfO_2–TiN interface (Torii et al., 2003).

Extensive work based on frequency-dependent charge pumping indicates that the defects are related to oxygen vacancy in HfO_2. Charge pumping measurements suggest the presence of trap levels at energies $Ev – Ec$ HfO_2~1.4 eV and additional levels at $Ev – Ec$ HfO_2 <1.2 eV, where Ec HfO_2 is the energy of the bottom of the HfO_2 conduction band, and Ec [Ev] is the energy of the bottom [top] of the Si conduction [valance] band as shown in Figure 1.8a.

These numbers fall well within the range of calculated electron energy levels for the various charge states of the oxygen vacancy in HfO_2 (Robertson, 2006). Since the electron trapping is due to tunneling of the carrier through the IL, increasing the thickness of IL reduces the magnitude of positive bias temperature instability (PBTI) while increasing the thickness of high-K material increases PBTI. The shift in NMOS device characteristics after PBTI-like stress is shown in Figure 1.8b.

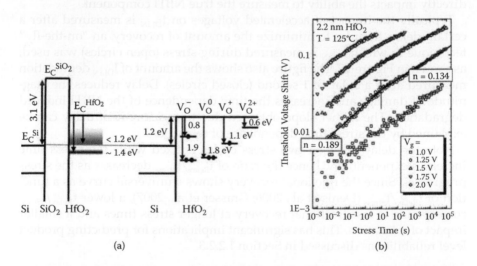

(a)   (b)

FIGURE 1.8
(a) Left: band diagram containing the energy level position of the electron traps as derived from spectroscopic ACP experiment. Right: calculated energy levels for oxygen vacancies in various charge states. (From Cartier, E. et al. 2006. *Proceedings of International Electronics Development Meeting*, IEEE, San Francisco, 321–324; Robertson, J. 2006. *Rep. Progr. Phys.*, 69, 327–396. With permission.) (b) Threshold voltage shift during PBTI stress at 125°C for 2.2-nm thick HfO_2 layer. (From Kerber, A. et al. 2008. *IEEE Trans. Electron. Dev.*, 55, 3175–3183. With permission.)

1.2.2.3 Relaxation in Bias Temperature Instability

It has been shown that BTI shows significant relaxation after stress is removed (Chen et al., 2002; Kerber and Cartier, 2009). Depending on the temperature and stress time, a V_{TH} recovery up to 80% is possible. Logarithmic time dependence is observed during relaxation for both NBTI and PBTI (see Figure 1.9a) unlike the stress phase in which a power law is observed.

If the fractional recovery is plotted as a function of relaxation time over stress time (t_R/t_S) a universal curve is obtained (Graser et al., 2007). For NBTI, the recovery is attributed to the re-passivation of the generated interface states (Alam, 2003; Chakravarthi et al., 2004) during the zero bias state and/or hole de-trapping (Huard et al., 2006) upon removal of stress. For PBTI, recovery is attributed to the de-trapping of trapped electrons. This implies that for a circuit level stress (such as a ring oscillator) in which a device is continuously switching, a lower BTI is expected and measured as shown in Figure 1.9b (Nigam, 2008).

1.2.3 Hot Carrier Injection

Hot carrier injection (HCI) occurs during the on-state condition with a high voltage on the drain that leads to a non-uniform field across the oxide. As with BTI, HCI also causes a shift in FET parameters such as threshold voltage (V_{TH}), saturation regime drain current (I_{DSAT}), and linear regime drain current (I_{DLIN}). HCI modeling and data have changed as the technology has scaled.

Hot carriers are generated by a high lateral electric field in the channel. When the mean kinetic energy of the carrier is higher than the lattice

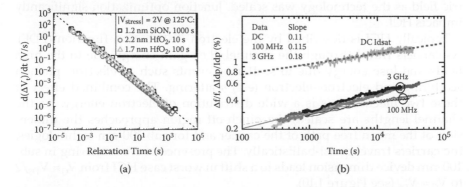

(a)   (b)

FIGURE 1.9
(See color insert.) (a) Threshold voltage recovery for NMOS and PMOS with 2.2-nm and 1.7-nm HfO2 gate dielectric and 1.2 nm SiON. The recovery follows a log(t) dependence for both NBTI and PBTI. (From: Kerber, A. et al. 2008. *IEEE Trans. Electron. Dev.*, 55, 3175–3183. With permission.) (b) Impact of AC stress using RO. Ring oscillator frequency degradation at 25OC along with I_{dsat} degradation for DC stress on PMOS transistor. Lower degradation is measured for RO as compared to DC stress. (From Nigam, T. 2008. *IEEE Trans. Dev. Mater. Reliability*, 8, 72–78. With permission.)

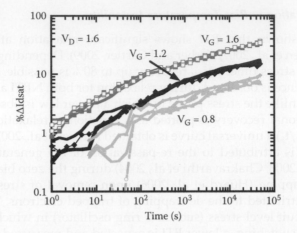

FIGURE 1.10
HCI degradation increases as a function of V_{GS} for a fixed V_{DS}. Channel length for these NMOS devices is 40 nm.

temperature, a carrier is "hot." The generated hot carriers can be injected into the oxide causing bulk defect generation or charge trapping. They can also lead to N_{IT} generation near the drain. Typically the damage due to HCI is highly localized. HCI increases as the channel length (L_G) is reduced. To reduce the HCI-induced device parameter shift, the lightly doped drain (LDD) was introduced. The main goal was to reduce the peak lateral electric field as the technology was scaled. Junction optimization significantly impacts HCI.

Typically HCI is described by the electron distribution function (EDF). As carriers travel through the channel, they gain energy due to the high fields and lose energy due to scattering events such as electron–phonon scattering and electron–electron (e-e) scattering. The combined effect of these two mechanisms is a wide distribution of electron energy. As the channel lengths are scaled, the pinch-off region approaches the dimension of the mean free path of the carrier and for sub-100-nm technologies, the carriers travel quasi-ballistically. The presence of e-e scattering in sub-100-nm device dimension leads to a shift in worst case HCI from $V_G = V_{DS}/2$ to $V_G = V_{DS}$ (see Figure 1.10).

At low V_{DS}, e-e scattering broadens the EDF tail as compared to the thermally distributed tail and exhibits a strong dependence on I_{DSAT}. In recent years, HCI-related degradation has received less attention due to the reductions of operating voltages, and it is expected that HCI damage will reduce as well. During circuit operation, HCI occurs only during switching. Therefore significant relief is expected as compared to the DC degradation (Nigam et al., 2009).

1.3 Back End of Line: Interconnect Technology

Advanced integrated circuits (ICs) require elaborate wiring (or interconnect) systems to distribute power, grounding, and various clock and input and output (I/O) signals to and from transistor devices. To maintain the cost and performance benefits associated with reduced transistor feature size and higher on-chip device density, the interconnect architecture must correspondingly increase in complexity and density; this is achieved by reducing the geometrical dimensions of the wirings and increasing the number of interconnect layers. The downsides to the "interconnect scaling," however, are an increase in wiring resistance (due to the smaller cross-sectional area) and the parasitic capacitance of wires that exert serious impacts on dynamic power consumption,[*] self-heating, and signal propagation speeds in the form of increased resistance–capacitance (RC) delay.

It was generally recognized that interconnect RC delay would effectively limit further gains in transistor speeds at the 0.25-µm technology node if no new interconnect materials with lower resistance and permittivity (K value) were used to replace the traditional Al and SiO_2 (Ho et al., 2002). To meet future performance requirements, the National Technology Roadmap for Semiconductors (NTRS) as early as 1994 began to predict the accelerated adoption of low-K interconnect materials. However, the commercialization of low-K materials to be more challenging than anticipated—only at the 0.13-µm technology node (in 2002) were higher conductivity Cu and low-K materials introduced. Even then, the implementation of even lower K materials (with nanometer porosities) has been slow and problematic, in part due to the reliability and yield issues associated with the integration of these materials. This delayed the production predictions of the International Technology Roadmap for Semiconductors (ITRS) by many years.

1.4 Evolution of Interconnect Materials and Patterning Schemes

Three metal candidates (Cu, Ag, and Au) exhibit lower bulk resistivities than Al, and the only practical option in terms of cost and manufacturability is Cu. Due to the high diffusivity of Cu into the dielectrics, a thin conformal barrier (liner) on the sidewalls and trench bottom of the interconnect wire is required. Since Cu does not form a volatile by-product, it is very difficult to

---

[*] The equation for dynamic power consumption P is generally $P = \alpha CfV^2$, where α is the wire activity, f is the transmission frequency, V is the power supply voltage, and C is the cumulative capacitances of the devices and interconnect wirings (Maex et al., 2003).

TABLE 1.2

Spun-On and Chemical Vapor-Deposited Low-K Dielectric Materials

Spun-On	Chemical Vapor-Deposited
B-staged polymers (CYCLOTENE™, SiLK™)	Fluorine-doped oxide (FSG)
Hydrogen silsesquioxane (HSQ)	Carbon-doped oxide (SiOCH)
Polybenzoxazole-based oxazole (OxD)	Parylene-N, parylene-F
Methyl silsesquioxane (MSQ) (LKD™)	
Divinylsiloxane-benzocyclobutene (BCB)	
Poly-tetrafluoroethylene (PTFE) (Teflon™)	
Fluorinated poly(arylene ether) (FLARE™)	

etch. This led to the development and adoption of the damascene approach whereby the wiring pattern is created in the dielectric by dry etching before filling with Cu and planarized to remove excess metal on top of the dielectric. This is in contrast to the traditional subtractive approach by which dielectric (SiO_2) is deposited on patterned Al metal and planarized.

The choice of low-K dielectric was less clear. Throughout the 2000s, research focused on different materials and deposition techniques for low-K materials, and a proliferation of spun-on and chemical vapor-deposited (CVD) dielectrics with bulk permittivities (K values) of 2.5 to 3.7 surfaced (Table 1.2). For these materials, the low K values (compared to ~4.2 for SiO_2) are achieved by incorporating atoms and bonds with lower polarizability, reducing the atomic and bond density, or both. Since 2005, the industry gradually gravitated toward CVD SiOCH because of its close resemblance to SiO_2 in terms of chemical, thermal, and mechanical properties.

While the further lowering of K values below 2.5 is possible by introducing nanometer porosity or macroscopic air gaps into the low-K dielectrics, these schemes create additional integration issues related to moisture uptake, damages arising from etch and chemical-mechanical polishing (CMP), and insufficient mechanical strength to withstand the forces during dicing, packaging, and assembly.

1.5 Challenges in Interconnect Reliability

More than three decades of continual CMOS scaling have now pushed existing interconnect materials to their reliability limits. The damascene fabrication process and materials fundamentally changed the stress states and interfaces of Cu wires compared to the subtractive patterning process for Al. In addition, the minimum thickness of low-K dielectrics between wires have scaled to deep submicron regimes (<50nm). Although this is nearly 50 times the gate oxide thickness, low-K dielectrics are far from the quality of gate oxides in terms of electrical, thermal and mechanical characteristics.

One major interconnect challenge associated with the adoption of new Cu and low-K materials is electromigration (EM). In Al interconnects, momentum exchange between the current-carrying electrons and host metal lattice can cause the Al ions to drift in the direction of the electron current, causing the formation of a void at flux divergent sites (i.e., regions of tensile stress). At locations of high compressive stress, extrusions and hillocks can also form, causing electrical shorts to neighboring wires. EM reliability has been shown to depend on the grain size and crystallographic texture of the Al (Vaidya and Sinha, 1981; Campbell et al., 1993; Toyoda et al., 1994; Knorr et al., 1996; Rodbell et al., 1996).

Cu has a higher melting point than Al, and at operating conditions is supposed to offer a 40× increase in reliability (Besser et al., 2000). However, while Al readily forms a stable native oxide that acts as a slow diffusion site, Cu requires a barrier (at the side walls and trench bottom) and a dielectric capping layer (at the top) to prevent diffusion into the low-K dielectric; these interfaces of Cu to barrier and capping layer provide a fast diffusion pathway for Cu to electromigrate (Hu et al., 2004).

Although the fundamental failure mechanism and processing of Al and Cu metallizations are different, their effective atomic diffusivity D_{eff} follows a similar equation (below) and is the sum of the individual diffusion coefficients along grain boundaries (D_{gb}), barrier interface (D_b), and capping layer interface (D_c). The diffusivities are functions of grain size (d) and the width (w) and height (h) of the wire; δ_{gb}, δ_b, and δ_c are the dimensions of the grain boundary and interface layers.

$$D_{eff} = D_{gb}\left(\frac{\delta_{gb}}{d}\right) + D_b\delta_b\left(\frac{2}{w} + \frac{1}{h}\right) + D_c\left(\frac{\delta_c}{h}\right) \qquad (1.5)$$

With geometrical scaling, EM performance is expected to degrade due to the relative increase in Cu interface area-to-volume ratio. The lower modulus low-K dielectrics also provide less backflow stress (i.e., reduced Blech effect) that may increase the risk of electromigration failure. Optimizing Cu interconnect reliability therefore includes, but is not limited to, optimizing the interfaces and process temperatures, using alternative capping materials such as CoWP, CoSnP and Pd (Hu et al., 2002), and doping the Cu with alloy elements (such as Al). A very thin (~10 nm) CoWP layer is reported to be sufficient in reducing most interfacial diffusion (Hu et al., 2003).

Another failure mechanism associated with interconnect metals is stress migration (SM) defined as the movements of metal atoms under the influence of mechanical stress gradients. Cu wires exist in a state of tensile stress due to the different materials and process temperatures of chip fabrication.

Metal atoms migrate from regions of low to high stress, contributing to void growth; when a sufficiently large void is formed (typically as the vias land on wide lines), a substantial increase in resistance or even an electrical opening can occur and interfere with chip functionality.

Bulk Cu has greater elastic modulus, melting point, and yield strength compared to Al and should be more resilient to stress-induced voiding. However, electrodeposited Cu has a much greater initial void concentration and therefore shows similar vulnerability to stress migration (JEDEC, 2009). With continued scaling in via sizes, fewer vacancies will be required to cause an unacceptable increase in via resistance. The SM time-to-failure (TTF) is generally based on the Eyring model, and can be described (Rzepka et al., 2004) by the equation:

$$TTF \propto (T_0 - T)^{-N} \exp\left(\frac{-E_a}{k_B T}\right) \tag{1.6}$$

where T_0 is the stress-free temperature for the metal (roughly equivalent to the deposition temperature for Cu), N is the creep exponent, E_a is the diffusion activation energy along the Cu/capping interface, k_B is the Boltzmann constant, and T is the temperature in Kelvin. One way to improve SM performance is the optimization of the Cu–capping interface using an alternative capping layer such as CoWP (Gambino et al., 2006), which also improves EM performance.

The shrinking of interconnect metal dimensions requires a corresponding decrease in Cu barrier thickness in order to mitigate the degradation in effective metal resistivity. According to the ITRS (2007), barrier thicknesses smaller than 5 nm were required by that year and are expected to scale below 2 nm by 2015. Maintaining barrier integrity against Cu diffusion into the dielectric is proving to be challenging with conventional physical vapor deposition (PVD) techniques, and is even more so with the introduction of ultralow-K dielectrics with increased porosities (Tökei et al., 2004).

The aggressive reduction in interconnect pitch greatly increases the operating electric fields, and correspondingly the leakage currents between wires. In addition, low-K dielectrics have inferior breakdown strengths compared to SiO_2 (Ogawa et al., 2003). Ultralow-K dielectrics are also susceptible to soft-breakdown (Chen et al., 2009), even though the physical thicknesses are at least an order of magnitude greater than gate oxides. These factors contribute to long-term wearout risks in the inter layer dielectric (ILD), and time-dependent dielectric breakdown (TDDB) is increasingly limiting back end-of-line (BEOL) reliability.

With increasing porosity, the effect of Joule heating in the metals will become more pronounced, due to the poorer thermal conductivity in the insulating films. This has a negative impact on all the reliability mechanisms associated with the back end-of-line (BEOL) interconnects. The conventional roadmap of reducing the permittivity of low-K dielectrics by increasing the porosity no longer appears feasible.

1.6 Emerging Interconnect Materials and Architecture

Beyond the 22-nm technology node, the effective resistivity of Cu is anticipated to increase dramatically as feature size scales due to electron scattering
from grain boundaries and the Cu–liner and Cu–dielectric-barrier interfaces.
The traditional approach to drive down the K value (below 2.0) by incorporating increasing porosity in the dielectrics no longer appears manufactur-
able due to extremely poor thermal, mechanical, and electrical properties.

In lithography, the 193-nm water immersion technology would no longer be able to resolve the tight pitches with sufficient resolution and would
require the use of double patterning or exposure to split the tight pitches
into larger ones. This would drive up the lithography cost and increase the
wafer line edge roughness (LER) and critical dimension (CD) tolerance budget. Although extreme-UV (EUV) lithography with smaller wavelengths and
better resolution imaging is currently in development, it does not promise a
cost-effective solution. To keep up with Moore's law at the BEOL, a radical
overhaul of the interconnect architecture is inevitable.

One proposed alternative to Cu is the cylindrical carbon nanotube (CNT)—
essentially a rolled-up graphene sheets made of one-atom thick carbon layers.
CNTs have large current carrying capacity, high electrical and thermal conductivity, and unlike Cu, are resistant to electromigration. However, several
major challenges must be overcome before CNTs are commercially viable:
the growth of CNTs has to be compatible with the BEOL thermal budget and
the chirality, diameter, and accuracy of placement must be well controlled.

Another potential candidate for Cu replacement is the graphene nanorib-
bon (GNR)—an unrolled CNT that may be patterned using conventional
top-down lithography. However, unlike CNTs that have no geometrical
edges, GNRs can suffer from electron scattering at imperfect edges and
hence exhibit increased resistivity at narrow widths.

There is increasing interest in 3D interconnect architectures that can significantly reduce global wiring lengths and therefore circumvent some of
the limitations in signal propagation delay and power dissipation. One 3D
scheme is the use of through-silicon vias (TSVs) to electrically connect two
or more heterogeneous chips that are stacked vertically. This scheme can
effectively increase interconnect densities and reduce the overall chip size.

Although TSVs are not new and have been used commercially since 1976
(primarily for grounding in GaAs dies with coplanar RF circuits), several major issues must be resolved before they can be adopted widely for
advanced ICs. These issues include the slow etch rate of silicon that limits
throughput, the relatively high cost of the associated equipment, and the
additional processes and mechanical stresses that can significantly impact
device and interconnect reliability.

1.7 Reliability for Design Enablement

Reliability is a stochastic phenomenon. It can be specified only in terms of probability of failure in a given lifetime and is governed by the specific reliability criterion applied. No absolute standards exist. The products that are manufactured using a certain process technology exhibit reliability that is governed by the intrinsic reliability of the technology (as described in the previous sections) and also by their specific design attributes and complexities.

1.7.1 Product Reliability Models

Typically, technology reliability specifications are very generic. For example, a typical gate dielectric TDDB specification usually states a certain maximum operating voltage (V_{max}) of the technology to support 11.4 years of product operating lifetime at 125°C and a 1 part per million (ppm) failure rate for a 10 mm^2 of gate dielectric area. In reality, not all the circuit blocks and elements in a product operate at V_{max} and 125°C and the product gate dielectric area may be smaller or larger than 10 mm^2; different lifetime and failure rate requirements may exist as well. For these reasons, a generic technology specification cannot accurately project or describe the reliability of a product operating within its specific usage profile.

On one hand, the intrinsic technology specifications may be overly optimistic and can over-predict reliability and possibly create risks of product failures and field returns. Conversely, specifications may be overly conservative and hence punitive for product design. Any conservatism left on the table cannot fully enable design or performance.

These non-optimal reliability projections based on generic technology specifications require development of product reliability models. An "intelligent" product model should comprehend all the characteristics and complexities of the product. For example, in a typical high performance CPU/GPU-like product, certain attributes should be included in a product reliability model:

- Multiple cores and functional blocks with different activity factors
- Multiple operating voltages for functional blocks
- Non-uniform temperature profiles
- Complex usage profiles (sleep states, idle states, etc.)
- Market segment (server, client, desktop, mobile, etc.)

Special CAD tools must be developed to map out and extract the various product operating characteristics of a product reliability model. These tools help identify the critical design paths that push reliability harder to enhance performance and enable the concept of "budgeting" reliability appropriately where it matters.

The development of product reliability models and budgeting reliability optimally across the different functional blocks form the backbone of a smart design methodology that "designs in" reliability rather than considering it an afterthought. Reliability modeling and budgeting allow designers to selectively overdrive up to several tens or hundreds of millivolts over and above the nominal operating voltage of a technology to push for performance without compromising reliability. These procedures also help achieve better management of the power–performance–reliability trade-off. An example of the application of this methodology to electromigration is outlined briefly in the next section.

1.7.2 Statistical Electromigration Budgeting

As the technology is pushed to its limits and is overdriven beyond nominal operating voltages, the interconnect wires of the BEOL also reach the DC EM limits that often restrict designs. Usually only certain critical design blocks must be pushed harder than others. To fully enable designs, a new methodology called statistical electromigration budgeting (SEB) has evolved. It can exploit the statistical nature of EM reliability to selectively supersede fixed current density design rules for some interconnects, allowing increased chip performance while simultaneously quantifying chip-level EM reliability to directly assure design conformance to reliability requirements (Kitchin, 1995).

Historically designers have compared interconnect dc average current per unit width I_{eff} to a conservative fixed limit using a dc severity ratio:

$$S_{dc} = \text{actual } I_{eff}/\text{design limit } I_{eff} \qquad (1.7)$$

Reliable design for EM was $S_{dc} \leq 1$, while any interconnect with $S_{dc} > 1$ had to be redesigned. However, from a process reliability perspective, EM degradation is inherently statistical. Also, not all the interconnect wires and vias are stressed to the design limit. This provides a way to statistically budget the electromigration failure rates over the various interconnect classes of a chip. As a result, this approach provides tremendous flexibility to the designers to exceed the limits in some critical paths to push performance without compromising reliability, as long as appropriate budgeting of the reliability is done in non-critical paths.

This method relies on an elegant method of tracking the failure in time (FIT) contribution per wire segment on a chip given by:

$$FIT_j = \frac{10^9}{LT} \ln 1 \quad \ln\left(S_{DC}^n\left(LT/T_{50,SH}\right)\right) \qquad (1.8)$$

where LT is the lifetime requirement, n is the current density exponent, S_{dc} is the severity ratio, $T_{50,\,SH}$ is the median time to failure with self heating, σ

is the log standard deviation parameter of the lognormal statistical model, and Φ is the standard normal cumulative distribution function. Summing up over all interconnect classes K and each wire per class n_k, a total chip failure rate can computed as:

$$FIT_{total} = \sum_{i=1}^{k} \sum_{j=1}^{n_k} FIT(i, j) \tag{1.9}$$

Critical paths can thus be pushed harder as long as the total chip failure rate thus computed is always below the required product EM failure rate specification.

1.8 Conclusions

As the technology keeps scaling according to Moore's Law, variability, power dissipation, leakage, and reliability become challenging. Key wear-out mechanisms related to transistors, i.e., gate dielectric time-dependent dielectric breakdown (TDDB), bias temperature instability (BTI), and hot carrier injection (HCI) were described in detail and models for each mechanism were outlined. As technology evolved from poly Si- and SiON-based gate dielectrics to metal gates and high-K gate dielectrics, the impacts on these reliability mechanisms and models were also highlighted.

BEOL interconnect technology is subject to a continual push to go to metal interconnects with lower resistivity and inter-metal dielectrics with lower dielectric constant values to reduce RC delays. Key wear-out mechanisms in the interconnect technology (electromigration, stress migration, and back end TDDB) were outlined. As the interconnect technology has evolved to replace Al with lower resistivity Cu interconnects and SiO2 based dielectrics with lower K or ultra low-K (ULK) dielectrics, the impacts on interconnect reliability and technology scaling have been explained.

Finally, the importance of product reliability models to fully enable performance without compromising reliability was discussed. The development of these models and budgeting reliability optimally across the different functional blocks form the backbone of a smart design that enables reliability to be designed in and not treated as an afterthought. An example of statistical electromigration budgeting (SEB) methodology that exploits the statistical nature of EM reliability was explained. It can selectively supersede fixed current density design rules for some interconnects, allowing increased chip performance while simultaneously quantifying chip-level EM reliability to directly assure design conformance to reliability requirements.

Acknowledgments

The authors would like to acknowledge the support of their team members and colleagues at Global Foundries through useful discussions that were instrumental in this work. They would like to thank Fab1 Integration in Dresden for providing the silicon hardware used for reliability measurements and the Technology Reliability Development Laboratory technicians and test engineers for test support and data acquisition for characterizing and modeling the reliability mechanisms described in this work.

References

Alam, M.A. 2003. A critical examination of the mechanics of dynamic NBTI for PMOSFETs. *Proceedings of International Electron Device Meeting*. IEEE, Allentown, PA, pp. 346–349.

Asenov, A. Brown, A.R., Davies, J.H., Kaya, S. and Slavcheva. G. 2003. "Simulation of intrinsic parameter fluctuations in decananometer and nanometer-scale MOSFETs," *IEEE Trans. Electron Devices*, 50, 1837.

Bernstein, K., Frank, D.J., Gattiker, A.E. et al. 2006. High-performance CMOS variability in the 65-nm regime and beyond. *IBM J. Res. Dev.*, 50, 433–449.

Bersuker, G., Heh, D., Young, C. et al. 2008. Breakdown in the metal–high-k gate stack: identifying the "weak link" in the multilayer dielectric. *Proceedings of International Electron Device Meeting*. IEEE, San Francisco, pp. 791–794.

Besser, P., Marathe, A., Zhao, L. et al. 2000. Optimizing the electromigration performance of copper interconnects. *Proceedings of International Electron Devices Meeting*. IEEE, San Francisco, pp. 119–122.

Borkar, S. 1999. Design challenges of technology scaling. *IEEE Micro.*, 171, 23–29.

Campbell A.N., Mikawa R.E., and Knorr D.B. 1993. Relationship between texture and electromigration lifetime in sputtered Al–1% Si thin films. *J. Electron. Mater.*, 22, 589–596.

Cartier, E., Linder, B.P., Narayanan, V. et al. 2006. Fundamental understanding and optimization of PBTI in nFETs with SiO$_2$/HfO$_2$ gate stack. *Proceedings of International Electron Devices Meeting*. IEEE, San Francisco, pp. 321–324.

Casse, M., Thevenod, L., Guillaumot, B. et al. 2006. Carrier transport in HfO$_2$/metal gate MOSFETs: physical insight into critical parameters. *IEEE Trans. Electron Dev.*, 53, 759–768.

Chakravarthi, S., Krishnan, A.T., Reddy, V. et al. 2004. A comprehensive framework for predictive modeling of negative bias temperature instability. *42nd International Reliability Physics Symposium*. IEEE, Phoenix, pp. 273–282.

Chen, F., Shinosky, M., Li, B. et al., 2009. Critical ultra low-k TDDB reliability issues for advanced CMOS technologies. *47th International Reliability Physics Symposium*. IEEE, Montreal, pp. 464–475.

Chen, G., Li, M.F., Ang, C.H. et al. 2002. Dynamic NBTI of p-MOS transistors and its impact on MOSFET scaling. *IEEE Electron Dev. Ltr.*, 23, 734–736.

Degraeve, R., Groeseneken, G., Bellens, R. et al. 1998. New insights in the relation between electron trap generation and the statistical properties of oxide break-down. *IEEE Trans. Electron. Dev.*, 45, 904–911.

Degraeve, R., Kaczer, B., Houssa, M. et al,. 1999. Analysis of high voltage TDDB measurements in Ta_2O_5/SiO_2 stack. *Proceedings of International Electron Device Meeting*. IEEE, Washington, pp. 327–330.

Denis, M., Bravaix, A., Huard, V. et al. 2006. Paradigm shift for NBTI characterization in ultra-scaled CMOS technologies. *Proceedings of 44th International Reliability Physics Symposium*. IEEE, San Jose, CA, pp. 735–736.

Frank, D.J., Dennard, R., Nowak, E., Solomon, P., Taur, Y., and Wong, H.-S.P. 2001. "Device scaling limits of Si MOSFETs and their application dependencies," *Proc. IEEE*, 89, 259–288.

Gambino, J.P., Johnson, C.L., Therrien, J.E. et al. 2006. Stress migration lifetime for Cu interconnects with CoWP-only cap. *IEEE Trans. Dev. Mater. Reliability*, 6, 197–202.

Ghai, T., Mistry, K., Packan, P. et al. 2000. Scaling challenges and device design requirements for high performance sub-50-nm gate length planar CMOS transistors. *Proceedings of Symposium on VLSI Circuits*. IEEE, Honolulu, pp. 174–175.

Grasser, T., Gos, W., Sverdlov, V. et al. 2007. The universality of NBTI relaxation and its implications for modeling and characterization. *Proceedings of 45th International Reliability Physics Symposium*. IEEE, Phoenix, pp. 268–280.

Gusev, E.P., Ed. 2006. *Defects in High-K Dielectrics*. Springer, New York.

Heh, D., Young, C.D., Brown, G.A. et al. 2006. Spatial distributions of trapping centers in HfO_2/SiO_2 gate stacks. *Appl. Phys. Ltr.*, 88, 152.

Ho, P.S., Leu, J., and Lee, W.W., Eds. 2002. *Low Dielectric Constant Materials for IC Applications*. Springer, New York.

Hu, C.K., Canaperi, D., Chen, S.T. et al. 2004. Effects of overlayers on electromigra-tion reliability improvement for Cu/low-*k* interconnects. *Proceedings of 42nd International Reliability Physics Symposium*. IEEE, Phoenix, pp. 222–228.

Hu, C.K., Gignac, L., Rosenberg, R. et al. 2002. Reduced electromigration of Cu wires by surface coating. *Appl. Phys. Ltr.*, 81, 178–184.

Hu, C.K., Gignac, L., Rosenberg, R., et al. 2003. Reduced Cu interface diffusion by CoWP surface coating. *Microelectron. Eng.*, 70, 406–411.

Huard, V., Denais, M., and Parthasarathy, C. 2006. "NBTI degradation: From physical mechanisms to modelling," *Microelectronics Reliability*, 46, 1–23.

Huff, H.R. and Gilmer, D.C., Eds. 2004. *High Dielectric Constant Materials: VLSI MOSFET Applications*. Springer, New York.

International Technology Roadmap for Semiconductors. Interconnect, 2007.

JEDEC. 2009. *Failure Mechanisms and Models for Semiconductor Devices*, Publication JEP122E. Solid State Technology Association, Arlington, VA.

Jeppson, K.O. and Svensson, C.M. 1977. Negative bias of MOS devices at high electric fields and degradation of MNOS devices. *J. Appl. Phys.*, 48, 2004–2014.

Kerber, A. and Cartier, E.A. 2009. Reliability challenges for CMOS technology qualifi-cations with hafnium oxide/titanium nitride gate stacks. *IEEE Trans. Dev. Mater. Reliability*, 9, 147–162.

Kerber, A., Cartier, E., Pantisano, L. et al. 2003. Origin of the threshold voltage insta-bility in SiO_2/HfO_2 dual layer gate dielectrics. *IEEE Electron. Dev. Ltr.*, 24, 87–89.

Kerber, A., Maitra K., Majumdar, A. et al. 2008. Characterization of fast relaxation during BTI stress in conventional and advanced CMOS devices with HfO$_2$/TiN gate stacks. *IEEE Trans. Electron. Dev.*, 55, 3175–3183.

Kitchin, J. 1995. Statistical electromigration budgeting for reliable design and verification in a 300-MHz microprocessor. *Proceedings of Symposium on VLSI Circuits.* IEEE, Kyoto, pp. 115–116.

Knorr, D.B. and Rodbell, K.P. 1996. The role of texture in the electromigration behavior of pure aluminum lines. *J. Appl. Phys.*, 79, 2409–2417.

Krishnan, A. and Nicollian, P. 2007. Analytic extension of the cell-based oxide breakdown model to full percolation and it implications. *Proceedings of 44th International Reliability Physics Symposium.* IEEE, Phoenix, pp. 232–239.

Lo, S.-H., Buchanan, D.A., Taur, Y., and Wang, W. 1997. "Quantum-mechanical modeling of electron tunneling current from the inversion layer of ultra-thin-oxide nMOSFETs," *IEEE Electron Device Lett.*, 18 (May), 209.

Maex, K., Baklanov, M.R., Shamiryan, D. et al. 2003. Low dielectric constant materials for microelectronics. *J. Appl. Phys.*, 93, 8793–8841.

Mahapatra, S., Alam, M.A., Kumar, P.B. et al. 2004. Mechanism of negative bias temperature instability in CMOS devices: degradation, recovery and impact of nitrogen. *Proceedings of International Electron Device Meeting.* IEEE, San Francisco, pp. 105–108.

Massoud, H.Z., Poindexter, E.H., and Helms, C.R., Eds. 1996. *The Physics and Chemistry of SiO$_2$ and the Si–SiO$_2$ Interface*, Vol. 3. Electrochemical Society, Pennington, NJ, pp. 3–14.

Mitani, Y. 2004. "Influence of nitrogen in ultra-thin SiON on negative bias temperature instability," *International Electron Device Meeting Tech. Dig.*

Nigam, T. 2008. Pulse-stress dependence of NBTI degradation and its impact on circuits. *IEEE Trans. Dev. Mater. Reliability*, 8, 72–78.

Nigam, T., Kerber, A., and Peumans, P. 2009. Accurate model for time-dependent dielectric breakdown of high-K metal gate stacks. *Proceedings of 47th International Reliability Physics Symposium.* IEEE, Montreal, pp. 523–530.

Nigam, T., Parameshwaran, B., and Krause, G. 2009. Accurate product lifetime predictions based on device-level measurements. *Proceedings of 47th International Reliability Physics Symposium.* IEEE, Montreal, pp. 634–639.

Ogawa, E.T., Kim, J., Haase, G.S. et al. 2003. Leakage, breakdown and TDDB characteristics of porous low-*k* silica-based interconnect dielectrics. *Proceedings of 41st International Reliability Physics Symposium.* IEEE, Dallas, pp. 166–172.

Okada, K., Ota, H., Nabatame, T. et al. 2007. Dielectric breakdown in high-K gate dielectric: mechanism and lifetime assessment. *Proceedings of 45th International Reliability Physics Symposium.* IEEE, Phoenix, pp. 36–43.

Robertson, J. 2006. High dielectric constant gate oxides for metal oxide Si transistors. *Rep. Progr. Phys.*, 69, 327–396.

Rodbell, K.P., Hurd, J.L., and DeHaven, P.W. 1996. Blanket and local crystallographic texture determination in layered Al metallization. *Proceedings of Materials Research Society Symposium.* Materials Research Society, Pittsburgh, pp. 617–626.

Rzepka, S., Lepper, M., Böttcher, M. et al. 2004. Convergence and interaction of BEOL and BE reliability methodology. *Proceedings of 42nd International Reliability Physics Symposium.* IEEE, Phoenix, pp. 49–60.

Schroder, D. and Babcock, J.F. 2003. Negative bias temperature instability: road to cross in deep submicron silicon semiconductor manufacturing. *J. Appl. Phys.*, 94, 1–18.

Stathis, J.H. 1999. "Percolation models for gate oxide breakdown," *Journal of Applied Physics*, 86, (10), 5757–5766.

Stine, B.E., Boning, D.S., and Chung, J.E. 1997. Analysis and decomposition of spatial variation in integrated circuit processes and devices. *IEEE Trans. Semiconductor Mfg.*, 10, 24–41.

Strong, A.W., Wu, W.Y., Vollertsen, R.P., Rauch, S.E., III, and Sullivan, T.D. 2009. *Reliability Wearout Mechanisms in Advanced CMOS Technologies*. John Wiley & Sons, Hoboken, NJ.

Suñé, J. 2001. New physics-based analytic approach to thin-oxide breakdown statistics. *IEEE Electron. Dev. Ltr.*, 22, 296–298.

Taur, Y., Buchanan, D.A., Chen, W. et al. 1997. CMOS scaling into the nanometer regime. *Proc. IEEE*, 85, 486–504.

Taur, Y., and Ning, T.H. 1998. *Fundamentals of Modern VLSI Devices*. Cambridge University Press, New York.

Tökei, Zs., Sutcliffe, V., Demuynck, S. et al. 2004. Impact of barrier–dielectric interface quality on reliability of Cu porous-low-k interconnects. *Proceedings of 42nd International Reliability Physics Symposium*. IEEE, Phoenix, pp. 326–332.

Torii, K., Shimamoto, Y., Saito, S. et al. 2003. Effect of interfacial oxide on electron mobility in metal insulator semiconductor field effect transistors with Al_2O_3 gate dielectrics. *Microelectron. Eng.*, 65, 447–453.

Toyoda, H., Kawanoue, T., Hasunuma, M. et al. 1994. Improvement in the electromigration lifetime using hyper-textured aluminum formed on amorphous tantalum–aluminum underlayer. *Proceedings of 32nd International Reliability Physics Symposium*. IEEE, San Jose, pp. 178–184.

Vaidya, S. and Sinha, A.K. 1981. Effect of texture and grain structure on electromigration in Al–0.5% Cu thin films. *Thin Solid Films*, 75, 253–259.

Weir, B.E., Silverman, P.J., Monroe, D. et al. 1997. Ultrathin gate dielectrics: they break down, but do they fail? *Proceedings of International Electron Devices Meeting*. IEEE, Washington, pp. 73–76.

Wilk, G.D., Wallace, R.M., and Anthony, J.M. 2001. High-K gate dielectrics: current status and materials properties considerations. *J. Appl. Phys.*, 80, 5243–5275.

2

Scaling and Radiation Effects in Silicon Transistors

Allan H. Johnston and Leif Z. Scheick

CONTENTS

ABSTRACT   This chapter reviews the effects of space radiation on microelectronic devices, and shows how device scaling has affected their vulnerability to radiation. The most dramatic impact has been on single-particle effects, particularly those caused by galactic cosmic rays. New effects can occur at the microscopic level, including permanent damage from the interaction of a single energetic particle when it strikes the gate of a metal oxide semiconductor (MOS) transistor. Scaling also impacts device complexity, not only because advanced circuits use such large number of transistors but also because of the complex processing steps needed to design individual transistors that can function with such small feature sizes and meet the requirements necessary for competitive performance. Many of the practical problems that need to be addressed in space are related to the complex design methods used to fabricate scaled devices.

2.1 Introduction

2.1.1 Overview of Space Environments

Space missions have to withstand large integrated fluxes of electrons and protons from trapped radiation belts, along with small fluxes of highly energetic cosmic rays. (Stassinopoulos and Raymond, 1988) We can divide the effects of radiation on semiconductor devices into two categories: (1) integrated damage effects from large numbers of particles that accrue throughout a mission, and (2) the effect of a *single* energetic particle when it strikes a particularly sensitive region. The latter effect has become far more complicated for highly scaled devices, partly because very small amounts of energy are involved in their electronic behavior.

In addition to the particles described above, we must be concerned also with solar flares (more properly, coronal mass ejections) that produce bursts of radiation over periods of several hours. Solar flares produce significant fluxes of high-energy protons, along with much smaller fluxes of high-energy particles that are similar to galactic cosmic rays (GCRs) but have lower energy.

Although displacement damage can also occur, the dominant effect of the integral particle flux is usually ionization damage that produces electron-hole

TABLE 2.1

Total Dose Requirements for Representative Space Missions

Mission	Description	Duration (years)	Total Dose [krad (Si)]
Earth Orbiter (300 km)	Space Station	10	5
Earth Orbiter (705 km)	Weather satellite	5	20
Earth Orbiter (36,000 km)	Geosynchronous orbit	5	50
Mars Rover	Martian surface	5	10
Jupiter moon exploration	Deep space mission	10	100

pairs in semiconductors and insulators. The total dose from ionization is usually specified in rads of a specific material where a rad is defined as the absorption of 100 erg/g of a specific material. Representative total dose levels for several space missions are shown in Table 2.1. The requirements for earth-orbiting missions are higher than those for deep space missions because of the Earth's trapped radiation belts. Even more intense radiation belts are encountered near Jupiter, which is important for planetary exploration missions that travel in its vicinity.

A more thorough treatment of radiation environments is contained in Barth (2003). The characteristics of the orbit, the amount of shielding used, and the effect of radiation from solar flares all affect the requirements for specific missions.

The effects of single particles are quite different and must be treated statistically when we discuss their impact. This is usually done by means of a cross section representing the average number of particles that must strike a device before a response is observed. The number of electron-hole pairs along the path length of an energetic particle is proportional to linear energy transfer (LET), with units of megaelectron volts per square centimeter per milligram (MeV-cm^2/mg). The number of particles decreases rapidly with increasing LET. Figure 2.1 shows the distribution of galactic cosmic rays for three different space environments. The values correspond to solar minimum conditions when GCR flux is highest. Two earth orbits are shown with different altitudes and inclinations.

Although linear energy transfer is the normal unit used to discuss single-particle effects, it is more intuitive to use a different unit (picocoulomb per micrometer or pC/µm) in discussions of scaling. For silicon, the conversion factor between pC/µm and MeV-cm^2/mg is 0.0106, i.e., an LET of 10 MeV-cm^2/mg is approximately 0.1 pC/µm. Thus, for single-event effects (SEEs) we are concerned with linear energy transfer values between about 0.01 and 1 pC/µm.

2.1.2 Device Scaling Aspects Important for Radiation Effects

Some of the changes imposed on device design by scaling (Dennard et al., 1974) exert large effects on radiation sensitivity. For simplicity, we will discuss

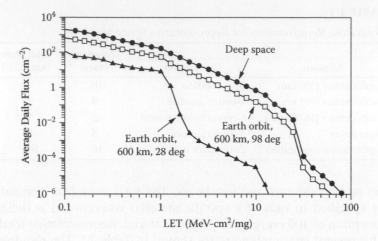

FIGURE 2.1
Linear energy transfer abundance for galactic cosmic rays (GCRs). The values are about four times larger during the solar quiet period (4 years) compared to the 7-year active period of the 11-year solar cycle. For the low inclination orbit, GCR particles with high atomic numbers are eliminated because of the effect of the Earth's magnetic field.

only moderate performance devices (Davari et al., 1995), avoiding the high-power, high-performance branch (where special cooling is required) and the low-power devices in which reduced noise margins restrict performance in high-reliability space systems.

For single-event effects, the most important features of scaling are reduction in cell area and reduced noise margin as power supply voltages are reduced. Recent scaling trends show that the power supply voltage levels off around 1 V (Frank et al., 2001), making the cell area the dominant factor in the future. Trends for gate length are shown in Figure 2.2, along with the approximate numbers of transistors per chip.

Different factors are important for total dose effects. The reduced gate oxide thickness exerts a powerful effect, eliminating charge trapping mechanisms in the core transistors that change gate threshold voltage. For scaled CMOS, total dose damage in isolation oxides that do not scale as dramatically as gate oxides has become the dominant mechanism.

2.2 Basic Mechanisms Producing Single-Event Effects

2.2.1 Particle Interactions

A heavy ion produces an intense charge track along its length when it interacts with a target material. This is illustrated in Figure 2.3 in which a particle

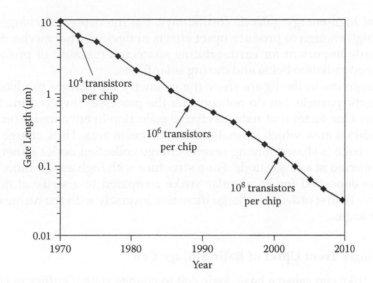

FIGURE 2.2
Gate length trends for scaled CMOS.

strikes a reverse-biased p-n junction. The track density with an initial diameter of about 0.1 µm is high enough to collapse the p-n junction. Ambipolar diffusion causes it to expand to about 1 µm approximately 1 ns after the particle strike.

Another mechanism can occur when a recoil atom from a high-energy proton interaction creates an ionization track. That process is shown at the right of Figure 2.3. The range of the recoil atom is much shorter than the

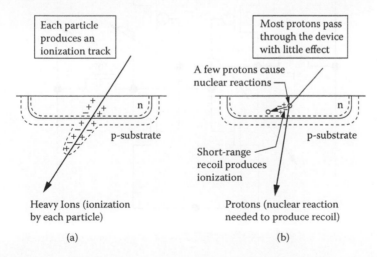

FIGURE 2.3
Direct and indirect processes for charge generation by energetic particles.

ranges of high-energy galactic cosmic rays, but the deposited charge may still be high enough to produce upset effects in devices. That mechanism is particularly important for earth-orbiting spacecraft (because of protons in the trapped radiation belts) and during solar flares.

The diagrams in the figure show the geometry for charges produced by an energetic particle, but do not consider the processes involved in charge collection. One factor that reduces charge collection in advanced structures is the reduced area, which is smaller than the beam area. Thus, charge from a single strike is shared among several charge collection nodes rather than being collected at a single node. For a structure with high aspect ratio, more charge is deposited for an angular strike compared to a strike at normal incidence. To first order, the charge increases inversely with the cosine of the incident angle.

2.2.2 Single-Event Upset of Basic Storage Cell

An ion strike can cause a basic logic cell to change state if sufficient charge is collected (Pickel and Blandford, 1980). The minimum charge to cause an upset is called the critical charge Q_c. The basic SRAM cell in Figure 2.4 can be upset from logic strikes in either of two locations. The critical charge is different for each of the internal transistors.

To produce an upset, charge collected from an ion strike must exceed the critical charge. An ion that produces just enough charge to cause an upset

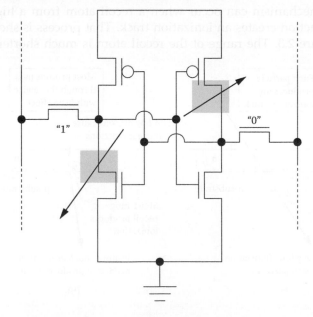

FIGURE 2.4
Upset of SRAM cell can take place in two different regions.

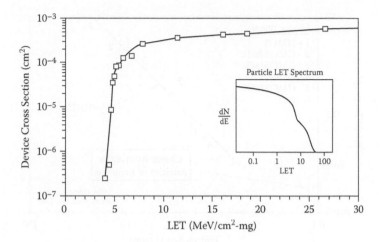

FIGURE 2.5
Typical upset cross section from an accelerator experiment. The device is continually evaluated during exposure to monitor the number of upset events that occur.

must strike very near the physical geometry of the transistor. Ions with more energy can pass through the structure further away and still cause upset because of charge diffusion. This causes the sensitive area to increase for ions with higher LET values, leading to an overall increase in cross section.

Single-event effects are usually evaluated by placing a device in the beam of a high-energy accelerator and measuring the cross section for upsets (observed when the beam is "on") for various ions. A typical measurement of cross section is shown in Figure 2.5. Although the cross section increases with LET, the abundance of GCR particles decreases with LET, as shown in the inset (as well as in Figure 2.1). The upset rate depends on the convolution of the cross section and ion abundance, as well as on the effective charge collection volume. The convolution process must consider the net effects of particles arriving at all angles. Computer programs (such as CREME96) can calculate the expected upset rate (Adams, 1986).

2.3 Effects of Device Scaling on Single-Event Upset

2.3.1 Basic Issues

Initial work on device scaling in 1982 showed that the critical charge decreased rapidly with feature size, as shown in Figure 2.6 (Petersen, et al., 1982). If we extrapolate this earlier result to the deep submicron feature sizes that are widely used today, it is apparent that the upset rate will increase by many orders of magnitude, making it an extremely severe problem. Other

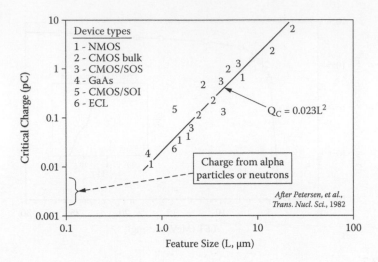

FIGURE 2.6
Decrease of critical charge with feature size (From Petersen, E.L. et al. 1982. *IEEE Trans. Nucl. Sci.*, 40, 1820–1830. With permission.) The charges from alpha particles or neutrons, important for modern devices, were added to the original figure.

factors that tend to reduce the severity of the effect, including charge collection efficiency must be considered in evaluating overall SEE sensitivity.

Another important factor is that modern devices must cope with the natural radiation environment that includes alpha particles from trace amounts of radionuclides in materials and packages and terrestrial neutrons produced by cosmic rays in the upper atmosphere (Baumann, 2005). The requirement to bound the responses of scaled devices to those effects places a lower bound on critical charge, as shown in the figure.

Although critical charge decreases with feature size, the cell dimensions have become so small that the size of an individual transistor is now much smaller than the effective diameter of the particle track. Consequently, only a fraction of the deposited charge is collected at an individual node. Figure 2.7 shows the results of a computer simulation for a 256-Mb dynamic random-access memory (DRAM) cell, showing how the collected charge decreases with junction area (Shin, 1999). This counteracts the decrease in critical charge from scaling, reducing the upset sensitivity compared to earlier predictions.

2.3.2 Changes in Device Technology

Scaling is affected by device technology as well as feature size. One important change during the last 10 years has been the emergence of silicon-on-insulator (SOI) complementary metal-oxide semiconductor (CMOS) technology. The underlying substrate is isolated by a special silicon dioxide structure instead of a reverse-biased p-n junction.

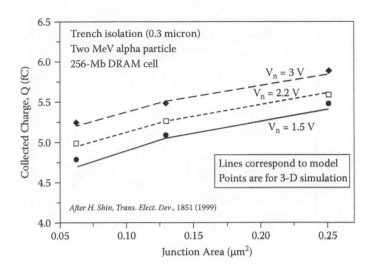

FIGURE 2.7
Collected charge from a 2-MeV alpha particle strike for DRAM cells with different junction areas. (From Shin, H. 1999. *IEEE Trans. Elect. Dev.*, 46, 1850–1857. With permission.)

A cross-sectional view of an SOI structure is shown on the right side of Figure 2.8 along with a junction-isolated CMOS structure at the left (the junction-isolated structure uses a low-resistivity substrate called the "epi" process). For the SOI structure, the oxide layer prevents charge collection from the substrate; the effective charge collection depth is more than a factor of ten lower than that of the junction-isolated structure (fabricated on an epi substrate to reduce charge collection depth).

At first glance, the SOI structure appears to be far better from the standpoint of single event upset (SEU) sensitivity. However, the compact structure shown in the figure has no electrical ties to the wells. Consequently the parasitic bipolar transistor between the closely spaced source and drain can amplify the (reduced) charge, eliminating much of the inherent advantage of the SOI structure. It is possible to reduce the bipolar effect by adding special body contacts, but that increases the cell area, in direct conflict with the purpose of scaling.

Figure 2.9 illustrates the importance of bipolar gain on the SEU rate. It shows the number of errors in an advanced commercial process from alpha

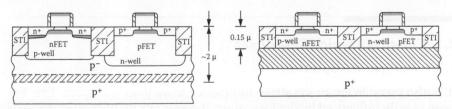

FIGURE 2.8
Comparison of charge collection depth of bulk and SOI CMOS structures.

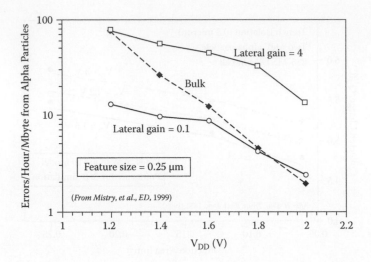

FIGURE 2.9
Influence of lateral bipolar gain in SOI NMOS transistor on alpha particle upset rate. (Adapted from Mistry, K.R. et al. 1999. *IEEE Trans. Electron. Dev.*, 46, 2201–2209. With permission.) Data for a bulk device with the same feature size is included for comparison.

particles generated by small amounts of radioactive impurities within the package and other local material (Mistry et al., 1999). Two SOI structures with different values of lateral gain are compared to a similar device fabricated in a bulk process. The high lateral gain of one of the structures makes the error rate much higher than that of a bulk device with the same feature size.

Figure 2.10 shows the practical impact of this limitation by comparing the heavy-ion cross sections of several commercial microprocessors from one manufacturer (Irom and Farmanesh, 2005). The threshold LET is nearly the same for several different generations, and is not significantly different for the SOI version because of the parasitic bipolar effect. However, the saturation cross section of the SOI version is about one order of magnitude below that of a bulk processor with nearly the same feature size because lateral charge collection cannot occur from the substrate in the SOI process.

2.3 Complex Device Responses

2.3.1 Multiple-Bit Upset

It is also possible for an ion strike to produce more than one upset. This has been heavily influenced by device scaling; the area of a highly scaled device is now much lower than the diameter of the ionization wake from a heavy ion, and a single ion strike can produce hundreds of upsets in modern devices.

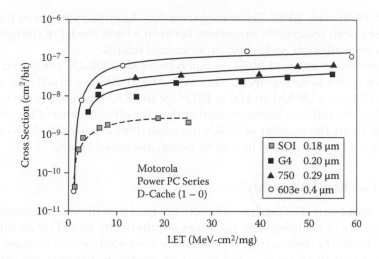

FIGURE 2.10
Effect of scaling on SEU response of internal cache registers in advanced microprocessors. (From Irom, F. and Farmanesh, F. 2005. *IEEE Trans. Nucl. Sci.*, 52, 1524–1529. With permission.)

Mulitple-bit upset (MBU) is important because it can interfere with error correction methods that are often used to mitigate SEU effects. The number of MBUs depends on the cell area relative to the effective area of the particle beam, which is about 0.1 μm. It also depends on the angle of incidence. An example is shown in Figure 2.11 for a 512-Mb SRAM with a feature size of

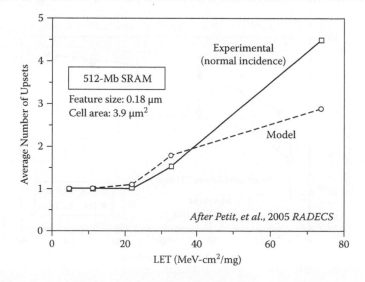

FIGURE 2.11
Average number of multiple-bit upsets versus LET for a 512-Mb SRAM. (From Petit. S. et al. 2006. *IEEE Trans. Nucl. Sci.*, 53, 1787–1793. With permission.)

0.13 μm (Petit et al., 2006). The average number of upsets increases as the LET increases, with reasonable agreement between a basic model of charge sharing between adjacent nodes and experimental results.

The average number of MBUs is much larger for a DRAM compared to an SRAM with the same feature size because of the difference in cell size, which is about 8 F^2 for a DRAM and 50 to 100 F^2 for an SRAM, where F is the feature size. Once the cell area becomes smaller than the effective area of the particle beam (0.1 μm), the number of MBUs increases dramatically. The MBU rate is also affected by ion strikes in control nodes, discussed below.

2.3.2 Functional Interrupts

An ion strike can also produce a functional interrupt in which internal control registers or complex state machines are altered by an SEU at an internal node. This has become a critical problem for advanced devices because it can place a device in an unexpected operating condition. For example, such an event can force a memory into a special test mode, allowing large parts of the memory to be corrupted.

Single-event functional interrupt (SEFI) events are particularly troublesome in microprocessors, where they can produce a wide range of malfunctions. It is difficult to establish effective ways to identify and correct these effects, and in some cases normal operation can be restored only by temporarily removing power, and re-initializing the processor.

Figure 2.12 compares the cross sections of "hangs" in two modern power PC processors. One is fabricated with the SOI process, resulting in a smaller

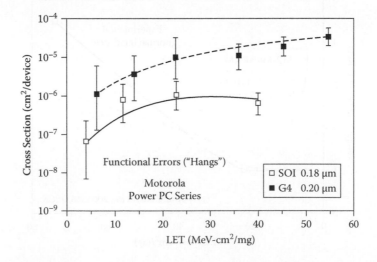

FIGURE 2.12
Cross section for "hang" events in a modern microprocessor. The feature size of the processor is shown in the inset.

cross section compared to a similar processor fabricated with bulk CMOS from the same manufacturer. The cross section for hangs is several orders of magnitude lower than the cross section for registers because only a small number of internal storage elements are involved.

One of the main problems with SEFIs is detecting them during testing. The cross section is low, and a certain amount of latency is involved between the time that the event takes place and the specific time in the diagnostics when the event is detected. Careful planning is essential for evaluating SEFI responses. For a complex circuit, it is possible to miss some events, particularly if the active operation during the test does not exercise all of the basic functional operations.

2.3.3 Trends for Scaled Devices

Several factors affect the way a highly scaled device responds to fundamental SEE effects. A decrease in switching energy introduces the possibility that secondary particles from interactions with highly energetic cosmic rays can also cause upsets. Those interactions can occur in other device regions including metallization and contacts (Dodds et al., 2009). This increases the upset rate and may also cause conditions that require examination of an entire spacecraft assembly when SEE rates are analyzed. Regions within device structures with higher atomic numbers (such as contacts and metallization) may also be sources of secondary particles.

Angular effects can also change for highly scaled devices, creating a scenario in which the upset rate is significantly higher for ion strikes at an angle compared to the effects of angle strikes from simple rectangular geometries. Recent work has discussed such mechanisms in considerable detail (Reed et al., 2002); they affect radiation testing and calculations of the upset rate.

The most important trend concerns device complexity. Although the designer of a complex device may understand the nuances of design in sufficient detail, in most cases complex devices are purchased from mainstream suppliers and end users have little or no knowledge of device design. Special methods may be installed to improve yield, for example, incorporating internal error detection and correction; storing hard codes to remove "bad" regions, replacing them with extra regions that are deliberately built into the device to allow such post-fabrication corrections; and introducing highly complex machines that correct for subtle differences in timing or other characteristics. Although such steps may operate seamlessly in a normal environment, internal ion strikes can alter conditions and subject parts to conditions in which architecture is no longer effective.

Successful operation in space requires the ability to detect such conditions and initiate corrective measures. This is a challenging problem that cannot always be solved by radiation testing. It is likely that the architectural problems observed for microprocessors and complex memories over the last 10 years will increase for those components and may appear in various guises in any circuit that contains large numbers of internal transistors with feature sizes below 250 nm (circuits with more than 100 million transistors are common today).

2.4 Single-Event Transients

2.4.1 Digital Circuits

An ion strike can cause internal transients in digital circuits that may affect logic chains, even when no storage elements are present. This effect has become very important as feature sizes have been reduced. In order for a transient to propagate through several logic elements, it must have a minimum amplitude and pulse width. Figure 2.13 shows test results of special structures designed to allow the effective pulse width to be measured in a particle beam test (Dodd et al., 2004). The critical pulse width decreases rapidly as the feature size is reduced, in reasonable agreement with computer modeling results. The maximum operating frequency of the technology node is also shown in the figure.

One reason that internal transients are important is that they can interfere with logic designed to suppress (or eliminate) SEU effects. This is an active topic for current research and is particularly important for the design of hardened circuits. The details depend on processing, architecture, and internal noise margin. The sensitivity increases for lower core logic voltage and increasing operating frequency.

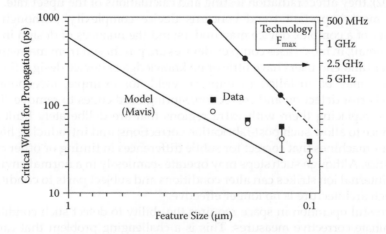

FIGURE 2.13
Minimum width of transient from a logic cell to allow it to propagate to other cells within the logic chain. (From Dodd, P.E. et al. 2004. *IEEE Trans. Nucl. Sci.*, 50, 3278–3284. With permission.)

2.4.2 Analog and Mixed-Signal Circuits

Spurious charges from heavy ions or protons can also be amplified by the gain stages contained within operational amplifiers, comparators, or mixed-signal circuits. If the charges are large enough, they can produce transients at the output of a circuit. In most cases, the earlier gain stages are saturated by the internal transient, producing an output signal that is initially unaffected by feedback. In general, mechanisms at the input are more important than those involved at later stages.

Peak transient voltages in op-amps and comparators often approach the supply rail levels and may persist for 10 µs or longer. In most cases, they produce only short "glitches" with minimal consequences. However, comparators in asynchronous applications (such as power supply monitoring circuits) can produce spurious signals that may produce false indications, including the possibility of shutting down power systems.

Transients in voltage regulators are more serious, because they may be large enough to exceed the voltage ratings of other circuits that are powered by a regulator affected by a transient. Figure 2.14 shows an example of transients from a point-to-point regulator. The initial value of the transient depends on the load resistance and load capacitance, and is larger for reduced loads.

Scaling provides little advantage for analog circuits, partly because most commercial circuits are designed to withstand circuit voltages of 40 V or more (this ensures wide applicability). However, analog circuits have become far more complex, increasing the difficulty of evaluating SET responses. Scaling in digital circuits has an indirect effect on power circuits; high-efficiency low-dropout regulators are often required to provide high currents with

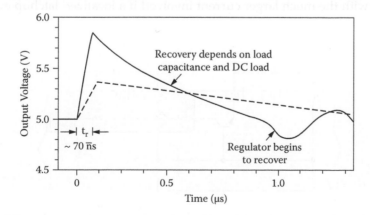

FIGURE 2.14
SET waveforms from voltage regulator. The dashed line corresponds to a circuit application with a load capacitance three times larger than the waveform with the solid line. The amplitude is also affected by the value of the unregulated input voltage.

voltages between 1 and 2.5 V for processors and other complex digital parts with less tolerance for voltage transients.

2.5 Catastrophic Single-Event Effects

2.5.1 Latchup

One of the most important catastrophic single-event effects is latchup (Soliman and Nichols, 1983). It can take place in most CMOS circuits because of four-layer structures formed by parasitic bipolar regions that are present because reverse-biased junctions are used to isolate the n- and p-channel transistors required in CMOS technology. Figure 2.15 shows the parasitic structure for a p-well CMOS process (similar parasitic junctions are also present in twin-well CMOS devices but not in SOIs). A heavy ion strike in the p-well or the n-substrate can trigger the four-layer structure, producing a low-resistance path between the power supply and ground that can cause circuit failure.

After such a structure is triggered, it remains in a highly conducting state until power is removed. Large-scale circuits typically have many potential latchup paths. The current through such a path may be only a small fraction of the normal power supply current. However, it is concentrated in a highly localized region and may cause damage even at currents on the order of 10 mA.

Failures from latchup tend to be of two types. Breaks or voids in metallization may occur, particularly in first or second level metallization where the cross section is small, consistent with low currents during normal operation, but not with the much larger current involved if a localized latchup event is

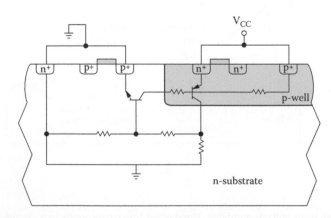

FIGURE 2.15
Parasitic four-layer structure in bulk CMOS.

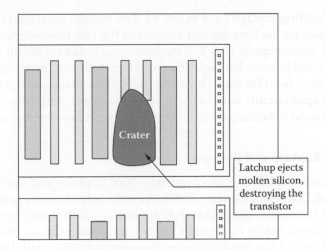

FIGURE 2.16
Void within silicon region of a CMOS circuit after failure from latchup.

triggered (Becker et al., 2002). Voids occur when molten metal is ejected, leaving only a small conducting region. The second type of catastrophic behavior occurs when the current is high enough to melt the silicon. Figure 2.16 shows an example of this effect; a large crater formed after silicon ejection.

Although manufacturers establish design and layout rules that prevent electrically induced latchup, they do not consider radiation-induced latchup (except for special radiation-hardened circuits). Energetic ions in space can inject currents in interior regions that can trigger latchup, and thousands of different latchup paths may exist within a complex large-scale integrated circuit.

Not all CMOS circuits are sensitive to latchup. Most high-performance microprocessors are free from latchup, partly because they are fabricated on low-resistivity substrates (with thin epitaxial layers), as shown in Figure 2.8. Although using such a structure does not guarantee latchup immunity, the low resistivity increases the current required to initiate latchup.

Many circuits are fabricated without using low resistivity substrates, partly because of lower cost, but also because substrates with higher resistivity reduce noise coupling. Most memory circuits and analog-to-digital converters use high-resistivity substrates and are typically very sensitive to latchup from radiation.

The effect of scaling on latchup is more difficult to analyze compared to other radiation issues, partly because the properties of the four-layer structure that affect it depend on the topology of a specific circuit element, including the locations of ties to the well and substrate regions. Latchup remains a critical problem for devices on bulk substrates and will likely continue to be a difficult issue.

One aspect of scaling that affects latchup is the power supply voltage. In order to sustain latchup, the voltage across the four-layer structure must exceed the holding voltage (~0.4 to 0.8 V). The voltage must also be capable of forward biasing the base emitter regions of the two transistors within the structure. If the voltage is <0.8 V, it becomes much more difficult to reach a sustainable condition for latchup. Thus, devices with low operating voltages may be immune from latchup. However, most low voltage technologies have input and output circuits with higher applied voltages, and those structures can still be prone to latchup even for processes with low core voltage

2.5.2 Gate Rupture and Burnout

Two other catastrophic SEE mechanisms—gate rupture and burnout—can occur in CMOS devices (Sexton, 2003). We will only consider gate rupture; single-event burnout is important only for devices with voltage applications above the normal range of digital CMOS logic. Gate rupture occurs when the charge from an ion produces a temporary conduction path through the gate that can produce filamentary damage. A great deal of concern surrounded gate rupture when the gate oxide thickness of mainstream CMOS was reduced below 150 nm, partly because the electric field was much higher than those in devices with larger feature sizes.

Work done in the mid-1990s showed that the critical field for gate rupture was actually higher for oxides in scaled devices, partly because of improved oxide quality. Furthermore, the breakdown characteristics changed from abrupt failure seen in thick oxides to small leakage currents (soft breakdown). Later work on hafnium oxide showed that the critical field for advanced gate materials was also above the normal electric fields for those structures as well (Massengill, 2001). Thus, gate rupture does not appear to be a significant problem for CMOS, with the possible exception of individual transistors at input or output stages that are connected to external nodes with higher voltages.

2.6 Total Dose Damage Fundamentals

2.6.1 Charge Generation and Trapping

When discussing total dose damage, the usual assumption is that such damage is caused by the integrated effect of large numbers of individual particles—typically electrons and/or protons—that interact with a device. We will initially discuss total dose damage from that perspective.

Microdose, which is caused by a large excess carrier density along the path of only one particle, will be discussed later. Microdose from a single galactic cosmic ray or proton can produce enough damage in a highly scaled device to cause it to fail. The end result is failure of individual transistors (in random locations), with a statistical probability similar to that of the SEE effects

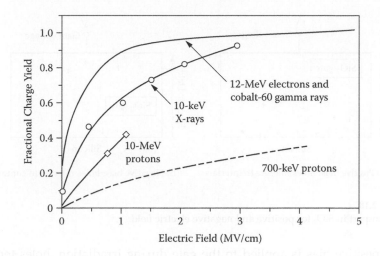

FIGURE 2.17
Charge yield versus electric field for various particles. (From Oldham, T.R. and McLean, B. 2003. *IEEE Trans. Nucl. Sci.*, 50, 483–489. With permission.)

discussed earlier. This contrasts with integrated total dose damage in which all the individual transistors within a circuit are damaged in roughly the same way.

Energy loss from ionization produces electron-hole pairs in SiO_2. Even though silicon-dioxide is a very high quality insulator, excess carriers generated by radiation will move after they are created. Large differences exist in the transport properties of the two types of carriers. Electron mobility is more than four orders of magnitude greater than hole mobility. Consequently, the electrons are quickly swept out, leaving relatively immobile holes that gradually migrate to interface regions where they can be trapped. If the trap density is large enough, it can affect device characteristics.

Although the electron mobility is very high, it is possible for some of the electron-hole pairs to recombine immediately after they are created if the oxide has only a weak electric field. The fraction of holes that recombine depends on the field and the density of electron-hole pairs that are initially created. Figure 2.17 shows how charge yield depends on electric field for various particles (Oldham and McLean, 2003).

Less recombination takes place for high-energy electrons and cobalt-60 gamma rays than for other particles, but at low fields more than one-half of the charge recombines before transport. In interpreting this curve, note that the electric field on the gate of a CMOS process with ¼ micron feature size (with a core voltage of 1.8 V) is about 0.5 MV/cm. Higher fields may be present at input and output terminals. As devices are scaled to smaller dimensions, the fields tend to increase because of the difficulty of meeting noise margin requirements if the core voltage is below 1 V.

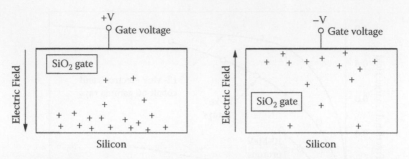

(a) Positive bias: holes migrate to interface (b) Negative bias: holes migrate to contact

FIGURE 2.18
Hole transport in SiO_2 for positive and negative electric field.

If a positive bias is applied to the gate during irradiation, holes tend to migrate toward the silicon–silicon dioxide interface, where some of them are trapped. For negative applied bias, the holes migrate toward the contact, where little or no charge trapping takes place, as illustrated in Figure 2.18. Thus, the sign and magnitude of the applied field are both important in establishing the number of holes trapped after transport.

2.6.2 Effects of Charge Traps on Gate Threshold Voltage

Charges trapped at the interface between a gate and channel region will alter the gate threshold voltage of an MOS transistor. Hole traps are formed near the interface immediately after transport. Interface traps take time to develop after the charge is transported. The present model for interface traps requires the initial formation of hole traps before the interface traps are formed (Oldham, 1999).

The effect of charge trapping on the subthreshold characteristics of an n-channel MOSFET after irradiation at high dose rate is shown in Figure 2.19 (Winokur et al., 1984). The hole traps produce a negative shift in gate threshold voltage. Interface traps change the slope of the subthreshold current and also produce a threshold shift in n-channel transistors that is opposite in sign to that of the hole traps. The dashed parameters show the regions of the subthreshold current relationship involved in decomposing the net result into contributions from hole and interface traps. In this example, the gate oxide thickness is 50 nm, a state-of-the-art technology in the early 1980s (note that the drain source voltage was 10 V, far higher than the drain source voltage of modern devices).

The hole traps gradually recover after irradiation at a high dose rate, with a time dependence that is approximately logarithmic. At room temperature, the time constant associated with the recovery is on the order of 10^6 s. Consequently, the interface trap component dominates in most space applications because the average dose rate is very low. The net threshold shift is positive (for n-channel devices), rather than negative, under low dose rate conditions.

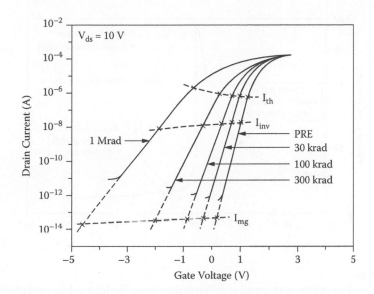

FIGURE 2.19
Effect of total dose on subthreshold characteristics of n-channel MOSFET. (From Winokur, P. et al. 1984. *IEEE Trans. Nucl. Sci.*, 31, 1453–1460. With permission.)

The interplay between the two types of traps has been taken into account by allowing laboratory radiation tests to be done at high dose rates and then annealing the damage at a nominal temperature of 100°C. That temperature is high enough to remove the hole traps, leaving only the interface traps, approximating the net effect of a device in space.

It is less straightforward to evaluate effects in CMOS circuits. Two factors must be considered. First, different failure mechanisms may occur when irradiation is performed at high and low dose rates. Second, the partially compensating effect of the two trap components may result in a region where a circuit will function at a much higher radiation level, leading to possible overestimation of radiation hardness. Figure 2.20 illustrates this for a microprocessor produced with relatively thick oxides in the mid-1980s (Johnston, 1994). The experimental failure level first increases and then decreases as the dose rate used for testing is reduced.

2.7 Total Dose Effects in Scaled Devices

2.7.1 Effects in Gate Oxides

To achieve the expected benefit from scaling, it is necessary to reduce the gate oxide thickness (Davari, Dennard and Shahidi, 1995). The effective volume

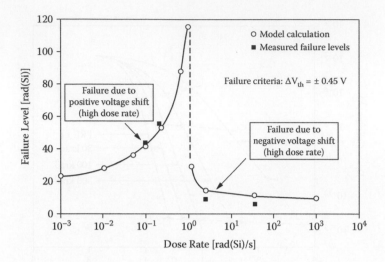

FIGURE 2.20
Failure level of microprocessor tested at different dose rates. The high failure level predicted at about 1 rad (Si)/s is due to partial compensation of the two types of traps. (From Johnston, A.H. 1994, *IEEE Trans. Nucl. Sci.*, 31, 1427–1433. With permission.)

for production of excess charge within an oxide then becomes smaller. If we consider the relationship of device parameters and doping levels, the maximum change in threshold voltage ΔV_T is related to oxide thickness t_{ox} by

$$\Delta V_T = -\frac{q}{\varepsilon_{ox}}t_{ox}^2 D \tag{2.1}$$

where D is the total dose. This is the limiting value of ΔV_T; it does not consider initial recombination or the fraction of transported holes that are trapped at the interface. The important point is the dependence of threshold voltage shifts on the square of the oxide thickness. For an older process (or a modern power MOSFET) with a gate oxide thickness of 70 nm, the threshold shift at 10 krad (SiO_2) is –2.3 V. For a scaled device with a gate oxide thickness of 10 nm, the threshold shift at that same radiation level is only 0.05 V.

As devices are scaled further, the gate becomes so thin that charges are no longer trapped because of tunneling, further reducing the importance of charge trapping in scaled devices, as illustrated in Figure 2.21. Thus, we can ignore conventional total dose effects due to charge trapping in very thin oxides.

Although gate oxide shifts are much lower for the core voltages of scaled CMOS devices, higher oxide thicknesses are still required in some cases. Examples include input and output circuits; charge pump circuits for flash memory, as well as the associated logic for erasing and writing; and CMOS circuits with larger feature size that are still used because of lower

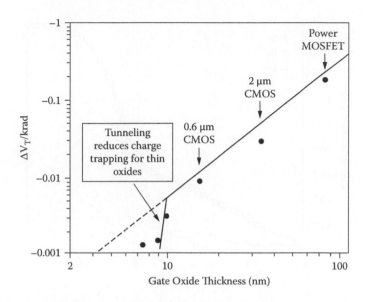

FIGURE 2.21
Dependence of maximum gate threshold shift (normalized to total dose) on gate oxide thickness. The sharp reduction for gate thicknesses below 10 nm is due to tunneling.

fabrication cost and compatibility with older logic voltages. Thus, changes in gate threshold voltage may still be important.

2.7.2 Microdose Damage

The importance of microdose damage in advanced devices was first studied by Oldham et al. (1993). They analyzed the distribution of the total number of charges produced by heavy ions in older dynamic memories and showed that a sufficient number were collected over the gate length to produce a change in threshold voltage large enough to increase leakage current in the pass transistor. However, they predicted that changes in DRAM technology would decrease the importance of the effect because of the reduction in the volume of the gate due to scaling.

Recent work has concentrated on flash memories in which one of the internal gates must withstand voltages of about 18 V for programming and erasure. Very few electrons are stored on advanced flash devices and this increases the importance of microdose effects on that basic technology.

Understanding of an internal architecture makes it possible to measure the distribution of individual cell threshold voltages for an entire flash memory array. Figure 2.22 shows how the cumulative threshold distribution of a NAND memory array with 90-nm feature size is affected by irradiation with heavy ions (Cellere et al., 2006). Larger changes take place for ions with higher LET values.

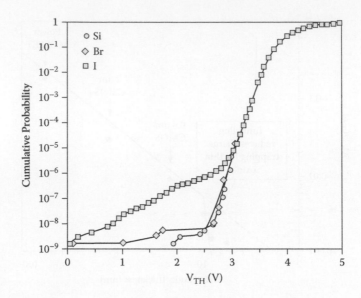

FIGURE 2.22

Change in threshold voltage distribution for storage cell transistors in advanced flash memory after irradiation with heavy ions. (From Cellere, C. et al. 2006. *IEEE Trans. Nucl. Sci.*, 53, 1813–1818. With permission.)

Although the fluence used for these experiments was many orders of magnitude higher than the fluences encountered in space, the results show that some of the storage cells within a large memory array will be affected by radiation, affecting the overall performance. Smaller threshold voltage changes were observed on similar devices with a feature size of 120 nm, leading to the conclusion that the charge loss effect becomes worse with scaling.

2.7.3 Effects in Isolation Regions

As discussed earlier, the reduced oxide thickness of modern devices has reduced the importance of gate oxide threshold shifts to the point that other mechanisms dominate total dose damage. For highly scaled CMOS devices, total dose damage in the isolation regions that surround individual transistors has become the dominant mechanism. This is illustrated for a device with shallow trench isolation in Figure 2.23 (note that the trench completely surrounds the transistor).

If sufficient charge trapping takes place at the interface between the isolation oxide and an n-dope well (or body) region, an inversion layer that produces a lateral leakage path between the drain and source region of an NMOS transistor will occur. The inversion charge increases with doping level, which generally increases with decreasing feature size.

One of the first studies of this effect was done by Shaneyfelt et al. (1998) using a CMOS technology with a core voltage of 5 V. For those devices, inversion

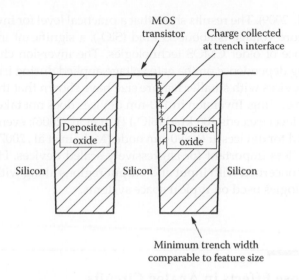

FIGURE 2.23
Location of drain source leakage path of NMOS transistor with trench isolation.

became important at a total dose level of about 10 krad (SiO_2), a very low value. One reason was the relatively low doping level of the silicon body region. As scaling has progressed, doping levels have increased, which in turn caused an increase in the radiation level for inversion. Figure 2.24 shows more recent results for a transistor with a gate length of 180 nm and a core voltage of 1.8 V

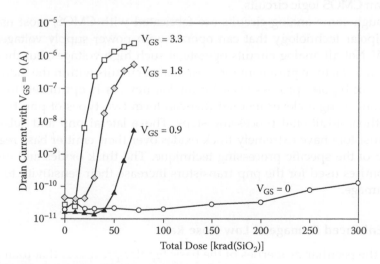

FIGURE 2.24
Leakage current from inversion in the trench isolation region for NMOS transistor with gate length of 180 nm. (From Johnston, A.H. et al. 2009. *IEEE Trans. Nucl. Sci.*, 56, 1535–1539. With permission.)

(Johnston et al., 2009). The results show that a practical level for inversion with the nominal core voltage is about 35 krad (SiO_2), a significant increase over the performance of older CMOS technologies. The inversion characteristics exhibit a strong dependence on the gate voltage applied during irradiation.

Studies of devices with smaller feature size have shown that the total dose increases with scaling. Inversion in 130-nm devices does not take place until the total dose level exceeds 100 krad (SiO_2) (Jun et al., 2006); even higher levels are required for devices at the 90-nm node (McLain et al., 2007). Thus, this mechanism is less important for aggressively scaled devices. However, we must remain concerned with input and output regions along with the mix of circuit technologies used on typical space systems.

2.8 Total Dose Effects in Analog Circuits

2.8.1 Basic Considerations

Scaling has far less benefit for analog integrated circuits for two reasons. First, analog circuit contain only a few elements (typically fewer than 200 transistors); and second, the area required for the output circuitry is so large that little direct benefit is obtained by scaling the other regions of the circuit. In addition, most analog circuits are required to operate at much higher voltages than CMOS logic circuits.

Although some analog circuits are fabricated with CMOS, most use an older bipolar technology that can operate with power supply voltages up to ±18 V. Not all analog circuits operate at such high voltages, but this has become a de facto requirement for most analog circuits within the industry.

The basic bipolar process used for analog devices is optimized for npn transistors, using a clever method that can form two types of pnn transistors without additional processing steps. Those lateral pnp and substrate pnp transistors have extremely thick oxides over their emitter base regions because of the specific processing technique. The thick oxides and design compromises used for the pnp transistors increase their sensitivity to total dose damage.

2.8.2 Enhanced Damage at Low Dose Rate

One of the peculiar properties of the basic bipolar process is that total dose damage in the transistors is actually greater under the low dose rate conditions of space than under the high dose-rate conditions common during laboratory evaluations. That phenomenon was first reported in 1991 (Enlow et al., 1991), and has been studied extensively during the last two decades.

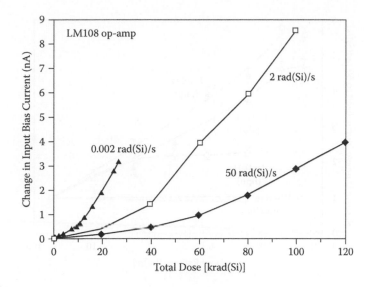

FIGURE 2.25
Degradation of input bias current in operational amplifier illustrating increased damage from
irradiation at low dose rate.

Figure 2.25 illustrates this effect for an operational amplifier in which the
input bias current (which is essentially proportional to the inverse of the
gain of the input transistor) caused damage. Changes in that parameter are
much higher at low dose rates due to the increased damage that occurs. One
way to deal with the enhanced damage at low dose rate (ELDR) effect is to
simply irradiate at very low dose rates. However, this is difficult in practice
because 6 months or more may be required to complete the irradiations.

Another potential approach is to perform the tests at a convenient high
dose rate, and apply de-rating factors to account for the additional damage.
Although that can be done for individual transistors, it is less effective for
analog circuits because multiple types of transistors are involved in the
internal circuit blocks.

Figure 2.26 shows the dependence of damage on dose rate for the three
basic types of transistors used in typical analog circuits (Johnston et al.,
1994). The damage is normalized to the value measured at a dose rate of 50
rad (Si)/s, so that the vertical axis is effectively an enhanced damage value.
First, note that typical discrete transistors do not exhibit enhanced damage;
annealing actually causes less damage at low dose rates. Second, for npn
transistors that have much thinner oxides over their emitter-base regions,
the dependence of damage on dose rate levels around 1 rad (Si)/s. Third, the
dependence of damage on dose rate is not only much larger for the two types
of pnp transistors, but it continues to increase until the dose rate is below
approximately 0.005 rad (Si)/s.

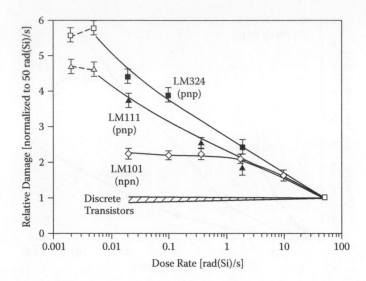

FIGURE 2.26
Comparison of enhanced damage in three types of internal transistors found in typical bipolar integrated circuits. (From Johnston, A.H. et al. 1994. *IEEE Trans. Nucl. Sci.*, 31, 2427–2436. With permission.)

These results have several consequences. In order to directly simulate the effects of damage in the low dose rate space environment, radiation tests must be carried out at a dose rate of 0.005 rad (Si)/s, requiring about 1 month to achieve a total dose of 12 krad (Si). From a practical standpoint, this is very inconvenient. Several attempts have been made to see whether alternative testing methods can be adopted but they have not worked satisfactorily.

We also have to be concerned about the dependence of damage on dose rate for the different types of transistors, all of which are used in a typical analog circuit. Because the ratio of the damage does not "track" with dose rate, it is not possible to add de-rating factors to high dose rate data; different circuit failure mechanisms may occur at high and low dose rates, depending on the specific circuit design.

2.8.3 Selecting Analog Device Technologies for Space

Several manufacturers produce analog circuits that are specifically hardened for space applications. In the past, the qualification approach has not included the ELDR phenomenon, but most manufacturers now include that effect as well. Another approach is to use analog circuits designed with lower voltage processes compared to the older commercial processes with high voltage ratings. The lower voltage rating allows devices to be made with thinner oxides and higher doping levels that increase radiation hardness and decrease the sensitivity to the ELDR effect.

CMOS analog circuits are also available, but are not necessarily improvements over their bipolar counterparts. The work of Pease (2003) should be consulted for more details on bipolar total dose damage.

2.9 Compound Semiconductors

Most transistors used in compound semiconductors have very different characteristics compared to mainstream CMOS devices fabricated with silicon. III–V semiconductors have poor surface properties. Advanced devices rely on heterostructures to isolate active regions from the surface and form new types of systems such as modulation-doped field-effect transistors in which the active region occurs well below the surface. The inherent differences in materials and device structures allow most compound semiconductor devices to withstand total dose levels of 1 Mrad or more. Thus, the main issue for most compound semiconductors is single-event upset.

Except for silicon–germanium devices that are similar to mainstream CMOS and high-speed bipolar devices in silicon, the integration density of compound semiconductor devices is extremely low. This makes SEE effects much easier to handle by reducing the overall cross section and eliminating the difficult problem of complexity that has become so important for mainstream CMOS devices.

For circuits, extremely high speed is often the justification for selecting a compound semiconductor technology (it has to compete with silicon). High speed reduces critical charge, and most high-speed circuits are highly susceptible to single-event upset. However, in most cases, the circuits are very small compared to the types of circuits that we consider for large-scale silicon, resulting in relatively small cross sections and low overall error rates that can be accommodated in subsystem design. A recent reference (Johnston, 2010) provides more detail on radiation effects in compound semiconductors.

2.10 Summary and Future Trends

The dramatic decrease in feature sizes of mainstream semiconductor devices during the last 40 years has also affected their sensitivity to space radiation. Initially the main problems faced by devices in space were changes in gain (bipolar technology) or threshold voltage (MOS technology) from the integrated effects of ionization damage. New mechanisms associated with the dense ionization track from a single energetic particle were first observed in the late 1970s, and have become the dominant technical problems for scaled devices in space.

The basic reason for this is the reduction in switching energy that has accompanied device scaling, which is now far lower than the energy deposited by galactic cosmic rays. Initially the spurious pulses from those particles affected only individual storage elements, but a plethora of new effects have emerged as devices have continued to advance, including multiple-bit upset, and functional interrupt.

Despite the increased concern for single-event effects, the overall upset rate is usually low enough to be accommodated at the system or circuit level using fault protection or error correcting techniques. For example, memory error rates in deep space are on the order of 10^{-7} per day, which is low enough to be handled by basic error detection and correction methods. The key to applying scaled devices is to have sufficient understanding of how they are affected and ensure that complex responses (such as functional interrupt or multiple-bit errors) and the higher event rate expected during an intense solar flare are accounted for. Permanent errors from microdose add an additional complication, but can be handled by fault-tolerant designs.

Based on the experiences of the past 10 years, the most important impact of scaling will be on device complexity rather than differences in basic response mechanisms. This will increase the cost of characterizing radiation responses and will also exact a penalty at the system design level.

Fortunately, total dose effects decrease with scaling, particularly for devices with feature sizes below 100 nm. The main concern for that environment surrounds interface circuits that must withstand higher voltages rather than effects on highly scaled logic devices. Special technologies such as flash memories must operate at much higher voltages during programming and erase cycles than mainstream logic devices.

References

Adams, J., Jr. 1986. Cosmic ray effects on microelectronics. Part IV. Naval Research Laboratory Memorandum Report 5901.

Barth, J., Dyer, C.S., and Stassinopoulos, E.G. 2003. Space, terrestrial and atmospheric radiation environments. *IEEE Trans. Nucl. Sci.*, 50, 466–482.

Baumann, R. 2005. Radiation-induced soft errors in advanced semiconductor technologies. *IEEE Trans. Dev. Mater. Reliability*, 5, 305–316.

Becker, H.N., Miyahira, T.F., and Johnston, A.H. 2002. Latent damage in CMOS devices from single-event latchup. *IEEE Trans. Nucl. Sci*, 49, 3009–3025.

Cellere, G., Paccagnella, A., Visconti, A. et al. 2006. Single event effects in NAND flash memory arrays. *IEEE Trans. Nucl. Sci.*, 53, 1813–1818.

Davari, B., Dennard, R.H., and Shahidi, G.G. 1995. CMOS scaling for high performance and low power: the next ten years. *Proc. IEEE*, 83, 595–606.

Dennard, R.H., Gaensslen, F.H., Rideout, V.L. et al. 1974. Design of ion-implanted MOSFETs with very small dimensions. *IEEE J. Solid-State Circuits*, 9, 256–268.

Dodd, P.E., Shaneyfelt, M.R., Felix, J.A. et al. 2004. Production and propagation of single-event transients in high-speed digital logic ICs. *IEEE Trans. Nucl. Sci.*, 50, 3278–3284.

Dodds, N.A., Reed, R.A., Mendenhall, M.H. et al. 2009. Charge generation by secondary particles from nuclear reactions in BEOL materials. *IEEE Trans. Nucl. Sci.*, 56, 3172–3179.

Enlow, E.W., Pease, R.L., Combs, W. et al. 1991. Response of advanced bipolar processes to ionizing radiation. *IEEE Trans. Nucl. Sci.*, 38, 1535–1539.

Frank, D.J., Dennard, R.H., Nowak, E. et al., 2001. Device scaling limits of Si MOSFETs and their application dependencies., *Proc. IEEE*, 89, 259–288.

Irom, F. and Farmanesh, F. 2005. Single-event upset in highly scaled commercial silicon-on-insulator PowerPC microprocessors. *IEEE Trans. Nucl. Sci.*, 52, 1524–1529.

Johnston, A.H. 1994. Super recovery of total dose damage in MOS devices. *IEEE Trans. Nucl. Sci.*, 31, 1427–1433.

Johnston, A.H., Swift, G.M., and Rax, B.G. 1994. Total dose effects in conventional bipolar transistors and integrated circuits. *IEEE Trans. Nucl. Sci.*, 41, 2427–2436.

Johnston, A.H., Swimm, R., and Miyahira, T.F. 2009. Total dose effects in CMOS trench isolation regions. *IEEE Trans. Nucl. Sci.*, 56, 1535–1539.

Johnston, A.H. 2010. *Reliability and Radiation Effects in Semiconductors*, World Scientific, Singapore.

Jun, B., Diestelhorst, R.M., Bellini, M. et al. 2006. Temperature dependence of off-state leakage current in x-ray irradiated 130-nm CMOS devices. *IEEE Trans. Nucl. Sci.*, 53, 3203–3209.

McLain, M., Barnaby, H.J., Holbert, K.E. et al. 2007. Enhanced TID susceptibility in sub-100 nm bulk CMOS I/O transistors and circuits. *IEEE Trans. Nucl. Sci.*, 54, 2210–2217.

Massengill, L.W., Choi, B.K., Fleetwood, D.M. et al. 2001. Heavy ion-induced breakdown in ultrathin gate oxides and high-k dielectrics. *IEEE Trans. Nucl. Sci.*, 48, 1904–1912.

Mistry, K.R., Sleight, J.W., Grula, G. et al. 1999. Parasitic bipolar gain reduction and the optimization of 0.25-μm partially depleted SOI MOSFET's. *IEEE Trans. Electron. Dev.*, 46, 2201–2209.

Oldham, T.R., Bennett, K.W., Beaucour, J. et al. 1993. Total dose failures in advanced electronics from heavy ions. *IEEE Trans. Nucl. Sci.*, 40, 1820–1830.

Oldham, T.R. 1999. *Ionizing Radiation Effects in MOS Oxides*, World Scientific, Singapore.

Oldham T. and McLean, B. 2003. Total dose effects in MOS oxides and devices. *IEEE Trans. Nucl. Sci.*, 50, 483–489.

Pease, R. 2003. Total dose effects in bipolar devices and circuits. *IEEE Trans. Nucl. Sci.*, 50, 539–551.

Petersen, E.L., Shapiro, P., Adams, J.H. et al. 1982. Calculation of cosmic ray-induced soft upset and scaling in VLSI devices. *IEEE Trans. Nucl. Sci.*, 29, 2055–2063.

Petit, S., David, J.P., Falguere, D. et al. 2006. Memories response to MBU and semi-empirical approach to SEE rate calculation. *IEEE Trans. Nucl. Sci.*, 53, 1787–1793.

Pickel, J. and Blandford, T. 1980. Cosmic ray-induced errors in MOS devices. *IEEE Trans. Nucl. Sci.*, 27, 1005–1015.

Reed, R.A., Marshall, P.W., Kim, H.S. et al. 2002. Evidence for angular effects in proton-induced singe-event upsets. *IEEE Trans. Nucl. Sci.*, 49, 3038–3044.

Sexton, F. 2003. Destructive single-event effects in semiconductor devices and ICs. *IEEE Trans. Nucl. Sci.*, 5, 603–621.

Shaneyfelt M., Dodd, P.E., Draper, B.L. et al. 1998. Challenges in hardening technologies using shallow trench isolation. *IEEE Trans. Nucl. Sci.*, 45, 2584–2592.

Shin, H. 1999. Modeling of alpha particle soft error rate in a DRAM. *IEEE Trans. Elect. Dev.*, 46, 1850–1857.

Soliman, K. and Nichols, D. 1983. Latchup in CMOS from heavy ions. *IEEE Trans. Nucl. Sci.*, 30, 4514–4519.

Stassinopoulos, E. and Raymond, J. 1986. The space radiation environment for electronics. *Proc. IEEE*, 76, 1423–1442.

Winokur, P., Schwank, J.R., McWhorter, P.J. et al. 1984. Correlating the radiation response of MOS capacitors and transistors. *IEEE Trans. Nucl. Sci.*, 31, 1453–1460.

3

Hewlett-Packard's MEMS Technology: Thermal Inkjet Printing and Beyond

James Stasiak, Susan Richards, and Paul Benning

CONTENTS

ABSTRACT In 1984, HP introduced the ThinkJet—the first low-cost, mass-produced thermal inkjet printer. Providing a reasonable alternative to noisy dot matrix printers, ThinkJet set the stage for subsequent generations of HP thermal inkjet (TIJ) technology. With each new generation, HP TIJ products provided new standards for print quality, color, and an unprecedented cost-to-performance ratio. Regarded as the first and most successful commercial MEMS technology, the development of HP's TIJ printheads required multi-disciplinary innovation in microfluidics, bulk and surface micromachining, large-scale integration of electronics, packaging, and high volume MEMS manufacturing. HP's current TIJ printhead products combine Pentium-class addressing circuitry, high voltage mixed-signal driver electronics, dense electrical interconnects, and up to 3900 high-precision microfluidic drop ejectors—all on a single silicon chip. In this chapter, we will provide a brief history of HP's TIJ technology and then discuss how the unique capabilities required to advance the state of the art of TIJ printing technology now provide platforms for the development of new MEMS devices and systems.

3.1 Introduction

Over the past two decades, the field of micro-electrical-mechanical systems (MEMS) has evolved from an area of fundamental research largely restricted to academic, government, and industrial laboratories to a mature technology that is enabling a wide variety of practical commercial applications. The list of successful MEMS-based commercial successes continues to grow and includes pressure sensors, accelerometers and gyroscopes, Texas Instruments' digital light processing (DLP) technology, microfluidic devices and bio-chips, and numerous applications in communications (Peeters, 1997; Ko, 2007).

However, the most notable commercial success has been the MEMS-based thermal inkjet (TIJ) printhead technology introduced by Hewlett-Packard in 1984. In this chapter the evolution of this technology is discussed from a MEMS perspective, highlighting the key technical accomplishments, and providing some examples of extending HP's MEMS design, processing, and manufacturing capabilities to develop new MEMS technologies and products.

3.2 Inkjet Printing Technology

Inkjet printing is a technology that uses small drops of ink to form an image on a medium such as paper. The two types of inkjet technologies are continuous flow and drop-on-demand. Continuous flow inkjet printing uses electrostatic acceleration and deflection to select drops from a constant flow of ink to form the image. Drop-on-demand (DOD) inkjet printing relies on an impulsive force to physically eject a single droplet of ink through an orifice or nozzle to form an image.

Although different physical processes may be used to eject droplets on demand (mechanical, thermal, electrostatic, and acoustic), the most common processes are mechanical and thermal. In mechanically driven inkjet processes, an impulsive force is delivered using the rapid displacement of a piezoelectric-coupled membrane to eject the droplet. The TIJ process relies on the use of a rapid, high energy heat pulse to vaporize a thin layer of ink that forms a bubble that expels the droplet. In Sections 3.2.1 and 3.2.2, the architecture of thermal inkjet devices and the droplet ejection process are discussed in more detail.

3.2.1 Thermal Inkjet Device Architecture

A cross section of an idealized thermal inkjet print cartridge is shown in Figure 3.1. In this example, the print cartridge integrates a local ink supply and the ink delivery components. These components (that generally are internal or external to the print cartridge) regulate ink pressure and supply ink to the printhead. The printhead is the MEMS device that provides the electro-mechanical functionality needed to eject ink onto the print medium.

The TIJ printhead architecture consists of a series of tiny ink firing chambers, each of which has a thin-film resistor on its ceiling that is aligned directly above a nozzle (Figure 3.2a). These architectural components form the inkjet drop ejector structure. Ink flows into the firing chamber as a result of capillary forces that draw it from the ink supply through the standpipe and around the ink-feed edge. Depending upon the printhead design, the ink-feed edge is either on a slot in the substrate or along the substrate edge. Figure 3.2b is a photograph of a typical firing chamber. The orifice plate has been removed to show the resistor and flow features.

Note that the current generation of thermal inkjet ejector architectures is considerably more sophisticated with highly engineered fluidic elements and flow features. In HP's most recent products, the microfluidics, thin-film resistors, and the CMOS control and addressing circuitry are fully integrated onto a common substrate using state-of-the-art semiconductor and MEMS bulk and surface micromachining processes.

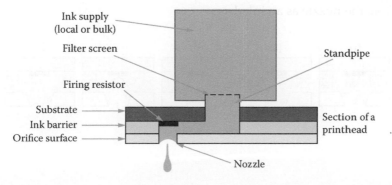

FIGURE 3.1
(See color insert.) Cross section of thermal inkjet print cartridge.

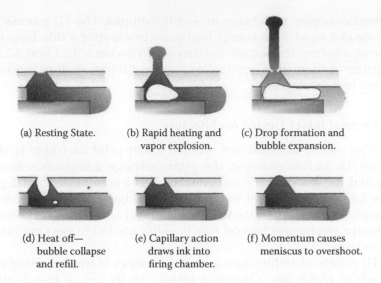

(a) Resting State.   (b) Rapid heating and   (c) Drop formation and
                                vapor explosion.   bubble expansion.

(d) Heat off—   (e) Capillary action   (f) Momentum causes
    bubble collapse   draws ink into   meniscus to overshoot.
    and refill.   firing chamber.

FIGURE 3.2
(a) Cross section of thermal inkjet firing chamber. (b) Photograph of early generation TIJ firing
chamber with orifice plate removed.

3.2.2 Thermal Inkjet Drop Ejection

The TIJ drop ejection process is illustrated in Figure 3.3. In Step 1, at the
resting state, the surface tension of the ink's meniscus is balanced by the
backpressure inside the print cartridge to prevent ink from leaking out of
the printhead. In Step 2, a short energy pulse is applied to the firing thin-
film resistor, resulting in rapid heating of the resistor surface. The fluid ink
at the thin-film resistor surface is quickly super-heated and forms small ink
vapor bubbles—a process known as nucleation. In Step 3, the nucleated bubble expands and acts to force the small volume of ink in the firing chamber
through the orifice. In Steps 4 and 5, the volume of ink is ejected and breaks
free from the nozzle as an ink drop.

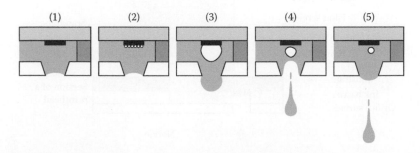

(1)   (2)   (3)   (4)   (5)

FIGURE 3.3
(See color insert.) Thermal inkjet drop ejection process.

With the ink drop ejected, the bubble collapses as it cools and condenses. The firing chamber refills due to the collapse of the bubble and capillary and meniscus forces that draw more ink through the delivery plumbing.

3.3 HP's Thermal Inkjet Printing History

In 1984, HP's calculator division introduced the first TIJ printer and print cartridge. The ThinkJet printer and cartridge were designed for use with personal computers and in small battery powered systems. The print cartridge used a glass substrate (instead of silicon), a nickel orifice plate, and a rubber bladder as the ink supply container. The next generation printers and printheads were designed for desktop applications.

The DeskJet designs became increasingly more sophisticated and capable. The glass substrates were replaced with silicon and new methods of fabricating the orifice plates and nozzles enabled higher print resolutions and performance. The development of new black and color inks paralleled the evolution of the printers and printheads to provide products that produced prints that rivaled photographic realism and approached the archival properties of silver halide. Inkjet ink formulation is a sophisticated science and is an HP core competency. HP's chemistry expertise, system reliability engineering, and chemical industry partnerships are key elements that have built its competitive advantage in color printing.

Over the past decade, HP engineers have focused on ever-increasing integration of the printhead functions to provide higher performance, print quality, and lower cost. A timeline of HP's TIJ technology development is shown in Figure 3.4.

By leveraging and adapting advances in thin film semiconductor processing and driving progress in silicon bulk and surface micromachining, and semiconductor packaging, it has been possible to integrate both the power electronics and per-nozzle interconnects to enable higher throughput and color resolution without increasing cost or size. The development of compact firing chamber architectures enabled smaller ejected drop volumes, higher nozzle packing densities, and higher throughput. The smaller drops provided more colors per dot, lighter tones, and photo quality printing on a wide variety of media. Smaller drops also required less firing energy per drop for increased frequency. Higher firing frequency along with higher nozzle packing density and integrating more nozzles on a single head enabled higher throughput printers.

In HP's latest generation of printhead technology, ink chambers, fluidic features, and nozzles are formed monolithically using photo-imageable thick-film polymers. All-lithographic patterning and new advances in silicon surface and bulk micromachining techniques eliminate the need for

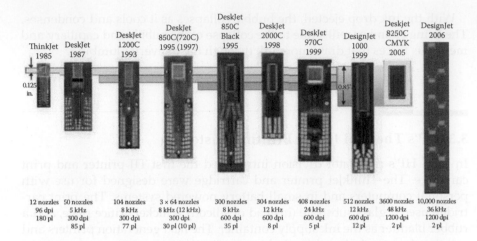

FIGURE 3.4
(See color insert.) Evolution of Hewlett-Packard's thermal Inkjet technology.

mechanical alignment processes and result in fluidic features that have the precision and integration scale of semiconductor devices and circuits. Figure 3.5 shows examples of lithographically defined fluidic features.

HP's new scalable print technology (SPT) provides a new paradigm in TIJ printhead manufacturing. SPT-based printheads combine Pentium-class addressing circuitry, high-voltage mixed-signal driver electronics, dense electrical interconnects, and up to 3900 high-precision microfluidic devices. These systems operate at high temperatures in the presence of aggressive inks and are expected to last more than 5 years.

Over the past 25 years, HP has continually pushed the limits of TIJ printhead performance. Moving from twelve nozzles in 1984 to thousands of nozzles in the current generation of products, HP roughly doubles the number of

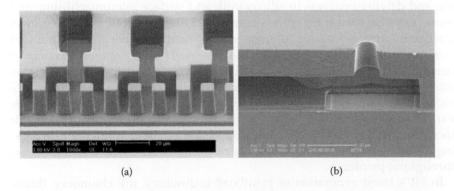

(a)   (b)

FIGURE 3.5
(a) Photomicrographs of TIJ fluidic features. (b) TIJ firing chamber fabricated using thick-film photo-imageable polymer materials.

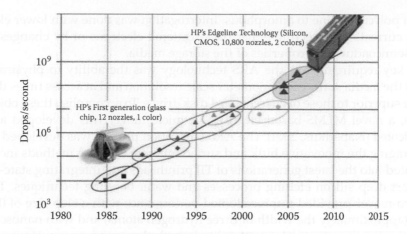

FIGURE 3.6
(See color insert.) HP's variation on Moore's law.

drops per second ejected from a printhead every 18 months, resulting in an analog of the semiconductor industry's Moore's law illustrated in Figure 3.6. The drop-per-second value is calculated as the product of the drop rate and the number of nozzles on a single printhead.

3.4 From TIJ to Atomic Resolution Storage and Beyond

As HP's TIJ printhead technology has evolved, new processing, fabrication, and manufacturing methods have been required. Furthermore, the extreme operating conditions of the TIJ ejection process required the development of new robust materials and innovative packaging technologies. By addressing these demands, HP established a unique expertise in integrating MEMS, microfluidics, novel materials, and semiconductor devices and circuitry. Over the past decade, HP has been leveraging these new capabilities and expertise to enable the invention and development of new technologies.

3.4.1 Atomic Resolution Storage

In the late 1990's, HP launched a program to develop a new, high-density probe-based memory technology called atomic resolution storage (ARS). Borrowing from advances in atomic probe microscopy, ARS was a thumbnail-size device with storage densities greater than one terabit per square inch (Toigo, 2000). The system used a dense array of electron field emitters to bombard the surface of a novel phase change material. The electrons locally modified the medium to reversibly change its state (in a bit-wise fashion) from polycrystalline to amorphous. Interrogation was done with lower electron currents by detecting either back-scattered electrons or by changes in the semiconducting properties of the storage media.

A key requirement of the ARS technology was the ability to physically scan the media surface at nanometer scale resolution and at access times that were superior to those of current hard disk drives. To accomplish these objectives, a novel MEMS-based electrostatic micro-mover was developed and fabricated (Naberhuis, 2001). The ARS fabrication process was developed by leveraging the innovative bulk and surface micromachining methods incorporated into the latest generations of TIJ printheads and integrating state-of-the-art deep silicon etching processes and wafer bonding techniques. The micro-mover provided unprecedented performance with a step size of 0.17 nm (approximately the width of three hydrogen atoms) and with nanoscale positioning accuracy and repeatability in both the x- and y-axes. Figure 3.7 shows the micro-mover plate and high-aspect-ratio plate flexures.

In addition to the revolutionary micro-mover device, two classes of high-precision MEMS-based electron emitters and focusing optics were also developed as part of the ARS program. Spindt tip emitters and integrated, multi-level electron optics were fabricated with characteristic feature sizes below 400 nm. A combination of computational modeling

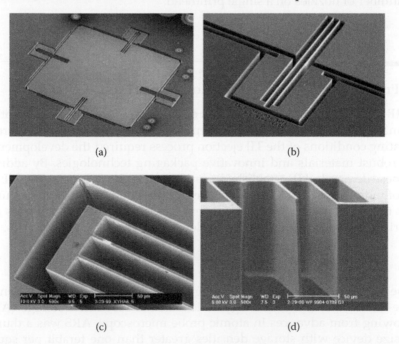

(a)   (b)

(c)   (d)

FIGURE 3.7
(a) Photograph of HP's MEMS micro-mover plate. (b)–(d) SEM micrographs of high-aspect-ratio flexures.

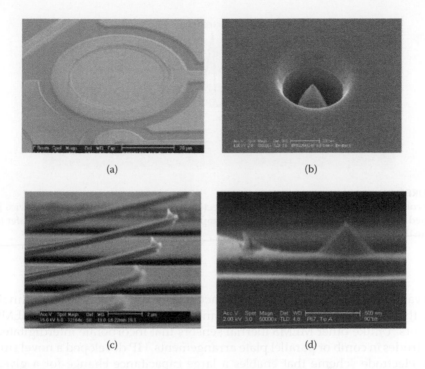

(a)

(b)

(c)

(d)

FIGURE 3.8
ARS emitters. (a) Metal insulator-metal flat emitter. (b) Single-level spindt tip emitter (multi-layer electron optics not shown). (c) and (d) MEMS cantilevers and expanded view of mechanical tip at end of one cantilever.

leveraged from HP's TIJ capability and empirical results demonstrated the ability of the devices to focus electron beams into spots with diameters smaller than 35 nm.

A second class of electron emitters with flat architectures and metal-insulator metal stacks provided breakthrough processing and integration potential but showed lower focusing performance. An alternative read–write approach that did not depend on electronic phase-change material was pursued and resulted in the demonstration of MEMS cantilever arrays with integrated nanomechanical tips. This approach used massive arrays of scanning probes to alter and probe the surface of a mechanical read–write media surface. Figure 3.8 illustrates both emitter types and the MEMS cantilevers.

3.4.2 MEMS Accelerometers and Gyroscopes

More recently, HP adapted the MEMS-based electrostatic micro-mover developed for ARS into a new mechanical sensor technology platform. Exploiting the nanoscale precision of the ARS electrostatic mover, researchers at HP Laboratories and the HP Technology Development Operations in

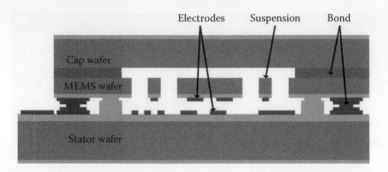

FIGURE 3.9
(See color insert.) MEMS accelerometer cross section (not to scale). The body of the device is formed by bonding three silicon wafers and completely etching through the middle wafer to create the proof mass and flexures.

Corvallis, Oregon developed a compact chip-scale accelerometer technology that sets new performance and manufacturability standards for MEMS sensors. Unlike other MEMS inertial sensors that incorporate interdigitated electrodes in comb or parallel plate arrangements, HP developed a novel surface electrode scheme that enables a large capacitance change for a given input and also provides a large dynamic range.

The in-plane accelerometer is constructed by bonding three silicon wafers to form the body of the device (Figure 3.9). The proof mass and structural elements are created by etching completely through the thinned middle wafer. The unique through-wafer etching process makes it possible to create proof mass elements with 1000× larger mass than typical surface micro-machined MEMS sensors.

Wafer bonding also enables wafer-scale vacuum packaging to maintain the required mechanical damping conditions. With the exception of thin films used for the wiring and electrodes, the bonded wafer scheme produces a device that is a nearly homogeneous block of single crystal silicon, greatly enhancing the thermal stability of sensor. Finally, this approach provides a path to on-chip electronics integration (Walmsley et al., 2009).

Figure 3.10a shows a close-up SEM image of the MEMS wafer used in a two-axis accelerometer. The image shows two perpendicular groups of surface electrodes patterned on the proof mass and two high-aspect-ratio flexures. Typical flexure width is 3 μ with an aspect ratio of 40:1. The two-axis accelerometer uses the same proof mass for both and x- and y-directions. Figure 3.10b shows a fully packaged two-axis accelerometer.

HP's MEMS processing capabilities that were originally developed for the ARS technology provide a platform that will make it possible to manufacture complex, high-performance inertial measurement technologies and will enable a new class of ultrasensitive, low-power MEMS inertial sensors.

(a)   (b)

FIGURE 3.10
(a) Perpendicular groups of surface electrodes patterned on proof mass. (b) Packaged two-axis accelerometer.

3.4.3 Optical MEMS

In 2000, HP began to extend its digital imaging expertise in printing, scanning, and digital photography into the digital projector and television markets. Continuing to push the state of the art of MEMS fabrication capabilities and leveraging its extensive color science and image processing competencies, HP began to develop highly innovative optical MEMS devices and systems. By adapting the MEMS fabrication processes and materials developed for TIJ, ARS, and accelerometer technologies such as deep silicon etching, bulk and surface micromachining, wafer bonding, and packaging, HP developed several movable pixel technologies. These devices strongly leveraged and optimized the architectural elements used in the fabrication of micromovers such as flexures, posts, and suspensions.

New capabilities for depositing high quality optical thin films and for texturing optical surfaces were developed and integrated into the MEMS device fabrication processes. These new capabilities made it possible to engineer new interferometric pixel technologies and color light modulators with exceptional color and image fidelity. The integration of sophisticated electrostatic pixel actuation electronics provided exceptional image processing, control, and fast refresh rates. Figures 3.11a and b show micrographs of two MEMS-based pixel devices. Figure 3.11c illustrates an interferometric-based pixel array producing primary colors and black.

Figure 3.11d is a photograph of a novel fully integrated Fabry-Perot color spectrophotometer used for color measurement and sensing. The device consists of an array of Fabry-Perot etalon filters that required the development of high quality partial reflectors and precisely controlled oxide spacer layers. By leveraging and extending semiconductor processing methods developed for TIJ printhead electronics, novel implant methods were developed to integrate highly responsive photodiodes below each etalon filter as shown in the figure.

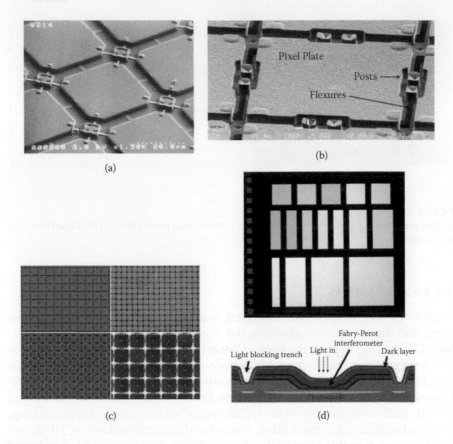

FIGURE 3.11
(See color insert.) (a) MEMS-based optical devices. (b) MEMS pixel plate devices. (c) Interferometric pixel array. (d) Fabry-Perot spectrophotometer chip and device cross section.

3.4.4 Silicon Photonics

A significant market focus for HP involves high performance computing and computing services. By targeting the development of high-performance technologies, groups in HPs laboratories and R&D areas have maintained a continuous cycle of innovation enables the company to lead the overall evolution of computing.

Over the past 10 years, HP has made significant investments in the development of silicon photonics, optical interconnection hardware, and new optical data bus architectures. In response to demands for increased performance, engineers and scientists at HP's laboratories and at its Technology Development Operations in Corvallis, Oregon focused on developing new micro- and nanofabrication processes, materials and new devices such as lasers, waveguides, microlenses, and beam splitters to help remove

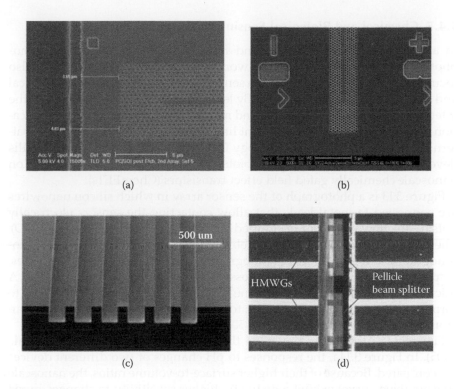

FIGURE 3.12
(a) and (b) Photonic waveguides fabricated using nanoimprint lithography. (c) Hollow metal wave guide. (d) HMWG and pellicle beam splitter. (From Yeo, J.S., Bicknell, R.N., Mathai, S. et al. 2009. *Appl. Phys. A*, 95, 1073–1077; Bicknell, R., King, L., Otis, C.E. et al. 2009. *Appl. Phys. A*, 95, 1059–1066.)

bottlenecks associated with interconnections and conventional data bus hardware (Tan et al., 2009).

The addition of new fabrication and processing capabilities such as nanoimprint lithography (Tong et al., 2005), focused ion and electron beam patterning, Langmuir–Blodgett coating, and atomic layer deposition made it possible to design and prototype sophisticated, nanoscale structures such as photonic band gap devices and photonic and hollow metal wave guides.

Micrographs of silicon photonic waveguides fabricated using nanoimprint lithography are shown in Figures 3.12a and 3.12b. Figure 3.12c shows an example of a silicon hollow metal wave guide (HMWG) fabricated using various patterning processes including sawing, laser machining, milling, wet and dry etching, and thin film deposition of metals and dielectrics. Figure 3.12d is a photograph of a 1 × 4 HMWG with a pellicle beam splitter inserted at 45 degrees. The beam splitter was fabricated using wafer level bonding of silicon and glass substrates, laser micromachining, wet and dry etching, and a mechanical polishing step to thin the glass substrate to 10 µm.

3.4.5 Chemical and Biological Sensing

In addition to the development and fabrication of the micro- and nano-photonic devices discussed above, work in HP's R&D laboratories has also focused on the development of high sensitivity and high selectivity chemical and biological sensing technology. By leveraging extensive experience in the integration of complex electronics and microfluidics and adding new fabrication capabilities such as nanoimprint lithography and focused ion beam patterning, researchers at the Technology Development Operations in Corvallis developed a novel chemical and biological sensing technology based on nanoscale chemically gated field effect transistors (ChemFETs).

Figure 3.13 is a photograph of the sensor array in which silicon nanowires are connected to metal leads that allow contacting the sensors electrically outside the flow channel area (dotted lines A, B, C) using a probe card (D). The silicon nanowire sensors were fabricated on SOI substrates using photolithography and electron beam lithography for micro- and nanoscale wires, respectively. The inset shows a cross section of one of the nanowires. A micro-molded poly dimethyl siloxane (PDMS, Sylgard 184) gasket is used to form the microfluid flow channel on top of the sensor chip. Figure 3.13b is an atomic force microscopy image of a 50-nm wide nanowire.

Figure 3.14 provides examples of the ChemFET operation (Moller et al., 2005). In Figure 3.14a, the responses to pH changes of two different devices are compared. Because of their higher surface-to-volume ratios, the nanoscale sensors (blue curve) exhibit a distinctly higher sensitivity to changes in pH than micro-scale devices (red curve). This is better illustrated in the inset in which the relative change of the sensor response ($\Delta I_{DS}/I_{DS}$) as a function of

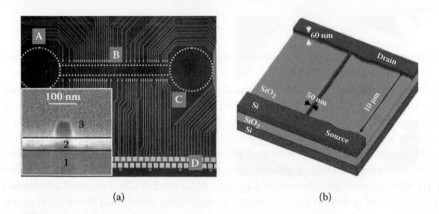

(a)   (b)

FIGURE 3.13
(See color insert.) (a) Optical micrograph of sensor array. The outline of the flow channel is shown as dotted line (A, B, C). The inset shows a SEM micrograph nanoscale silicon wire. Wires are insulated from the silicon substrate (1) by the buried oxide (2) and a thermally grown gate oxide with a thickness of ~3 nm acts as a gate dielectric for ChemFET operation of the silicon wires (3). (b) Atomic force microscopy image of 50-nm Si nanowire (with substrate).

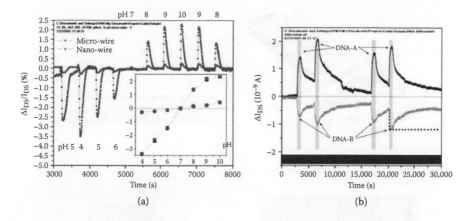

(a)   (b)

FIGURE 3.14

(See color insert.) (a) Response of drain-source current of 2-μ and 50-μ wide silicon wires to injection of analyte solutions with different pH. In the inset, the response of nanoscale wires to a change in pH is up to one order of magnitude higher than that of micro-scale wires.(b) Differential sensor signal for injections of two different DNA sequences (A and B).

pH for 50 nm and 2 μ wide wires is plotted. The response of nanoscale wires to a change in pH is up to one order of magnitude higher.

In a companion study, the specific binding of DNA target and probe molecules was characterized. By immobilizing DNA probe molecules on the silicon wire surface, the ChemFET devices are rendered specific to the sequence of the DNA probes. As shown in Figure 3.14b, at a concentration of C_{DNA} = 10 μM, two different single-stranded 24-bp DNA oligonucleotides could be distinguished in the sensor response. Details of the experiment and analysis are provided in Moller et al. (2005).

3.4.6 Expanding HP's Printing Portfolio: Piezoelectric Inkjet Technology

HP has leveraged its core capabilities in MEMS, microfluidics, thin film processing, packaging, and nanofabrication to explore new technologies and applications. HP also continues to leverage these capabilities to further expand its inkjet printing technology portfolio.

Recently, researchers and engineers in HP's printhead development laboratories in Corvallis, Oregon and Netanya, Israel have focused on the design, fabrication, and production of a new piezoelectric inkjet printhead platform. The new printhead technology enables the use of a broader range of inks including ultraviolet (UV) inks for industrial print markets. The fundamental operating mode of the piezo inkjet uses a mechanical impulse to eject a droplet of ink as described above in Section 3.2.

Fabrication of HP's X2 piezoelectric inkjet printhead uses much of the same equipment and processes developed for TIJ and the MEMS devices described above (Figure 3.15). Fabrication of the X2 printhead combines precise depth

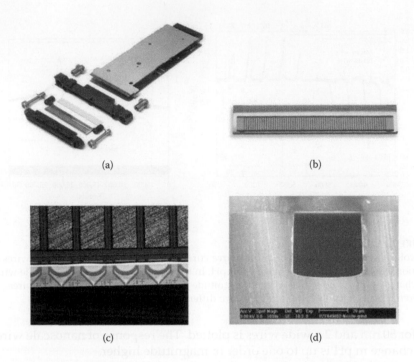

(a)

(b)

(c)

(d)

FIGURE 3.15
(See color insert.) (a) Major parts assembled to produce the HP X2 piezoelectric printhead assembly. (b) Closer view of silicon and glass printhead. (c) Four pumping chambers. (d) Edge-on view of a single X2 inkjet nozzle.

and feature control of deep silicon etch with wafer level glass bonding, die-level piezoelectric material bonding, and photo and mechanical patterning processes. Pumping chambers are fabricated on both sides of a thick silicon wafer to double the nozzles available on each printhead (128 nozzles on each 1.28-inch X2 head). The combination of precision processes enables the necessary control to produce 50-μ ink orifice features, 2000-μ long pumping chambers and fluidic architectures needed to produce consistent, high quality droplets.

3.5 Conclusions and Future Directions

In this chapter, we have provided a brief history of the evolution of HP's TIJ printhead technology from a MEMS perspective. We outlined the basic thermal process, discussed the fundamental architectural elements of the printhead, and highlighted key technical challenges. We discussed the evolution of HP's TIJ technology and described how new processing, fabrication, and manufacturing methods were developed to address increasing demands for higher performance, print quality, and lower costs.

Furthermore, to develop products that would operate reliably under the extreme operating conditions of the TIJ process, we discussed how the development of new thermally and mechanically robust materials, adhesives, thin film coatings, wafer bonding techniques, circuit technologies, and hermetic packaging were required. In Section 3.4, we provided examples of new MEMS, photonics, and nanoelectronic technologies developed by leveraging and extending the core competencies, knowledge, and capabilities acquired over the past 25 years of TIJ technology development and commercialization.

HP scientists and engineers continue to work at the leading edges of MEMS, microfluidics, electronics, materials science, and manufacturing to extend its printing technology portfolio and continue to develop new innovative emerging technologies and products. Current efforts focus on developing technologies that will enable HP's Central Nervous System for the Earth (CeNSE) vision (Kirkpatrick, 2009; Mims et al., 2009). CeNSE is a company-wide project focused on developing technology to enable a network of billions of macro-, meso-, micro-, and nanoscale sensors that will connect with computing systems, software, and services to share real-time and archived data on environmental, biological, and structural health to facilitate situational awareness, control and optimization of infrastructure, disaster prevention, and intelligence collection.

Acknowledgments

The work presented in this chapter represents the efforts of a large number of HP engineers and scientists who, for the past 25 years, have been responsible for the invention, innovation, development, commercialization, and evolution of HP's MEMS technologies. The authors gratefully acknowledge these contributions. We would also like to thank the following people for technical material presented in the chapter and for useful discussions and comments: Sam Angelos, Tim Weber, Don Ouchida, Pete Hartwell, Dave Erickson, Don Milligan, David Champion, John Yeo, Pavel Kornilovich, Robert Bicknell, Jenny Wu, and Kevin Peters.

References

Bicknell, R., King, L., Otis, C.E. et al. 2009. Fabrication of pellicle beam splitters for optical bus application, *Appl. Phys. A*, 95, 1059–1066.

Kirkpatrick, D., HP's grand vision: measure everything. 2009. http://money.cnn.com/2008/07/18/technology/kirkpatrick_nano.fortune/index.htm.

Ko, W. 2007. Trends and frontiers of MEMS. *Sens. Actuat. A*, 136, 62–67.

Mims, C., Schupak, A., Moyer, M. et al. 2009. World changing ideas: 20 ways to build a cleaner, healthier, smarter world. *Scient. Amer.* December, 50–61.

Moller, S., Hinch, G.D., Duda, K.J. et al. 2005. Nanoscale chemical and biological sensors. In Islam, M. and Dutta, Eds., *Proceedings of SPIE 6008. Nanosensing: Materials and Devices II*, SPIE, Boston, pp. 87–98..

Naberhuis, S. 2001. Probe-based recording technology: A MEMS perspective. *Microelectromechanical Systems Conference*, pp. 33–35.

Peeters, E. 1997. Challenges in commercializing MEMS. *IEEE Comp. Sci. Eng.*, 4, 44–48.

Tan, M., Rosenberg, P., Yeo, J.S. et al. 2009. A high-speed optical multi-drop bus for computer interconnections. *Appl. Phys. A*, 95, 945–953.

Toigo, J.W. 2000. Avoiding the data crunch. *Scient. Amer.*, May, 58–71.

Tong, W.M, Hector, S.D., Jung, G.Y. et al. 2005. Nanoimprint lithography: the path toward high-tech, low-cost devices. In Mackay, R. and Scott, R., Eds., *Proceedings of SPIE 5751. Emerging Lithographic Technologies IX*, SPIE, San Jose, pp. 46–55.

Walmsley, R.G., Kiyama, L.K., Milligan, D.M. et al. 2009. Micro-G silicon accelerometer using surface electrodes. In French, P. and Mukhopadhyay, S., Eds., *Sensors 2009*, IEEE, Christchurch, NZ, pp. 971–974.

Yeo, J.S., Bicknell, R.N., Mathai, S. 2009. Compensation of beam walk-off in hollow metal waveguides. *Appl. Phys. A*, 95, 1073–1077.

4

Silicon MEMS Resonators for Timing Applications

Bongsang Kim, Matthew A. Hopcroft, and Rob N. Candler

CONTENTS

ABSTRACT   This chapter includes a review of MEMS resonators for use as frequency references and filters. Basic modeling and transduction techniques for resonators are described with emphasis on capacitive transduction. A brief history of resonator architectures is given, beginning with early devices consisting of linear comb drive actuators and flexural beams to later devices utilizing bulk resonant modes and parallel plate actuation. The underlying causes of frequency instability are described, with emphasis on the dependence of frequency upon temperature. Descriptions are provided for the various energy dissipation mechanisms that are central to determining the quality factors of resonators. The concept of coupled resonators for narrow-band frequency filters is also covered.

4.1 Introduction

The need for timing is not new. From early Egyptian obelisks and sundials to precision timekeeping for navigation in the eighteenth century (Newton 1714), timing has been a large part of our technological development. Timing is ever-present in the modern era, from the obvious watches on our wrists to the unseen timing references used for wireless communication that fills the airwaves around us. The twentieth century saw the expansion of timing from simple calendars and sailing ship navigation devices to the basis of worldwide communications and commerce, and the quartz crystal was at the heart of the expansion.

Quartz crystals have been workhorses for communications since the mid-twentieth century. More than one billion quartz crystals are produced for timing and communications applications every year (Vig 1992).

More recently, MEMS resonators have been considered as replacements for quartz crystals. The potential advantages of MEMS resonators are the following:

- Reduced cost from batch fabrication processes for silicon that were largely developed by the semiconductor industry
- Smaller size enabled by precision fabrication methods
- Integration of silicon resonators with circuitry on the same wafer

The adoption of MEMS resonators will not come without effort. The challenges that must be overcome include (1) packaging of resonators for long-term stability, and (2) achieving efficient transduction in miniaturized devices. Many people believe that the challenges facing MEMS resonators are not fundamental. However, quartz is well entrenched in the marketplace, making the challenge of earning trust in MEMS resonators a real issue.

Some of the earliest MEMS resonators achieved resonant frequencies in the kilohertz (kHz) range using electrostatic comb-drive actuation and sensing (Tang et al., 1989). More recently, the linearity of comb drives has been relinquished to achieve increased actuation forces and greater sensing currents enabled by parallel plate actuation. Rapid progress has been made to increase the resonance frequency from kilohertz to gigahertz (GHz) devices (Jing et al., 2004; Weinstein and Bhave, 2009) by moving toward stiffer, bulk mode structures to attain higher frequencies, and reducing the electrostatic gap to detect the small displacements of the stiff structures. Indeed, some electrostatic gaps have been partially or completely filled with high-k dielectric materials to improve transduction.

While the focus of this chapter is on electrostatically actuated devices, other transduction mechanisms must also be identified as valid techniques. The incorporation of piezoelectric materials for actuation and sensing of MEMS resonators and filters has already led to commercial success (Ruby et al., 2004b), and development of gigahertz devices from piezoelectric materials is an active area of research (Chengjie et al., 2010; Abdolvand et al., 2008). Generally speaking, piezoelectric materials provide much improved signal transduction but require the use of non-standard materials.

What does the future of MEMS resonators hold? A number of avenues may be pursued. New resonator materials could be investigated to enable pushing the limits of performance with regard to frequency and energy dissipation. New transduction mechanisms or new materials for existing transduction mechanisms could be pursued to improve the integration of these devices into oscillator and sensor systems. New packaging techniques could be developed to enable the inclusion of MEMS resonators in nearly every electronic system, no matter how small or thin. Self-calibration techniques could help lower prices; at present most of the cost of an oscillator system is a result of testing and packaging costs. Finally, the performance of MEMS resonators could be improved to enable their adoption into systems that currently require more expensive timing references (e.g., ovenized quartz resonators and atomic clocks). MEMS resonators will likely represent active areas of research and industry for many years to come.

The following sections cover the basic characteristics of resonators, the common transduction mechanisms used to actuate and sense resonators, the frequency stability of resonators, and various energy dissipation mechanisms.

4.1.1 Resonator Characteristics

We have many ways to transmit energy to and from a MEMS resonator, but a resonator is a mechanical device at its core. The simplest representation of a MEMS resonator is the lumped element model shown in Figure 4.1. This model captures the basic nature of the device as a second order system that contains two energy storage elements (a mass and a spring), and one energy dissipation element (a dashpot). While this model requires further

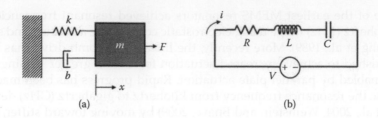

(a)   (b)

FIGURE 4.1
(See color insert.) (a) Simple mass spring damper lumped element mechanical resonator. (b) Resistor–capacitor–inductor electrical resonator.

refinement in some cases (e.g., to ensure non-linear behavior or shunt capacitance is included for electrostatically actuated resonators), this model successfully captures the fundamental behavior of the devices.

This mechanical lumped element model can be converted to an electrical lumped element model based on the analogous terms listed in Table 4.1. Just as the mechanical model has two energy storage elements and one dissipation element, so does the electrical model (energy storage in the capacitor and inductor and energy dissipation in the resistor). Also, the electrical model can be expanded to include the transduction phenomena from electrical to mechanical energy and back.

Resonator vibration can be measured in the time or frequency domain. Besides resonance frequency f, the quality factor Q is another main parameter describing resonator performance. As can be seen in Figure 4.2, the Q of a resonator is a measure of the amount of energy stored as compared to the amount of energy dissipated. The Q can be determined experimentally in the time or frequency domain, also shown in Figure 4.2. Because Q is in part a function of frequency, resonators are sometimes compared on the basis of their Q-f product.

TABLE 4.1

Comparison of Analogous Properties of Mechanical and Electrical Resonators

Mechanical	Electrical
Force, F	Voltage, V
Velocity, $\dot{x}$	Current, I
Spring constant, k	Capacitance, C
Mass, m	Inductance, L
Dashpot (damper), b	Resistor, R
Resonant frequency, $f = \dfrac{1}{2\pi}\sqrt{\dfrac{k}{m}}$	Resonant frequency, $f = \dfrac{1}{2\pi}\sqrt{\dfrac{1}{LC}}$

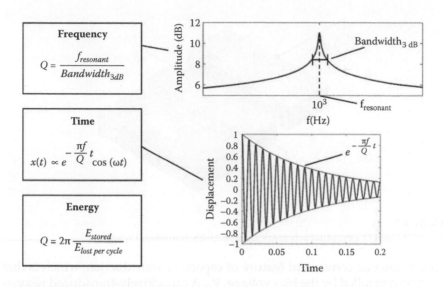

FIGURE 4.2
Resonator quality factor (Q) that may be measured in the time or frequency domain.

4.2 Transduction Mechanisms in MEMS Resonators

A MEMS resonator is a mechanical device, but electrical signals are required for connection to modern electronics. The choice of the transduction mechanism which converts energy between the mechanical and electrical domains is fundamental to the design of a MEMS resonator. While there are a large number of transduction mechanisms, capacitive- and piezoelectric-based designs are the most popular due to their superior performance and fabrication requirements.

4.2.1 Capacitive Transduction

Electrostatic transduction in MEMS resonators utilizes the capacitance change between the movable resonator structure and stationary electrodes as shown in Figure 4.3. The efficiency of the transduction between the mechanical force and electrical signal is the electrostatic coupling coefficient, η, which is defined for parallel plate capacitors as

$$\eta = V_P \frac{\partial C}{\partial x} \propto \frac{V_P \varepsilon_r A}{d^2} \tag{4.1}$$

where V_P is the DC bias voltage applied to the resonator, C is the capacitance between the resonator and the surrounding electrodes, ε_r is the relative permittivity, A is the overlap area, d is the capacitive gap size. The definition of

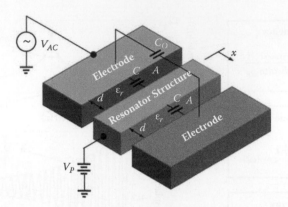

FIGURE 4.3
(See color insert.) Conceptual schematic of capacitively transduced resonator.

η illustrates one convenient feature of capacitive transduction, which is that it can be controlled by the bias voltage, V_P. A capacitively-transduced resonator can effectively be switched off by simply setting the bias voltage to zero.

One drawback of electrostatic transduction is the relatively poor electro-mechanical coupling compared to other options. Reducing the motional impedance of electrostatic resonators remains a challenge. This issue becomes even more serious for high frequency devices because the device size shrinks and the transducer overlap area decreases. To overcome this issue, researchers have introduced several techniques for enhancing electro-mechanical coupling in resonators (Figure 4.4 and Figure 4.5).

One approach is high-k dielectric gap filling. As can be seen in Equation (4.1), the electro-mechanical coupling of the capacitive transducer is proportional to ε_r, the relative permittivity of the gap; therefore, by filling this gap with high-k dielectric material, its coupling can be effectively enhanced. For example, resonator capacitive transduction could be improved 7.5 times by filling the gap with silicon nitride ($\varepsilon_r = 7.5$) or 18 times with hafnia ($\varepsilon_r = 18$), resulting in significant improvement in its motional impedance (Lin et al., 2005). Recently demonstrated internal transduction also utilizes these high-k dielectric materials (Weinstein and Bhave, 2007; Ziaei-Moayyed et al., 2009). However, the use of high-k dielectric materials raises another issue. While the electro-mechanical coupling increases due to the increase in relative permittivity, at the same time the feed-through capacitance C_o (see Figure 4.3) also increases in the same ratio, creating difficulties in practical circuit applications.

Another approach is reducing the gap spacing d in Equation (4.1). Several fabrication techniques have been developed to reduce the gap size beyond the capability of conventional lithography and etch. The most notable method is the deposition of a thin sacrificial layer between the structure and the electrodes. The layer is later etched out, leaving an extremely small gap. Silicon dioxide has been frequently used to achieve gap sizes as small as 100

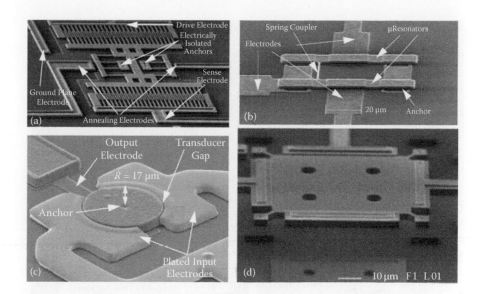

FIGURE 4.4
Various silicon MEMS resonator designs for signal processing applications. (a) Comb drives (From Wang et al. 1997). (b) Beam-type tuning forks. (From Bannon and Nguyen. 1996. With permission.) (c) Disk resonator. (From Clark, J.R., Abdelmoneum, M.A., and Nguyen, C.T.C. 2003. *Proceedings of Joint IEEE International Frequency Control Symposium and PDA Exhibition with 17th European Frequency and Time Forum*, pp. 802–809. With permission.) (d) Square resonator. (From Wojciechowski et al. 2007. With permission.)

nm (Pourkamali et al., 2003). Recently atomic layer deposition has been introduced to reduce this gap even further (Hung et al., 2008).

4.2.2 Piezoelectric Transduction

Piezoelectric transducers utilize the relationship between an internal electric field and mechanical strain in a piezoelectric material for transduction. This method is the almost ideal coupling of electrical and mechanical domains, and indeed piezoelectric devices typically exhibit very good electro-mechanical coupling. In traditional macroscale frequency control devices, piezoelectric transduction of quartz has been the dominant technology for almost 100 years. At the microscale, materials such as lead zirconium titanate (PZT), zinc oxide (ZnO), and aluminum nitride (AlN) have been investigated extensively in the MEMS community for resonator applications. AlN is the basis for the most successful MEMS resonator products of the past decade, Avago Technologies' FBAR (Ruby et al., 2004a).

However, piezoelectric devices require separate electrodes (in most cases metals) attached to the piezoelectric structure to pick up the electrical output, and the quality of a resonator is significantly diminished by the interface effects and/or the electrode material. Thus, piezoelectric MEMS devices have

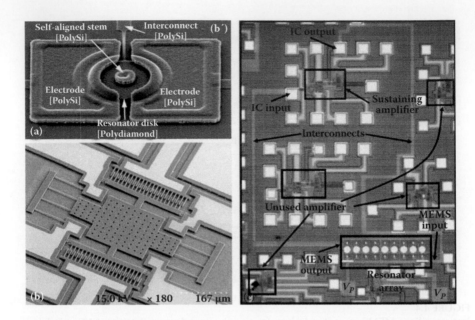

FIGURE 4.5
Electrostatic MEMS resonators in various materials. (a) Diamond resonator. (From Wang et al. 2004c. With permission.) (b) Silicon carbide (SiC) (From Roy et al. 2002. With permission.) (c) Nickel resonator. (From Huang, W.L., Ren, Z., Lin, Y.W. et al. 2008. *Proceedings of 21st IEEE International Conference on Microelectromechanical Systems.* Tucson, pp. 10–13. With permission.)

traditionally lagged behind other types, particularly electrostatic, in terms of their maximum quality factor and Q-f product. Because of the potential for high performance, efforts to achieve high quality factors in piezoelectric MEMS resonators are research topics of great interest, but piezoelectric MEMS resonators with Q-f products approaching the theoretical limits have not yet been demonstrated.

While silicon does not have piezoelectric properties, some efforts have been made to couple silicon with piezoelectric materials (Abdolvand et al., 2008; Lavasani et al., 2008). It is hoped that these combinations can utilize the best attributes of both materials (good electromechanical transduction and good material properties). However, the complexities of fabrication and material interface issues have limited the performance of these types of devices to date.

4.2.3 Other Transduction Techniques

Several other transduction techniques have been demonstrated in MEMS resonators, but most have limitations which make their use in electronic circuits impractical. Magnetomotive transducers utilize the Lorentz force. This transduction mechanism is extremely sensitive even with very small signals and displacements. However, it requires large magnetic fields and

FIGURE 4.6
Examples of use of magnetomotive transduction in micro- and nanomechanical resonators. (a) Silicon nanowire resonator. (From Cleland, A.N. and Roukes, M.L. 1996. *Applied Physics Letters*, 69, 2653–2655. With permission.) (b) Si_3N_4 cantilever resonator. (From Kemp, T. and Ward, M. 2005. *Sensors and Actuators A*, 123, 281–284. With permission.)

large electric currents. Therefore the overall size and power consumption of the system is typically very large and this option has only been shown to be favorable in some fundamental research demonstrations such as nanoscale resonators (Cleland and Roukes, 1996; Kemp and Ward, 2005). See Figure 4.6. Electro-thermal actuators use resistive Joule heating to produce mechanical forces through thermal expansion (Othman and Brunnschweiler, 1987; Rahafrooz and Pourkamali, 2011). While many functional devices have been demonstrated, they typically have lower frequencies than other resonator types and require high power for heating which limits their application in modern circuit design. Optical transduction for MEMS resonators based on coherent light (lasers) is also used, particularly in research applications where the introduction of magnetic, electrostatic, or other forces is not possible (e.g., Foulgoc, 2006). The additional complexity required for lasers and optical path management typically precludes the use of optical transduction techniques in electronic circuits. However, optical resonance and transduction techniques are sure to play an important role in the future as optical signals are increasingly used for high-speed communications.

4.3 Frequency Stability

Despite many breakthroughs in MEMS resonator development, the stability of the resonance frequency continues to be an active area of development for both research and commercial devices. Frequency stability, the fractional change of the resonance frequency, can be defined as:

$$\frac{f - f_0}{f_0} = \frac{\Delta f}{f_0}$$

(4.2)

TABLE 4.2

Typical Stability Requirements of Frequency References for Various Applications

Application	Typical Frequency Stability Requirement $\Delta f/f_0$ from 0 to 70°C
Consumer electronics, USB (universal serial bus)	100 ppm
Mobile phone radio	10 ppm
Mobile phone base station	<1 ppm
Navigation base station, military radar	<0.1 ppm

where f_0 is the nominal or expected frequency, and is typically given in parts-per-million (ppm units). Typical stability requirements for different applications are listed in Table 4.2.

A large number of sources of potential instability affect the resonance frequency of silicon-based micromechanical devices. In this section, the three most important stability issues: long-term stability (aging), temperature stability, and vibration sensitivity will be discussed.

4.3.1 Long-Term Stability of Silicon Resonators

4.3.1.1 Potential Aging Mechanisms

The long-term time dependence of frequency of timing devices is often called frequency aging. Several potential mechanisms may affect the aging performance of silicon-based MEMS resonators. The most important among them are discussed below.

Surface contamination and decontamination—Resonance frequency can drift by the mass change due to adsorption and desorption of contaminants on the resonator surface. As the size of a resonator structure shrinks, the surface-to-volume ratio increases, thus the role of surface contamination becomes more important. Typically, the adsorption or desorption of one atomic layer of contaminants can shift the resonance frequency of a microscale mechanical resonator by parts per million (ppm) levels. Often, both the mass addition and surface stress, induced or relieved by the reaction of the structure surface with adsorbed contaminants, can significantly affect resonance frequency stability.

Humidity and oxygen are the most common sources of surface contamination. Besides oxidation by humidity or oxygen, many other gases can also permanently or temporarily change the surface properties of a resonator structure. To minimize this contamination-induced aging, a resonator must be fabricated and packaged in a very clean environment, and must be completely isolated from the outside environment during its operation.

Material fatigue—Resonators vibrate at their resonance frequency during operation. In many applications, high frequency operation is desired, thus the devices experience a large number of cyclic loadings and material fatigue becomes a possible factor affecting long-term frequency stability. Fatigue is a progressive and localized structural damage that occurs when a material is subjected to a large number of cycles with nominal stresses smaller than its critical or yield stress.

While fatigue has been observed in silicon microstructures, fatigue of silicon at the micro- and nano-scales is not fully understood. In general, silicon does not exhibit plasticity or crack-tip shielding phenomena at room temperature, so its failure mechanisms due to cyclic loading cannot be interpreted with the traditional ductile or brittle material fatigue models.

After some researchers reported subcritical crack development under cyclic loading in micron-scale silicon structures (Connally and Brown, 1992), significant effort has been focused on analyzing fatigue in silicon (Muhlstein et al., 2001; Kahn et al., 2002). Among proposed models, the reaction layer fatigue mechanism model of Muhlstein et al. and the mechanically induced subcritical cracking model of Kahn et al. are attracting the most attention. Although the exact model for silicon fatigue remains controversial, a few concepts are commonly accepted.

First, all reported fatigue behavior is found only in the gigapascal range of large stress amplitudes that represent the tens of percent of the critical fracture stress. Second, although the exact roles of humidity and oxygen are still uncertain they play key roles in silicon fatigue. This has been recently confirmed by experiments conducted by Yamaji et al. (2007).

Stress effects—Mechanical resonating structures can encounter residual stresses after fabrication. Specifically, contaminant coatings and native oxide formations on surfaces are the main sources of surface stress. Also, a structure's mounting to the package enclosure and their thermal expansion mismatch cause significant residual stress. When these residual stresses are not stable and change over time, aging can result. These stress effects can be minimized by fabrication process control such as annealing or engineering designs such as smart anchoring.

Additional effects may affect the long-term frequency stability of mechanical resonating structures, for example, timed degradation of electrodes, diffusion of electrodes and resonating structures, and operating pressure changes due to package leakage.

4.3.1.2 Role of Packaging in Long-Term Resonator Stability

To achieve long-term resonance frequency stability of silicon-based MEMS resonators, all the issues above must be successfully addressed simultaneously. Losing control over any one of the potential aging mechanisms will result in frequency instability. One very important point is that all the potential aging mechanisms discussed are very closely related to the encapsulation of the resonator structure. For example, to minimize the contamination-induced aging, the resonator structure must be fabricated and hermetically packaged in an ultraclean high vacuum environment. Furthermore, the sealed cavity cleanliness must be maintained for long periods.

To avoid fatigue problems, a resonator must be designed to avoid excess stress concentration (up to gigapascal range) and the exposure of the silicon surface to humidity and oxygen must be prevented.

The main causes of residual surface stress generation, such as contaminants (including humidity and oxygen that can form native oxides) and residues, must be restricted by controlling the operating cavity and preventing their intrusion from the outer environment. Also, by managing the thermal expansion of the packaging materials and resonator structures, stress-induced aging can be minimized. Thus the long-term stability of MEMS resonators depends strongly on the quality of the packaging and if resonators can be packaged in an extremely clean environment with ultra-hermetic encapsulation, these possible aging factors can be effectively eliminated or at least minimized.

4.3.1.3 Frequency Stability of Silicon MEMS Resonators

Guckel et al. (1992) pioneered long-term stability investigation of silicon MEMS resonators. Polysilicon resonators were encapsulated by LPCVD (low pressure chemical vapor deposition)-deposited polysilicon cap layers. Their resonance frequency change was about 4,000 ppm after 7,000 hours of operation. Later, Koskenvuori et al. (2004) demonstrated ppm-level stability over 42 days of operation of silicon resonators packaged in vacuum using soldered brass caps. Similar resonators without encapsulation exhibited tens of ppm of frequency drift. Although the solder-cap package was not suitable for mass production and commercialization of MEMS resonators, it demonstrated the importance of long-term packaging stability.

Most notably, Kim et al. (2007) first demonstrated ppm levels of long-term stability of commercial grade silicon MEMS resonators. As shown in Figure 4.7a, a tuning-fork type silicon resonator was encapsulated by an epitaxially deposited silicon film package in a 1,000°C CMOS clean chamber, destroying any surface contaminants or native oxide on the resonator surface. Also, the hermeticity of this package provided an effective barrier to prevent atmospheric gas or environmental contaminant intrusion. The resonance frequency of these encapsulated silicon resonators showed less than a couple ppm long-term stability over more than 1 year of operation as shown in Figure 4.7b.

Similar experiments were also conducted by Kaajakari et al. (2006). Silicon MEMS resonators were encapsulated by anodic bonding of two wafers. Depending on the designs, the tested resonators exhibited 4 to 500 ppm/month of long-term drift. The beam-type resonator with one free end showed much better long-term stability (it was decoupled from external package stress), but its resonance frequency still increased by 4.2 ppm/month, limited by the cleanliness and hermeticity performance of anodic bonding encapsulation.

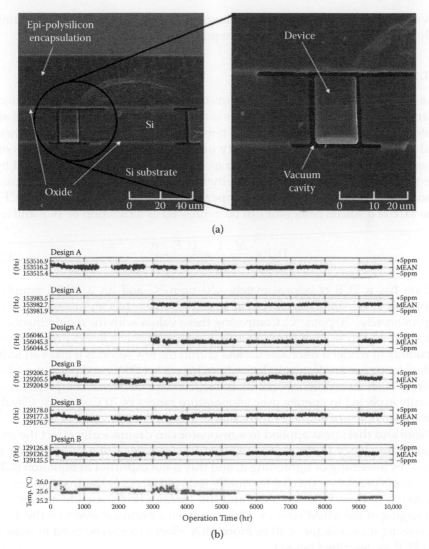

FIGURE 4.7
(See color insert.) Demonstration of long-term stability of silicon-based MEMS resonators. Silicon MEMS resonators are encapsulated in an epitaxially deposited wafer-level packaging as shown in the SEM images (a). Their resonance frequency has not changed more than a few ppm (within the measurement error range) during more than a year of operation (b). (From Kim, B., Candler, R.N, Hopcroft, M. et al. 2007. *Sensors and Actuators A*, 136, 125–131. With permission.)

Recent improvements in MEMS packaging technologies (e.g., Hsu, 2007 and Lutz et al., 2007) have dramatically advanced the long-term stability of MEMS resonators and the commercialization of silicon MEMS resonator-based oscillators has become possible. For example, SiTime, Inc. uses thin-film encapsulation and Discera, Inc. uses glass-frit bonding. Both companies have been able to control package hermeticity and cavity cleanliness, enabling volume production of commercial frequency references.

4.3.2 Temperature Stability

The most important aspect of medium-term frequency stability is the effect of ambient temperature. Temperature is a feature of all environments, and all material properties are functions of temperature. The most important thermal properties for frequency stability are expansion, described by the coefficient of thermal expansion (CTE or α), and elasticity change, described by the temperature coefficient of elasticity (TCE). The effect of temperature on resonator frequency is called the temperature coefficient of frequency (TCF). The TCF of a silicon resonator can be expressed as a function of TCE and α:

$$TCF = \frac{\Delta f}{\Delta T} \frac{1}{f_0} = \frac{TCE}{2} + \frac{\alpha}{2} \tag{4.3}$$

Silicon has a TCE of approximately −60 ppm/°C and an α of 2.6 ppm/°C (Hopcroft et al., 2010); thus a typical silicon resonator has an inherent TCF of approximately −30 ppm/°C. Comparing these values with the requirements in Table 4.2, we can see that a temperature variation of only a few degrees has the capability to render a resonator useless within the few seconds it takes an individual to remove an electronic device from his pocket (35°C) and put it on a table in front of an air conditioner (22°C). In fact, modern electronic components are usually expected to operate over temperature ranges of 70°C or more. Clearly some form of temperature compensation is required for practical devices!

Note also that stress applied to a resonator can dramatically affect the resonance frequency in the same way that the frequency of a guitar string is tuned by loosening or tightening it, and this is not captured in Equation (4.3). The most common source of thermally induced stress is thermal expansion mismatch between the resonator and the surrounding structure and/or the packaging of the MEMS device. This can have a dramatic effect on the TCF, increasing it by a factor of 10 or more (this effect can also be used to reduce the TCF, as described below).

The simplest remedy for thermally induced stress is to anchor the resonator to the surrounding structure at only one point, so that the resonator is free to expand or contract with temperature (for example, see Figure 4.4c). The approaches for temperature compensation of MEMS resonators typically involve one or more of the following:

1. Limiting the temperature change experienced by the resonator
2. Designing the resonator materials and geometries to reduce the effect of temperature change
3. Using an external circuit to change the resonator frequency as required

4.3.2.1 Limiting Temperature Change: Ovenizing

The first approach requires insulating the resonator from the ambient temperature changes and using a temperature control system to maintain the temperature at a set point. The result is an "ovenized" resonator. Typically, a heater is used to maintain a temperature set point above the maximum ambient temperature so that no refrigeration is required.

An ovenized system is somewhat complex and expensive because it requires an insulated enclosure (usually large) and a temperature control system that generally consumes large amounts of power. However, the efficacy of the approach is limited only by the ability of the temperature controller to maintain temperature inside the enclosure near the set point.

For this reason, this approach is used in traditional macroscale quartz-based precision timing devices capable of achieving frequency stabilities of 0.01 ppm/°C or better over large temperature ranges. Such devices typically have volumes of cubic centimeters and consume up to 3 W of power. While such a solution is clearly not appropriate for portable devices, the high frequency stability and low oscillator noise enabled by this approach make it attractive for many modern applications.

In microscale silicon resonators, "micro-ovenizing" has the potential to overcome some of the difficulties of macroscale implementations. The volumes required for insulation are small and the power required for heating is consequently reduced. Ovenized MEMS resonators have been investigated (Nguyen, 1995; Hopcroft et al., 2006; Pourkamali and Ayazi, 2006). Hopcroft et al. (2007) and Salvia et al. (2010) achieved sub-ppm stability levels inside cubic millimeter volumes while consuming less than 20 mW of power for heating. However, the complexities of measuring temperature inside small volumes and designing ovenized MEMS packages has limited the commercial applications of microscale ovenizing to date.

4.3.2.2 Temperature-Insensitive Design

Another approach to temperature stability of frequency involves temperature-insensitive design. This involves using a combination of materials with different thermal properties arranged so that the temperature of the structure exerts only a limited effect on the frequency of the resonator. For example, Melamud et al., 2005; and Melamud, 2008 deposited a layer of aluminum on top of a silicon resonator. As the temperature increased, the Al layer (α_{Al} ~25 ppm/°C) expanded more than the silicon resonator, putting tensile stress on the resonator. The resonance frequency of the resonator increased due to this tension (similar to a guitar string), and the increase in frequency counteracted the decrease due to temperature. Other examples of this type were devised by Stemme and Stemme (1992) and Jianqiang et al. (2005). Designs based on a thermal expansion mismatch that change the electrostatic gap d have also been demonstrated (Hsu et al., 2000; Hsu and Nguyen, 2002).

Temperature-insensitive designs have also been demonstrated using materials with different TCEs. Silicon dioxide is the best known material exhibiting a positive TCE (it becomes stiffer as temperature increases). Several researchers have demonstrated temperature-insensitive designs using a combination of silicon dioxide and silicon (Othman and Brunnschweiler, 1987; Shen et al., 2001; Abdolvand et al., 2008; Melamud et al., 2009).

Temperature-insensitive design has the advantage of being passive and automatic, but typically the total frequency variation over a useful temperature range will still be in the range of tens or hundreds of ppm, and the mismatch of materials, particularly metals and silicon, can cause issues related to aging and repeatability. However, the combination of silicon dioxide and silicon seems promising and will likely play a role in future resonator products.

The electronic doping of silicon can also be used in temperature-insensitive design. High levels of doping can modify the effective TCE of the silicon, reducing it to less than 10 ppm/°C (Csavinszky and Einspruch, 1963; Ono et al., 2000). This technique has only recently been demonstrated in practical devices (Samarao and Ayazi, 2009; Samarao et al., 2010) and appears promising.

4.3.2.3 Frequency Tuning

The third method of improving frequency stability is to use an external circuit to change or tune the resonance frequency as necessary in response to temperature changes. The most established technique in quartz resonators is inclusion of a tunable capacitor in the oscillator feedback loop. The capacitor is then adjusted in response to the measured temperature. In electrostatic MEMS designs, the frequency can be adjusted by changing either the gap spacing d or the bias voltage V_b (Sundaresan et al., 2007; Lee et al., 2008).

More recent techniques involve digital frequency adjustments using the oscillator circuit; for example a fractional-n phase-locked-loop circuit is used by SiTime, Inc., to insert or delete pulses in the oscillator output as necessary, maintaining the average output frequency (Lutz et al., 2007). The circuit-based frequency tuning methods are widely used in inexpensive commercial oscillators because they enable arbitrary frequency selection and tuning, but they introduce noise into the output signal, and are generally not suitable for precision frequency references (<10 ppm).

4.3.3 Vibration Sensitivity

Vibrations are everywhere in the environment. Vibration often affects the performance of electronic devices and this sensitivity must be minimized for practical applications. Resonators are also affected by environmental vibration. In particular, in frequency reference oscillators, external vibrations can introduce rapid changes in resonance frequency that modulates

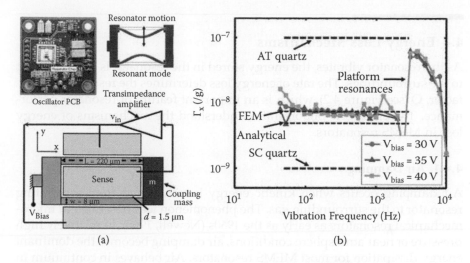

FIGURE 4.8
(See color insert.) Vibration sensitivity measurement of silicon MEMS resonators. The vibration sensitivity of MEMS resonators was shown to be on part with mid-grade quartz crystal resonator-based oscillators. (From Agarwal, M., Park, K.K., Chandorkar, S.A. et al. 2007. *Applied Physics Letters*, 90, 14–103. With permission.)

the oscillator output signal phase and the resulting phase noise degrades oscillator performance (Filler, 1988).

In silicon MEMS resonators, two main potential mechanisms cause resonance frequency shifts due to vibration-induced acceleration. First, external acceleration may induce stress on a resonator structure that alters the resonance frequency. Anchor location and design are critical for this kind of mechanism. Second, in case of capacitive transduction, when a resonator experiences acceleration, the structure is dislocated and the capacitive gap between the resonator and the surrounding electrode changes. This change alters the electrical spring softening of the capacitive transduction, resulting in a shift in resonance frequency (Figure 4.8).

Agarwal et al. (2007) investigated acceleration sensitivity of capacitively transduced MEMS resonators. A tuning fork-type silicon resonator-based oscillator was actuated by a magnetic vibration table and the sideband peaks in the output power spectrum were measured. The acceleration sensitivities of silicon MEMS resonators were shown to be on par with mid-grade quartz crystal-based oscillators, but still dominated by the induced stress due to the mass of a large coupling block. As the frequency of resonators increases, the structure size shrinks and the mass becomes smaller. Thus, improvements in resonator design and circuitry are expected to boost the acceleration sensitivity performance of MEMS-based oscillators far beyond the capabilities of current high-grade quartz-based oscillators.

4.4 Energy Loss Mechanisms

As the resonator vibrates, the energy stored in the resonator is gradually lost to the surroundings. The rate of energy loss determines the resonator quality factor, Q (see Figure 4.2), which is an important feature of resonator performance. Therefore, it is important to understand the mechanisms of energy loss in MEMS resonators.

4.4.1 Air Damping

Air damping occurs when kinetic energy is transferred from a vibrating resonator to the surrounding gas. The phenomenon was described in micromechanical resonators as early as the 1960s (Newell, 1968). In relatively high pressure or near atmosphere conditions, air damping becomes the dominant energy dissipation for most MEMS resonators. Air behaves in continuum in this regime, and viscous damping (or Couette flow damping) between the moving structure and substrate is often the primary source of air damping (Cho et al., 1994; Hosaka et al., 1995).

Most MEMS resonators are operated in low pressure environments, where air molecules behave instead as kinetic particles, to avoid air damping. In this low pressure regime, energy loss occurs due to collisions of kinetic gas particles with vibrating structures. This energy loss should be proportional to the number density n and mean velocity $\bar{v}$ of air molecules; thus,

$$E_{dissipated} \sim n \cdot \bar{v} \sim \frac{P}{k_b T} \sqrt{\frac{8 k_b T}{\pi m}} \sim \frac{P}{\sqrt{k_b T}} \tag{4.4}$$

where, m is molecular mass, k_b is the Boltzmann constant, P is pressure, and T is temperature.

Models for more complex conditions (squeeze film damping and torsional vibration) have been developed. While they do not typically match the experimental values perfectly, they capture the energy dissipation from air damping as proportional to the square root of temperature and inversely proportional to pressure. Thus, the observed air damping can usually be fit to Equation (4.4) with a constant C involving the surface area normal to the direction of the motion, the mode shape of the resonating structures, wall collision effects, etc.

$$Q_{air} = C \frac{\sqrt{k_b T}}{P} \tag{4.5}$$

4.4.2 Thermoelastic Dissipation

Thermoelastic dissipation occurs when mechanical energy in a resonator is irreversibly converted to thermal energy. When a resonator structure is deformed, a gradient of the strain rate is produced, leading to a temperature gradient. This induced temperature gradient yields heat flow to equilibrate the temperature across the resonator.

The heat flow is irreversible (the energy is not returned to the mechanical energy of the beam), which means that energy dissipation has occurred. This phenomenon was first analyzed by Zener in the 1930s for the rectangular cross-section beam-type geometry, using a one-dimensional approximation for the heat flow (Zener, 1937):

$$Q_{TED} = \left(\frac{f_M^2 + f_T^2}{f_M \cdot f_T} \right) \cdot \frac{\rho C_P}{\alpha^2 T E} \qquad (4.6)$$

where ρ is density, C_P is specific heat, α is thermal expansion coefficient, and E is Young's modulus. For characteristic frequencies, f_M is the mechanical resonance frequency and f_T is the thermal frequency, the inverse expression of the thermal decay time of heat flow across the flexing beam.

As shown in Equation (4.6), energy loss due to thermoelastic dissipation is determined by the relative magnitudes of two frequency values (mechanical resonance frequency and thermal frequency). As the period of mechanical resonance $1/f_M$ approaches thermal relaxation time $1/f_T$, this thermo-mechanical coupling increases, producing increased energy loss.

Zener's model has recently been extended to predict TED-limited Q for more complex geometries (Lifshitz and Roukes, 2000; Abdolvand et al., 2006; Candler et al., 2006). However, in these cases, Zener's simple expression in Equation (4.5) does not hold, and more general methods that include multiple thermal modes are required. For such analysis, numerical simulation tools are required to estimate exact energy loss (Duwel et al., 2006).

4.4.3 Anchor Loss

Anchor loss (or support damping) occurs when energy is transmitted through the support structure of a resonator and not returned. The vibrating structure launches elastic waves through the supports that attach the resonator to the substrate. Energy that is not reflected back into the resonator is lost to the surrounding substrate. Several research groups have recently reported investigations of this phenomenon (Hao et al., 2003; Park and Park, 2004; Bindel et al., 2005; Hao and Ayazi, 2005), and a predictive model for anchor loss will likely be available in the near future.

Several strategies for reducing anchor dissipation have been proposed, including impedance mismatch of the resonator and anchor structure (Wang et al., 2004b), isolation of devices on mesa structures (Pandey et a., 2009), and quarter-wave mechanical impedance transformers (Wang et al., 2000).

4.4.4 Surface Loss

Surface loss is most evident as resonators approach the nanoscale, creating a large surface area-to-volume ratio (Yang et al., 2000; Yasumura et al., 2000; Yang et al., 2002; Wang et al., 2004a; Seoanez et al., 2008). Resonator surfaces may contain impurities, lattice defects, adsorbates, or other imperfections that may lead to energy dissipation. Several researchers reported quality decreases as micromechanical structures are miniaturized (Ekinci and Roukes, 2005). Others investigated the impacts of thermal treatment and monolayer surface coatings (Yang et al., 2000; Wang et al., 2004a; Seoanez et al., 2008).

4.4.5 Material Loss

Material loss generally describes the attenuation of a wave traveling in a solid material. The supporting literature describes material loss in terms of phonon–phonon interactions such as the Akheiser and Landau–Rumer effects (Chandorkar et al., 2008; Tabrizian et al., 2009). While thermoelastic dissipation can be described by continuum mechanics and involves spatial transport of thermal energy, material losses involve localized phonon–phonon coupling.

In other words, thermoelastic dissipation has a spatial dependency, making it partially design-dependent, whereas material losses depend only on the material and frequency of vibration. Because material loss is a function only of the material properties of a resonator, it sets a theoretical upper bound on the Q-f product (Chadokar et al., 2008).

4.5 Applications of MEMS Resonators

While silicon MEMS resonators have been topics of research for over 20 years, technology of the first decade of the twenty-first century has advanced to the point that commercial products based on MEMS resonators have begun to appear. The potential applications of MEMS resonators to modern electronic design are significant.

4.5.1 Frequency Reference Oscillators

The most common use of microelectronic resonators is in frequency reference oscillators. Frequency references are fundamental components of every electronic device. In digital devices, a frequency reference provides the clock signal for a microprocessor and the timing signal for data transfer. In analog devices, the frequency reference provides the signal for radio transceiver frequencies, the output frequency for audio speakers, and so on. In short,

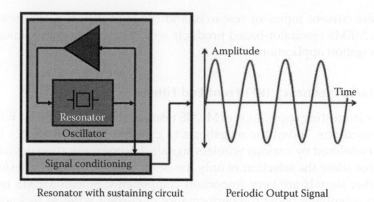

FIGURE 4.9
(See color insert.) Frequency reference composed of resonator with an oscillator circuit.

whenever we speak of an event "at some frequency" or "at some speed," a reference oscillator with a resonator at its core is involved.

A reasonably complex electronic device like a digital camera or television will require 5 to 20 separate frequency references for different functions. In current electronic design, each of these frequency references uses a separate quartz crystal that represents an additional component that must be sourced and assembled, increasing the cost and size of the final device. One future vision for silicon MEMS oscillators is direct integration with silicon integrated circuits (ICs), leading to cost and size reductions.

The resonator is the heart of the oscillator circuit (Figure 4.9) and the overall performance of the oscillator can be no better than that of the resonator. Silicon resonators offer the potential for very high performance with low power consumption in microscale form. Several companies have begun offering silicon frequency reference products in the last 5 years, including Discera, Inc., SiTime, Inc., and Silicon Labs. These products are typically in the 100-MHz range, with frequency stability of 50 ppm over a standard temperature range and power consumption of approximately 10 mW. They are targeted at consumer electronics applications such as USB keys and video game consoles.

A critical issue limiting the performance of silicon MEMS resonators in frequency reference applications is phase noise or very short-term frequency stability. One reason is that the temperature compensation methods for existing MEMS products are typically based on external circuit tuning methods. These methods are the least expensive to implement, but introduce phase noise into the output of the oscillator.

A second reason is that the power-handling capability of electrostatic MEMS resonators is much lower than comparable quartz (piezoelectric) resonators. Increasing the power through the MEMS resonator beyond a certain point increases the phase noise dramatically due to nonlinear effects. Both issues are current topics of research, and the expectation is that within a decade, MEMS resonator-based products will expand into communications and navigation applications.

4.5.2 Radio Frequency (RF) Front End Filters

Another interesting application of MEMS resonators is in RF filters for wireless communications. Today, the megahertz to gigahertz range frequency bands are overwhelmed by various wireless signals. In communication devices, RF filters that allow the selection of only the desired signal without interference from other signals are very important components. Silicon MEMS resonators can potentially boost the performance of current wireless technologies by providing RF filters with lower insertion losses and steeper filter skirts due to the 10 to 100 times higher quality factor than other current technologies such as ceramic filters and film bulk acoustic resonator (FBAR) filters. Also, as with frequency references, the potential of silicon MEMS RF components to be integrated with silicon ICs is very promising for future radio technologies.

Filters can be made by mechanically or electrically coupling two or more resonators. Figure 4.10a is a schematic of a basic filter consisting of two mechanically coupled resonators whose resonant frequencies are the same. When the two resonators are coupled, the resonator array exhibits two resonance modes: (1) an in-plane mode (two resonators resonate in the same

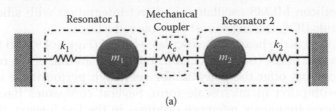

(a)

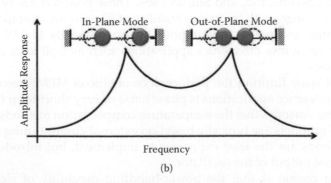

(b)

FIGURE 4.10
(a) Mechanically coupled resonator filter. (b) Amplitude responses of mechanically coupled resonators.

direction) and (2) an out-of-plane mode (two resonators resonate in opposite directions). The frequency response plot is illustrated in Figure 4.10b. The center frequency of this filter is the resonance frequency of the individual resonators and the width of passband is determined by the stiffness of the coupler between two resonators.

Recently, silicon MEMS resonator-based filters have demonstrated impressive characteristics (Li et al., 2007). Faster passband-to-stopband roll-offs from high-Q resonators enable channel-selecting RF front-end filter bank architectures in wireless communications. Power consumption and signal processing efficiency are expected to be greatly improved as described by Nguyen (2007).

Parallel arraying techniques have proven effective for overcoming the long-time bottleneck high impedance arising from capacitive coupled resonators (Figure 4.11; Demirci et al., 2003). Also, atomic layer deposition enabled partial high-k dielectric-filled gap-reduced resonator motional impedance without too much increase in feed-through capacitance (Hung et al., 2008; Akgul et al., 2009).

However, even with all the recent progress, several hurdles still remain for silicon MEMS resonator-based filters. First, even with techniques such as arraying or atomic layer deposition (ALD) partial gap filling, resonator motional impedance remains too high, requiring unwanted additional termination matching networks. Second, the operating frequencies of current silicon-based MEMS resonators are much lower than they need to be. Most modern wireless frequency bands are in gigahertz; only a few silicon MEMS resonators have been demonstrated at that frequency range.

Furthermore, they exhibited far from practical performance—too much parasitic loss or feed-through capacitance, excessive motional impedance, inadequate Q factor— (Weinstein and Bhave, 2009; Ziaei-Moayyed et al., 2009). While silicon MEMS resonators for filter applications still present

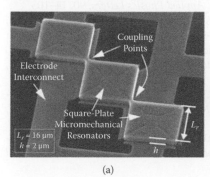
(a)

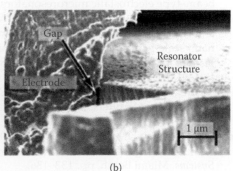
(b)

FIGURE 4.11
(a) Micromechanical resonator array. (From Demirci, M.U., Abdelmoneum, M.A., and Nguyen., C.T.C. 2003. *Proceedings of 12th International Conference on Solid State Sensors and Actuators.* Boston, pp. 955–958. With permission.) (b) ALD high-k dielectric partial gap filling of capacitively transduced MEMS resonators. (From Akgul et al. 2009. With permission.)

undeniable attractions, further research and development are needed to address these issues.

4.6 Conclusions and Future Directions

While quartz-based oscillators still dominate the timing market, MEMS resonators have emerged from the research phase in the past 15 years to become commercial competitors. Now the focus is on improving the understanding of resonator stability and energy dissipation mechanisms in order to target higher-performance applications. The main advances projected for MEMS resonators in the next 15 years remain to be seen and may include arrays of narrow band frequency filters, highly sensitive chemical and biological sensors, competition for small-scale atomic clocks, and other applications that have not yet been discovered.

References

Abdolvand, R., Johari, H., Ho, G.K. et al. 2006. Quality factor in trench-refilled poly-silicon beam resonators. *Journal of Microelectromechanical Systems*, 15, 471–478.

Abdolvand, R., Lavasani, H., Ho, G. et al. 2008. Thin-film piezoelectric-on-silicon resonators for high-frequency reference oscillator applications. *IEEE Transactions on Ultrasonics, Ferroelectrics and Frequency Control*, 55, 2596–2606.

Agarwal, M., Park, K.K., Chandorkar, S.A. et al. 2007. Acceleration sensitivity in beam–type electrostatic microresonators. *Applied Physics Letters*, 90, 14–103.

Akgul, M., Kim, B., Hung, L.-W., Yang, L., Li, W.-C., Hwang, W.-L., Gurin, I., Borna, A., and Nguyen, C. T. C., 2009. Oscillator far-from-carrier phase noise reduction via nano-scale gap tuning of micromechanical resonators. *TRANSDUCERS 2009, the 15th International Conference on Solid-State Sensors, Actuators and Microsystems*, June 21–25, 2009, Denver, Colorado.

Bannon, F.D., III, Clark, J.R., and Nguyen, C.T.C. 1996. High frequency microelectromechanical IF filters. *Electron Devices Meeting, 1996. IEDM '96., International*, pp. 773–776.

Bindel, D.S., Quévy, E., Koyama, T. et al. 2005. Anchor loss simulation in resonators. *Proceedings of 18th IEEE International Conference on Microelectromechanical Systems*. Miami Beach, pp. 133–136.

Candler, R.N., Duwel, A., Varghese, M. et al. 2006. Impact of geometry on thermoelastic dissipation in micromechanical resonant beams. *Journal of Microelectromechanical Systems*, 15, 927–934.

Chandorkar, S.A., Agarwal, M., Melamud, R. et al. 2008. Limits of quality factor in bulk–mode micromechanical resonators. *Proceedings of 21st IEEE International Conference on Microelectromechanical Systems*, pp. 74–77.

Chengjie, Z., Van Der Spiegel, J., and Piazza, G. 2010. 1.05-GHz CMOS oscillator based on lateral field-excited piezoelectric AIN contour mode MEMS resonators. *IEEE Transactions on Ultrasonics, Ferroelectrics and Frequency Control,* 57, 82–87.

Cho, Y.H., Pisano, A.P., and Howe, R.T. 1994. Viscous damping model laterally oscillating microstructures. *Journal of Microelectromechanical Systems,* 3, 81–87.

Clark, J.R., Abdelmoneum, M.A., and Nguyen, C.T.C. 2003. UHF high order radial contour mode disk resonators. *Proceedings of Joint IEEE International Frequency Control Symposium and PDA Exhibition with 17th European Frequency and Time Forum,* pp. 802–809.

Cleland, A.N. and Roukes, M.L. 1996. Fabrication of high frequency nanometer scale mechanical resonators from bulk Si crystals. *Applied Physics Letters,* 69, 2653–2655.

Connally, J.A. and Brown, S.B. 1992. Slow crack growth in single-crystal silicon. *Science,* 256, 1537–1539.

Csavinszky, P. and Einspruch, N.G. 1963. Effect of doping on elastic constants of silicon. *Physical Review,* 132, 24–34.

Demirci, M.U., Abdelmoneum, M.A., and Nguyen., C.T.C. 2003. Mechanically corner coupled square microresonator array for reduced series motional resistance. *Proceedings of 12th International Conference on Solid State Sensors and Actuators.* Boston, pp. 955–958.

Duwel, A., Candler, R.N., Kenny, T.W. et al. 2006. Engineering MEMS resonators with low thermoelastic damping. *Journal of Microelectromechanical Systems,* 15, 1437–1445.

Ekinci, K.L. and Roukes, M.L. 2005. Nanoelectromechanical systems. *Review of Scientific Instruments,* 76, 61–101.

Filler, R.L. 1988. The acceleration sensitivity of quartz crystal oscillators: a review. *IEEE Transactions on Ultrasonics, Ferroelectrics and Frequency Control,* 35, 297–305.

Foulgoc, B.L., et al. 2006. Highly decoupled single-crystal silicon resonators: An approach for the intrinsic quality factor. *Journal of Micromechanics and Microengineering,* 16, S45–S53.

Guckel, H., Rypstat, C., Nesnidal, M. et al. 1992. Polysilicon resonant microbeam technology for high performance sensor applications. *Proceedings of 5th IEEE Solid State Sensor and Actuator Workshop.* Hilton Head, SC, pp. 153–156.

Hao, Z. and Ayazi, F. 2005. Support loss In micromechanical disk resonators. *Proceedings of 18th IEEE International Conference on Microelectromechanical Systems.* Miami Beach, pp. 137–141.

Hao, Z., Erbil, A., and Ayazi, F. 2003. An analytical model for support loss in micromachined beam resonators with in-plane flexural vibrations. *Sensors and Actuators A,* 109, 156–164.

Hopcroft, M.A., Agarwal, M., Park, K.K. et al. 2006. Temperature compensation of a MEMS resonator using quality factor as a thermometer. *Proceedings of 19th IEEE International Conference on Microelectromechanical Systems.* Istanbul, pp. 222–225.

Hopcroft, M.A., Lee, H.K., Kim, B. et al. 2007. A high-stability MEMS frequency reference. *Proceedings of 14th IEEE International Conference on Solid State Sensors, Actuators and Microsystems.* Lyon, pp. 1307–1310.

Hopcroft, M.A., Nix, W.D., and Kenny, T.W. 2010. What is the Young's modulus of silicon? *Journal of Microelectromechanical Systems,* 19, 229–238.

Hosaka, H., Itao, K., and Kuroda, S. 1995. Damping characteristics of beam-shaped micro-oscillators. *Sensors and Actuators. A*, 49, 87–95.

Hsu, W.-T. 2007. Low cost packages for MEMS oscillators. *Proceedings of IEEE/CPMT 32nd International Electronic Manufacturing Technology Symposium*. San Jose, California, pp. 273–277.

Hsu, W.T., Clark, J.R., and Nguyen, C.T.C. 2000. Mechanically temperature compensated flexural mode micromechanical resonators. *Proceedings of International Electron Devices Meeting*. San Francisco, pp. 399–402.

Hsu, W.T. and Nguyen, C.T.C. 2002. Stiffness-compensated temperature-insensitive micromechanical resonators. *Proceedings of 15th Annual International Conference on Microelectromechanical Systems*. Las Vegas, pp. 731–734.

Huang, W.L., Ren, Z., Lin, Y.W. et al. 2008. Fully monolithic CMOS nickel micromechanical resonator oscillator. *Proceedings of 21st IEEE International Conference on Microelectromechanical Systems*. Tucson, pp. 10–13.

Hung, L.W., Jacobson, Z.A., Ren, Z. et al. 2008. Capacitive transducer strengthening via ALD-enabled partial-gap filling. *Proceedings of Solid State Sensors, Actuators, and Microsystems Workshop*. Hilton Head, SC, pp. 208–211.

Jianqiang, H., Changchun, Z., Junhua, L. et al. 2005. A novel temperature-compensating structure for micromechanical bridge resonators. *Journal of Micromechanics and Microengineering*, 15, 702–705.

Jing, W., Ren, Z., and Nguyen, C.T.C. 2004. 1.156-GHz self-aligned vibrating micromechanical disk resonator. *IEEE Transactions on Ultrasonics, Ferroelectrics and Frequency Control*, 51, 1607–1628.

Kaajakari, V., Kiihamaki, J., Oja, A. et al. 2006. Stability of wafer level vacuum encapsulated single crystal silicon resonators. *Sensors and Actuators. A*, 130, 42–47.

Kahn, H., Ballarini, R., Bellante, J.J. et al. 2002. Fatigue failure in polysilicon not due to simple stress corrosion cracking. *Science*, 298, 1215–1218.

Kemp, T. and Ward, M. 2005. Tunable response nano-mechanical beam resonator. *Sensors and Actuators A*, 123, 281–284.

Kim, B., Candler, R.N, Hopcroft, M. et al. 2007. Frequency stability of wafer-scale film-encapsulated silicon-based MEMS resonators. *Sensors and Actuators A*, 136, 125–131.

Koskenvuori, M., Mattila, T., Haara, A. et al. 2004. Long-term stability of single crystal silicon microresonators. *Sensors and Actuators A*, 115, 23–27.

Lavasani, H.M., Abdolvand, R., and Ayazi, F. 2008. Low phase-noise UHF thin film piezoelectric on substrate LBAR oscillators. *Proceedings of 21st IEEE International Conference on Microelectromechanical Systems*. Tucson, pp. 1012–1015

Lee, H.K, Hopcroft, M.A., Melamud, R.K. et al. 2008. Electrostatic tuning of hermetically encapsulated composite resonators. *Proceedings of IEEE Solid State Sensor, Actuator and Microsystems Workshop*. Hilton Head, SC, pp. 48–51.

Li, S.S., Lin, Y.W., Ren, Z. et al. 2007. An MSI micromechanical differential disk array filter. *Proceedings of International Conference on Solid State Sensors, Actuators and Microsystems*. Lyon, pp. 307–311.

Lifshitz, R. and Roukes, M.L. 2000. Thermoelastic damping in micro- and nanomechanical systems. *Physical Review B*, 61, 5600–5609.

Lin, Y.W., Li, S.S., Xie, Y. et al. 2005. Vibrating micromechanical resonators with solid dielectric capacitive transducer gaps. *Proceedings of IEEE International Frequency Control Symposium and Exposition*. Vancouver, pp. 128–134.

Lutz, M., Partridge, A., Gupta, P. et al. 2007. MEMS oscillators for high volume commercial applications. *Proceedings of 14th IEEE International Conference on Solid State Sensors, Actuators and Microsystems.* Lyon, pp. 49–52.

Melamud, R. 2008. Temperature-Insensitive Micromechanical Resonators. PhD Thesis, Stanford University, Palo Alto, CA.

Melamud, R., Chandorkar, S.A., Bongsang, K. et al. 2009. Temperature-insensitive composite micromechanical resonators. *Journal of Microelectromechanical Systems,* 18, 1409–1419.

Melamud, R., Hopcroft, M., Jha, C.M. et al. 2005. Effects of stress on the temperature coefficient of frequency in double clamped resonators. *Proceedings of 12th IEEE International Conference on Solid State Sensors, Actuators and Microsystems.* Seoul, pp. 392–395.

Muhlstein, C.L., Brown, S.B., and Ritchie, R.O. 2001. High cycle fatigue of single crystal silicon thin films. *Journal of Microelectromechanical Systems,* 10, 594–600.

Newell, W.E. 1968. Miniaturization of tuning forks. *Science,* 1320–1326.

Newton, I. 1714. To consider what encouragement was fit to be given to such as should find longitude at sea. *Journal of the House of Commons,* 677–678.

Nguyen, C.T.C. 2007. MEMS technology for timing and frequency control. *IEEE Transactions on Ultrasonics, Ferroelectrics, and Frequency Control,* 54, 251–270.

Nguyen, C.T.C. 1995. Micromechanical resonators for oscillators and filters. *IEEE Ultrasonics Symposium,* 481, 489–499.

Ono, N., Kitamura, K., Nakajima, K. et al. 2000. Measurement of Young's modulus of silicon single crystal at high temperature and its dependency on boron concentration using the flexural vibration method. *Japanese Journal of Applied Physics,* 39, 368.

Othman, M.B. and Brunnschweiler, A. 1987. Electrothermally excited silicon beam mechanical resonators. *IEEE Electronics Letters,* 23, 728–730.

Pandey, M., Reichenbach, R.B., Zehnder, A.T. et al. 2009. Reducing anchor loss in MEMS resonators using mesa isolation. *Journal of Microelectromechanical Systems,* 18, 836–844.

Park, Y.H. and Park, K.C. 2004. High fidelity modeling of MEMS resonators. I: Anchor loss mechanisms through substrate. *Journal of Microelectromechanical Systems,* 13, 238–247.

Pourkamali, S. and Ayazi, F. 2006. High frequency low impedance capacitive silicon bar structures. *Proceedings of IEEE Solid State Sensor, Actuator and Microsystems Workshop* Hilton Head, SC, pp. 284–287.

Pourkamali, S., Hashimura, A., Abdolvand, R. et al. 2003. High Q single crystal silicon HARPSS capacitive beam resonators with self-aligned sub-100-nm transduction gaps. *Journal of Microelectromechanical Systems,* 12, 487–496.

Rahafrooz, A., and Pourkamali, S. 2011. High-frequency thermally actuated electromechanical resonators with piezoresistive readout. *IEEE Transactions on Electron Devices,* 58, 1205–1214.

Roy, S., DeAnna, R.G., Zorman, C.A., and Mehregany, M. 2002. Fabrication and characterization of polycrystalline SiC resonators. *IEEE Transactions on Electron Devices,* 49 (12): 2323–2332.

Ruby, R., Bradley, P., Clark, D. et al. 2004. Acoustic FBAR for filters, duplexers and front end modules. *IEEE MTT–S International Microwave Symposium Digest,* 932, 931–934.

Salvia, J.C., Melamud, R., Chandorkar, S.A. et al. 2010. Real time temperature compensation of MEMS oscillators using an integrated micro-oven and a phase-locked loop. *Journal of Microelectromechanical Systems*, 19, 192–201.

Samarao, A.K. and Ayazi, F. 2009. Temperature compensation of silicon micromechanical resonators via degenerate doping. *IEEE International Electron Devices Meeting*. Baltimore, pp. 1–4.

Samarao, A.K., Casinovi, G., and Ayazi, F. 2010. Passive TCF compensation in high Q silicon micromechanical resonators. *Proceedings of 23nd IEEE International Conference on Microelectromechanical Systems*. Hong Kong, pp. 116–119.

Seoanez, C., Guinea, F., and Neto, A.H. 2008. Surface dissipation in nanoelectromechanical systems: unified description with standard tunneling model and effects of metallic electrodes. *Physical Review B*, 77, 125–127.

Shen, F., Lu, P., O'Shea, S.J. et al. 2001. Thermal effects on coated resonant microcantilevers. *Sensors and Actuators A*, 95, 17–23.

Stemme, E. and Stemme, G. 1992. A capacitively excited and detected resonant pressure sensor with temperature compensation. *Sensors and Actuators A*, 32, 639–647.

Sundaresan, K., Ho, G.K., Pourkamali, S. et al. 2007. Electronically temperature compensated silicon bulk acoustic resonator reference oscillators. *IEEE Journal of Solid State Circuits*, 42, 1425–1434.

Tabrizian, R., Rais–Zadeh, M., and Ayazi, F. 2009. Effect of phonon interactions on limiting the f-Q product of micromechanical resonators. *Proceedings of 14th IEEE International Conference on Solid State Sensors, Actuators and Microsystems*. Lyon, pp. 2131–2134.

Tang, W.C., Nguyen, C.T.C., and Howe, R.T. 1989. Laterally driven polysilicon resonant microstructures. *Sensors and Actuators*, 20, 25–32.

Vig, J.R. 1992. Introduction to quartz frequency standards. Technical Report SLCET-TR-92-1, Army Research Laboratory, Electronics and Power Sources Directorate.

Wang, D.F., Ono, T., and Esashi, M. 2004. Thermal treatments and gas adsorption influences on nanomechanics of ultra-thin silicon resonators for ultimate sensing. *Nanotechnology*, 15, 1851–1854.

Wang, J., Butler, J.E., Feygelson, T. et al. 2004a. 1.51-GHz nanocrystalline diamond micromechanical disk resonator with material-mismatched isolating support. *Proceedings of 18th IEEE International Conference on Microelectromechanical Systems*. Maastricht, pp. 641–644.

Wang, J., Ren, Z., and Nguyen, C.T.C. 2004b. 1.156-GHz Self-aligned vibrating micromechanical disk resonator. *IEEE Transactions on Ultrasonics Ferroelectric, and Frequency Control*, 51, 1607–1628.

Wang, K., Wong, A.C., and Nguyen, C.T.C. 2000. VHF free beam high Q micromechanical resonators. *Journal of Microelectromechanical Systems*, 9, 347–360.

Wang, K., Wong, A.-C., Hsu, W.-T., and Nguyen, C.T.C. 1997. Frequency trimming and Q-factor enhancement of micromechanical resonators via localized filament annealing. *Solid State Sensors and Actuators, 1997. Conference on TRANSDUCERS '97 Chicago., 1997 International*, vol. 101, pp. 109–112.

Weinstein, D. and Bhave, S.A. 2009. Internal dielectric transduction in bulk mode resonators. *Journal of Microelectromechanical Systems*, 18, 1401–1408.

Wojciechowski, K.E., Olsson, R.H., Baker, M.S. et. al., 2007. Low vibration sensitivity MEMS resonators. *IEEE Frequency Control Symposium*, May 29–June 1, Geneva, Switzerland, pp. 1220–1224.

Yamaji, Y., Sugano, K., Tabata, O. et al. 2007. Tensile mode fatigue tests and fatigue life predictions of single crystal silicon in humidity controlled environment. *Proceedings of 20th IEEE International Workshop on Microelectromechanical Systems*. Kobe, pp. 267–270.

Yang, J., Ono, T., and Esashi, M. 2000. Surface effects and high quality factors in ultra-thin single crystal silicon cantilevers. *Applied Physics Letters, 77*, 3860–3862.

Yang, J., Ono, T., and Esashi, M. 2002. Energy dissipation in submicrometer thick single crystal silicon cantilevers. *Journal of Microelectromechanical Systems, 11*, 775–783.

Yasumura, K.Y., Stowe, T.D., Chow, E.M. et al. 2000. Quality factors in micron and submicron thick cantilevers. *Journal of Microelectromechanical Systems, 9*, 117–125.

Zener, C. 1937. Internal friction in solids. II: General theory of thermoelastic internal friction. *Physical Review, 53*, 90–99.

Ziaei-Moayyed, M., Elata, D., Hsieh, J. et al. 2009. Fully differential internal electrostatic transduction of a Lamé mode resonator. *Proceedings of 22nd IEEE International Conference on Microelectromechanical Systems*. Sorrento, pp. 931–934.

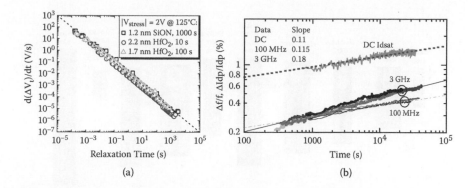

FIGURE 1.9

(a) Threshold voltage recovery for NMOS and PMOS with 2.2-nm and 1.7-nm HfO2 gate dielectric and 1.2 nm SiON. The recovery follows a log(t) dependence for both NBTI and PBTI. (From Kerber, A. et al. 2008. *IEEE Trans. Electron. Dev.*, 55, 3175–3183. With permission.) (b) Impact of AC stress using RO. Ring oscillator frequency degradation at 25OC along with I_{dsat} degradation for DC stress on PMOS transistor. Lower degradation is measured for RO as compared to DC stress. (From Nigam, T. 2008. *IEEE Trans. Dev. Mater. Reliability*, 8, 72–78. With permission.)

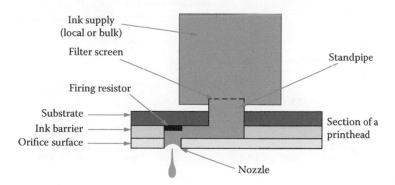

FIGURE 3.1

Cross section of thermal inkjet print cartridge.

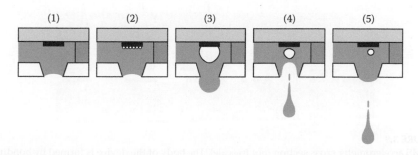

FIGURE 3.3

Thermal inkjet drop ejection process.

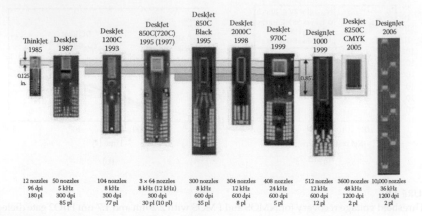

12 nozzles	50 nozzles	104 nozzles	3 × 64 nozzles	300 nozzles	304 nozzles	408 nozzles	512 nozzles	3600 nozzles	10,000 nozzles
96 dpi	5 kHz	8 kHz	8 kHz (12 kHz)	8 kHz	12 kHz	24 kHz	12 kHz	48 kHz	36 kHz
180 pl	300 dpi	300 dpi	300 dpi	600 dpi	600 dpi	600 dpi	600 dpi	1200 dpi	1200 dpi
85 pl		77 pl	30 pl (10 pl)	35 pl	8 pl	5 pl	12 pl	2 pl	2 pl

FIGURE 3.4
Evolution of Hewlett-Packard's thermal Inkjet technology.

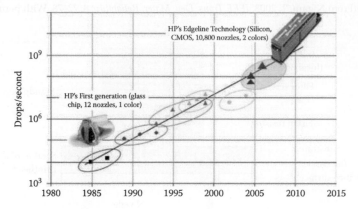

FIGURE 3.6
HP's variation on Moore's law.

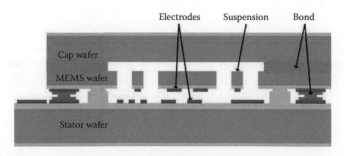

FIGURE 3.9
MEMS accelerometer cross section (not to scale). The body of the device is formed by bonding three silicon wafers and completely etching through the middle wafer to create the proof mass and flexures.

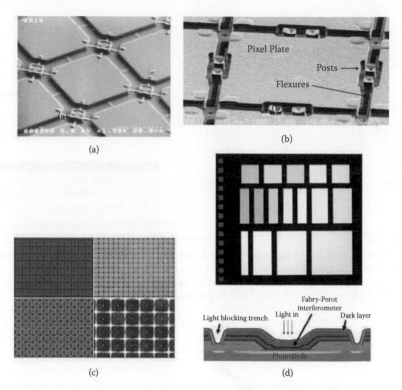

FIGURE 3.11
(a) MEMS-based optical devices. (b) MEMS pixel plate devices. (c) Interferometric pixel array.
(d) Fabry-Perot spectrophotometer chip and device cross section.

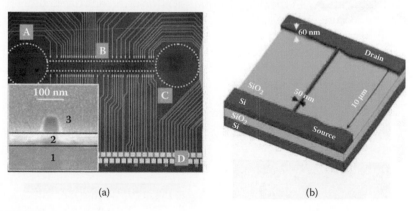

FIGURE 3.13
(a) Optical micrograph of sensor array. The outline of the flow channel is shown as dotted line
(A, B, C). The inset shows a SEM micrograph nanoscale silicon wire. Wires are insulated from
the silicon substrate (1) by the buried oxide (2) and a thermally grown gate oxide with a thickness of ~3 nm acts as a gate dielectric for ChemFET operation of the silicon wires (3). (b) Atomic
force microscopy image of 50-nm Si nanowire (with substrate).

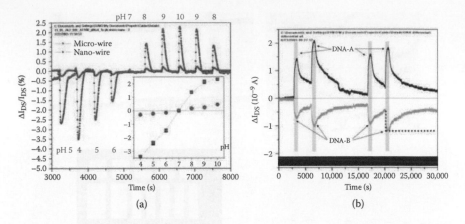

(a)   (b)

FIGURE 3.14
(a) Response of drain-source current of 2-μ and 50-μ wide silicon wires to injection of analyte solutions with different pH. In the inset, the response of nanoscale wires to a change in pH is up to one order of magnitude higher than that of micro-scale wires.(b) Differential sensor signal for injections of two different DNA sequences (A and B).

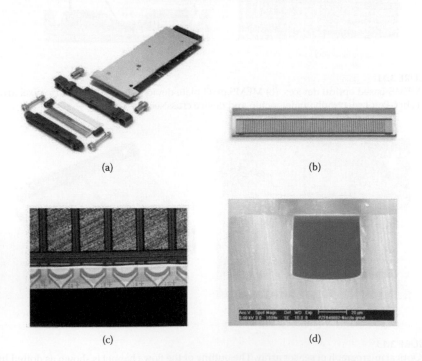

(a)   (b)

(c)   (d)

FIGURE 3.15
(a) Major parts assembled to produce the HP X2 piezoelectric printhead assembly. (b) Closer view of silicon and glass printhead. (c) Four pumping chambers. (d) Edge-on view of a single X2 inkjet nozzle.

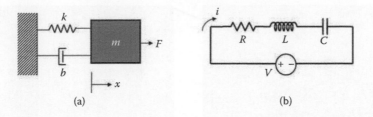

FIGURE 4.1
(a) Simple mass spring damper lumped element mechanical resonator. (b) Resistor–capacitor–inductor electrical resonator.

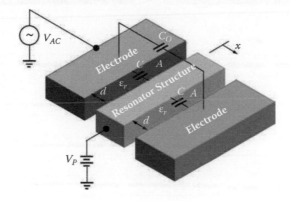

FIGURE 4.3
Conceptual schematic of capactively transduced resonator.

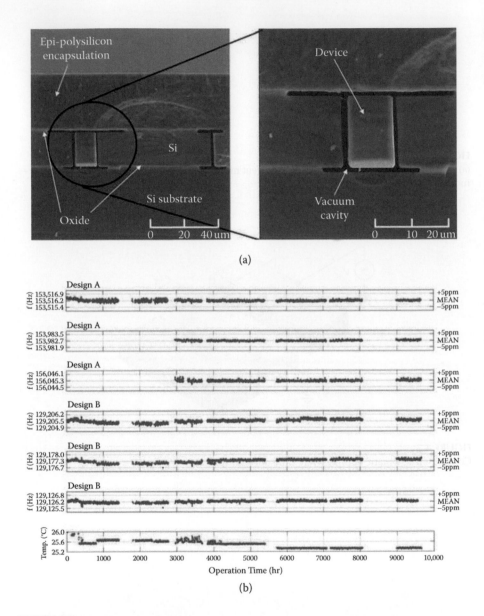

FIGURE 4.7
Demonstration of long-term stability of silicon-based MEMS resonators. Silicon MEMS resonators are encapsulated in an epitaxially deposited wafer-level packaging as shown in the SEM images (a). Their resonance frequency has not changed more than a few ppm (within the measurement error range) during more than a year of operation (b). (From Kim, B., Candler, R.N, Hopcroft, M. et al. 2007. *Sensors and Actuators A*, 136, 125–131. With permission.)

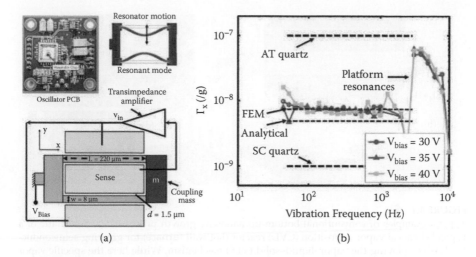

(a)   (b)

FIGURE 4.8
Vibration sensitivity measurement of silicon MEMS resonators. The vibration sensitivity of MEMS resonators was shown to be on part with mid-grade quartz crystal resonator-based oscillators. (From Agarwal, M., Park, K.K., Chandorkar, S.A. et al. 2007. *Applied Physics Letters*, 90, 14–103. With permission.)

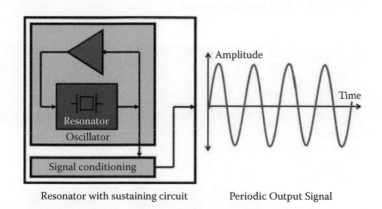

Resonator with sustaining circuit   Periodic Output Signal

FIGURE 4.9
Frequency reference composed of resonator with an oscillator circuit.

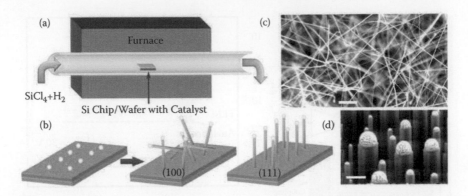

FIGURE 5.1
Typical examples of conventional bottom-up nanowire growth processes. (a) Schematic of a typical chemical vapor deposition (CVD) reactor (hot wall furnace) for growing semiconductor NWs employing the vapor–liquid–solid (VLS) mechanism. While here the specific vapor precursors are for Si NWs growth, the same thermal CVD setup can be used for other NWs by varying precursors. (b) Simplified schematics showing the VLS growth process, using Si NWs growth as an example, on Si (100) and (111) substrates, respectively. The gold color spheres are the catalyst nanoparticles, and the light blue rods highlight the Si NWs. (c) A typical SEM image of Si NWs grown on Si (100) substrate. Scale bar is 1μm. (d) A typical SEM image of vertically aligned Si NWs grown on Si (111) substrate. Scale bar is 300nm. (From Gao, D., et al., 2005, *J. Am. Chem. Soc.* 127, 4574–4575. With permission.)

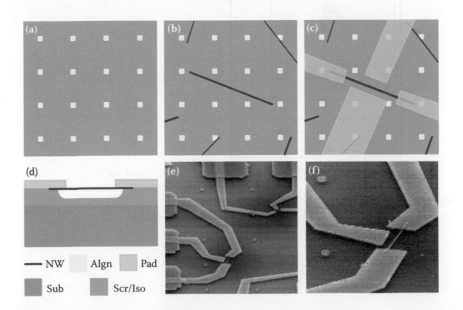

FIGURE 5.2
Conventional nanowire and other one-dimensional NEMS device fabrication processes. (a) EBL patterning markers. (b) NWs are distributed on the substrate and coordinates identified. (c) EBL patterning of electrodes. (d) Device suspension. (e) and (f) SEM images of devices.

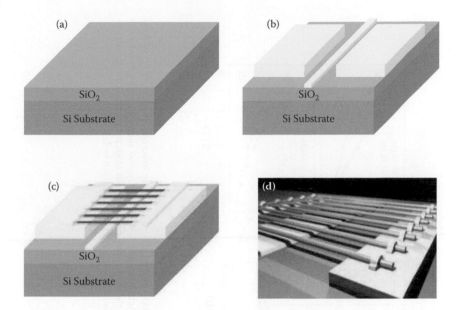

FIGURE 5.3
Recent development of new transfer print technique for making self-aligned nanowire NEMS devices and arrays.

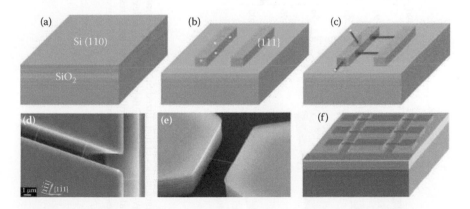

FIGURE 5.4
Recently developed novel hybrid bottom-up and top-down approach for making self-aligned suspended nanowire NEMS devices and arrays.

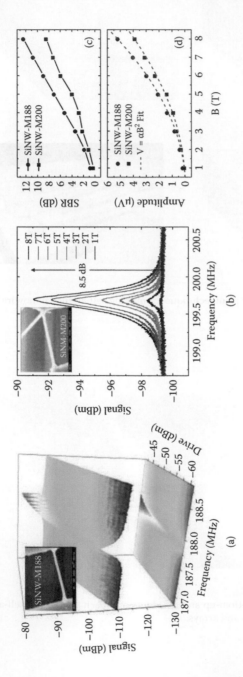

FIGURE 5.5

Measured resonance characteristics of metallized SiNW VHF resonators. (a) Magnetomotively transduced response [referred-to-input (RTI) of first-stage amplifier] from SiNW-M200 at varying *B* fields (RF drive to device is –46 dBm). (b) Achieved detection efficiency and signal-to-background ratio (SBR) in decibels. (c) Actual voltage signal amplitude of the SiNW resonances displaying the B^2 dependence expected from magnetomotive transduction. (d) Magnetomotive response (RTI) from SiNW-M188 as RF drive to the device sweeps from –61 to –41 dBm (at *B* = 6 tesla).

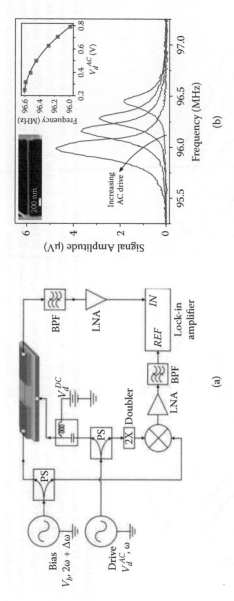

(a)

(b)

FIGURE 5.6

Very high frequency (VHF) Si nanowire resonators with fully integrated electrostatic actuation and piezoresistive self-detection. (a) Bias and drive circuitry. (b) Performance of 40 nm-thick, 96MHz nanowire resonator with quality (Q) factor ~550. It is 1.8 μm long and has a dc resistance of 80 kΩ. The AC drive is set at $V_{d,AC}$ = 0.50, 0.63, 0.71, and 0.79 V for the curves, respectively, with DC voltage fixed at $V_{d,DC}$ = 0.2 V. The left inset is an SEM image of the device. The right inset shows the drive dependence of the resonance frequency. (c) Performance of a 30 nm-thick, 75MHz Si nanowire resonator with Q ~700. Its length is 1.8 μm and its dc resistance is 300 kΩ. The curves are taken at different bias voltages with the same drive. The insets show the SEM image and the linear dependence on bias voltage. (d) Data taken at varied drives with the same bias. Inset shows quadratic dependence on drive voltage. (Continued)

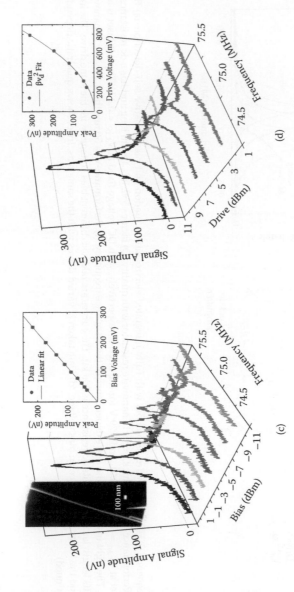

FIGURE 5.6 (CONTINUED)

Very high frequency (VHF) Si nanowire resonators with fully integrated electrostatic actuation and piezoresistive self-detection. (a) Bias and drive circuitry. (b) Performance of 40 nm-thick, 96MHz nanowire resonator with quality (Q) factor ~550. It is 1.8 μm long and has a dc resistance of 80 kΩ. The AC drive is set at $V_{d,AC}$ = 0.50, 0.63, 0.71, and 0.79 V for the curves, respectively, with DC voltage fixed at $V_{d,DC}$ = 0.2 V. The left inset is an SEM image of the device. The right inset shows the drive dependence of the resonance frequency. (c) Performance of a 30 nm-thick, 75MHz Si nanowire resonator with Q ~700. Its length is 1.8 μm and its dc resistance is 300 kΩ. The curves are taken at different bias voltages with the same drive. The insets show the SEM image and the linear dependence on bias voltage. (d) Data taken at varied drives with the same bias. Inset shows quadratic dependence on drive voltage.

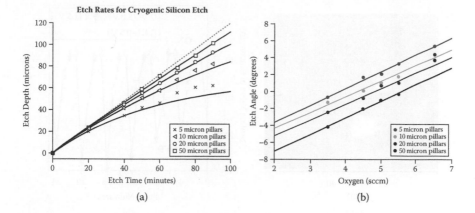

FIGURE 6.12
(a) Graph of theoretical and experimental results of cryogenically etching silicon that illustrates the aspect ratio-dependent etching behavior. (b) Graph of experimental data showing etch angle dependence on oxygen flow rate for cryogenic silicon etching.

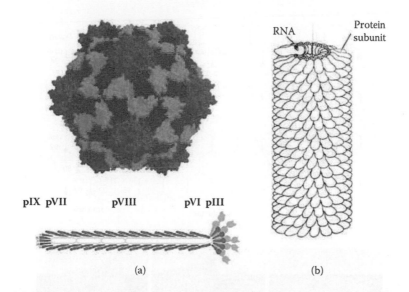

FIGURE 7.1
a) CPMV structure is shown at top. The CPMV is composed of 60 copies of an asymmetric unit made from a small subunit (blue) and a two-part large subunit (green and red). The M13 bacteriophage structure is shown at bottom. The five structural proteins of the M13 virus are labeled. (b) TMV structure. The 2130 identical structural proteins are arranged around the single-stranded RNA that holds its genetic information. (From Caspar, D.L.D. 1963. *Advances in Protein Chemistry*, 18, 37–121; Lin, T.W., Chen, Z.G., Usha, R. et al. 1999. *Virology*, 265, 20–34; Flynn, C.E., Lee, S.W., Peelle, B.R. et al. 2003. *Acta Materialia*, 51, 5867–5880. With permission.)

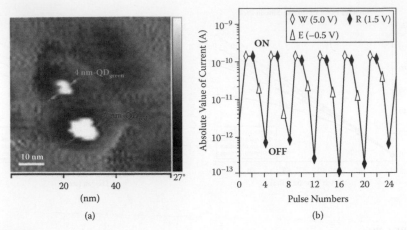

FIGURE 7.6
(a) AFM phase image of CPMV with one red and one green quantum dot bound to the CPMV surface. (b) Current measurements for several write-read-erase cycles applied to the CPMV-QD device. (From Portney, N.G., Martinez-Morales, A.A., and Ozkan, M. 2008. *ACS Nano*, 2, 191-196. With permission.)

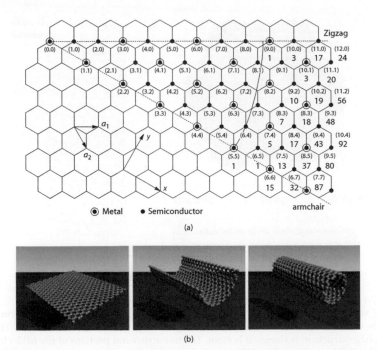

FIGURE 8.2
(a) Chiral orientations of CNTs represented by alignment of CNT axis with particular lattice points of graphene sheet that determine n,m index of CNT. (From Dresselhaus, M.S., Dresselhaus, G., and Saito, R. 1995. *Carbon*, 33, 883–891. With permission.) (b) Visualization of SWNT structure by wrapping of graphene sheet.

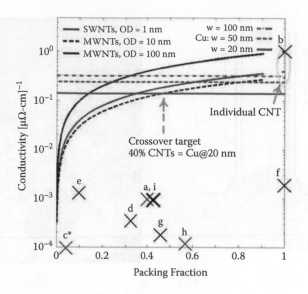

FIGURE 8.3
Previously published calculations of electrical conductivity along continuous CNTs as related to CNT packing fraction compared to previously measured electrical data for individual CNTs and CNT networks. The packing fraction is defined as the ratio of the CNT areal density (number of CNTs per square centimeter) to the areal density of hexagonally packed CNTs (modeled as cylinders) having the same outer diameter. Solid and dashed x marks are MWNTs and SWNTs, respectively. We assume that the conductivity of an aligned CNT network is linearly proportional to the number of CNTs per unit cross-sectional area and that the constant of proportionality is the conductivity of an individual CNT. (From (a): Fischer, J.E., Zhou, W., Vavro, J. et al. 2003. *Journal of Applied Physics*, 93, 2157–2163. (b): Li, Y.L., Kinloch, I.A., and Windle, A.H. 2004. *Science*, 304, 276–278. (c): Kang, S.J., Kocabas, C., Ozel, T. et al. 2007. *Nature Nanotechnology*, 2, 230–236. (d) Atkinson, K.R., Hawkins, S.C., Huynh, C. et al. 2007. *Physica B*, 394, 339–343. (e): Yokoyama, D., Iwasaki, T., Ishimaru, K. et al. 2008. *Japan Journal of Applied Physics*, 47, 1985–1990. (f): Close, G.F. and Wong, H.S.P. 2007. *Proceedings of IEEE International Electronic Devices Meeting*, pp. 203–206. (g): Wang, D., Song, P.C., Liu, C.H. et al. 2008. *Nanotechnology*, 19. (h): Hayamizu, Y., Yamada, T., Mizuno, K. et al. 2008. *Nature Nanotechnology*, 3, 289–294. i: Tawfick, S., O'Brien, K., and Hart, A.J. 2009. *Small*, 5, 2467–2473. With permission.)

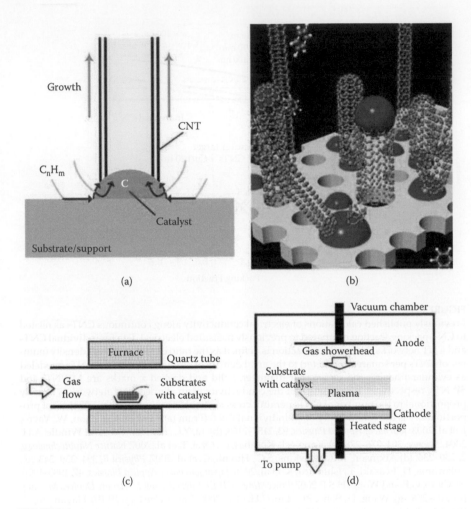

FIGURE 8.4
Individual substrate-bound CNTs growing by CVD. (a) Base growth of CNT. A gaseous carbon source diffuses at a metal catalyst particle that remains attached to the substrate and a CNT grows upward from the surface of the particle. (b) Base and tip growth of CNTs rooted in nanoporous (zeolite) substrate. (From Hayashi, T., Kim, Y.A., Matoba, T. et al. 2003. *Nano Letters*, 3, 887–889. With permission). Classical furnace designs for CVD synthesis of CNTs and like nanostructures. (From Teo, K.B.K., Singh, C., Chhowalla, M. et al. 2004. *Encyclopedia of Nanoscience and Nanotechnology*. With permission.) (c) Horizontal tube furnace with fixed catalyst. (d) Low pressure plasma-enhanced (PECVD) chamber with heated stage.

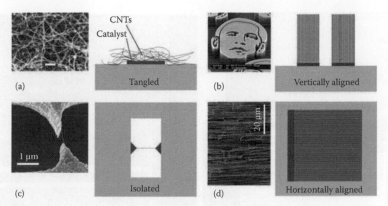

FIGURE 8.5
Classical morphologies of CNTs grown on substrates. (a) Tangled film that terminates around micrometer thickness due to steric hindrance among CNTs. (From Hart et al., 2006. *Carbon*, 44, 348–359. With permission). (b) Vertically aligned film that can grow uniformly to millimeter thickness. (www.nanobama.com). (c) Single CNT suspended over microfabricated channels. (From Jungen, A., Durrer, L., Stampfer, C. et al. 2007. *Physics Status Solidi B*, 244, 4323–4326. With permission.) (d) Horizontally aligned CNTs on substrate (From Kocabas, et al., 2005. *Small*, 1, 1110–1116. With permission.)

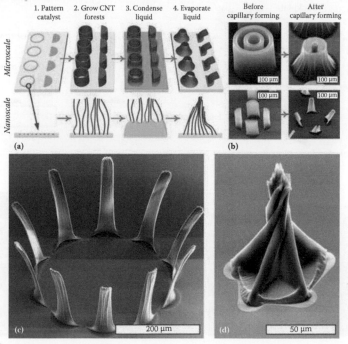

FIGURE 8.8
Diverse 3D microarchitectures made by capillary forming of carbon nanotubes. (a) Capillary forming process showing micro- and nanoscale detail. (b) Concentric wells and bending structures before and after capillary forming. (c) Blooming flower of bending structures. (d) Micro-helix made by combining contraction and bending operations. (From De Volder, M., Tawfick, S, Park, S.J. et al. 2010a. *Advanced Materials*, 22, 4384–4389. With permission.)

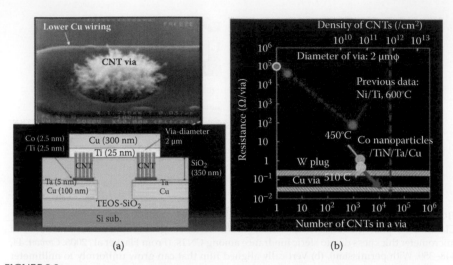

FIGURE 8.9
Integration and performance of vertical CNTs as microelectronic interconnects by Fujitsu.

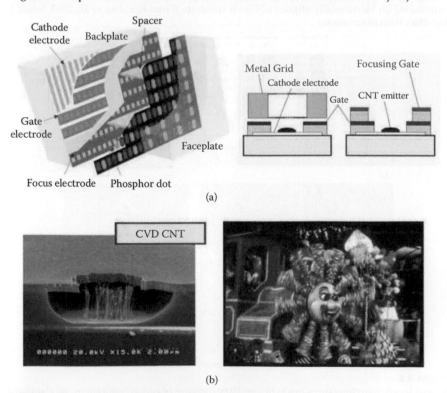

FIGURE 8.10
Samsung CNT field emission display. (a) Display structure and triode pixel architecture. (b) SEM image of individual pixel with aligned CNTs grown directly on substrate, and image of prototype 15-inch diagonal display.

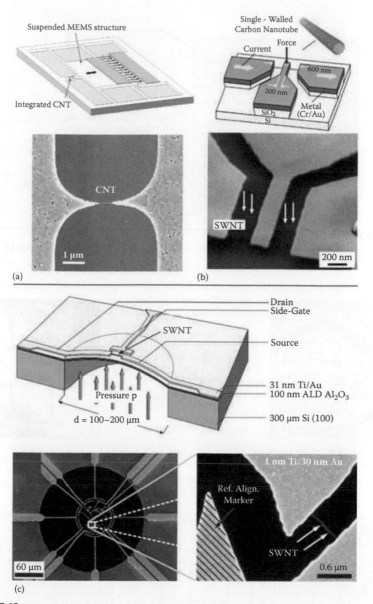

FIGURE 8.12

(a) NEMS with integrated SWNT by direct growth and comb drive for sensing and actuation. SEM image shows a free-standing SWNT bridging two poly-Si tips. (b) SWNT-based nanoscale sensor system that may be used for measuring electrical responses to mechanical deformations applied at the center cantilever via an AFM tip. SEM image shows SWNT bridging two electrodes with the center cantilever exerting a force downward on the SWNT. (c) CNT-based pressure sensor consisting of an ultrathin Al_2O_3 membrane with SWNT adhering to it with electrodes. SEM images of the device show electrodes extending onto the membrane (black circle) and electrically contacting SWNT (left image), which is shown at higher magnification in the right image.

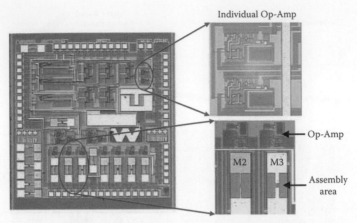

FIGURE 9.1
CMOS chip designed and fabricated using the AMI 0.5-μm CMOS process provided by MOSIS.

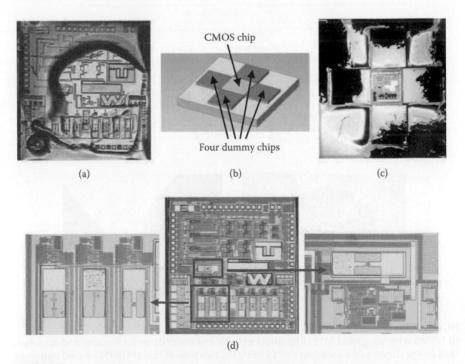

(a)   (b)   (c)

(d)

FIGURE 9.3
(a) Optical photograph of CMOS chip after spinning photoresist on top. Due to the edge bead problem, optical lithography on a CMOS chip is not possible. (b) Placing four dummy chips around the CMOS chip during photoresist application to solve the edge bead problem. (c) Optical photograph of CMOS chip surrounded by four dummy chips. (d) Optical photograph of CMOS chip after photolithography. Two inset images show close-ups of lithography patterns.

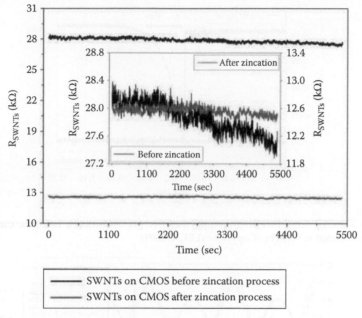

SWNTs on CMOS before zincation process
SWNTs on CMOS after zincation process

FIGURE 9.5
Comparison of resistances of SWNTs assembled onto metal electrodes before and after the zincation process. The inset shows magnified resistance measurements from the SWNTs versus time. Before zincation, the resistance of the SWNTs assembled onto the CMOS circuitry was around 27.86 ± 0.22 KΩ measured for 90 min. After zincations, the measured resistance of SWNTs assembled onto CMOS circuitry was about 12.56 ± 0.05 KΩ during the same period. Therefore, the zincation process improves the SWNT-to-electrode contacts by decreasing contact resistance by about 54.9% and stabilizes the value with a smaller standard deviation.

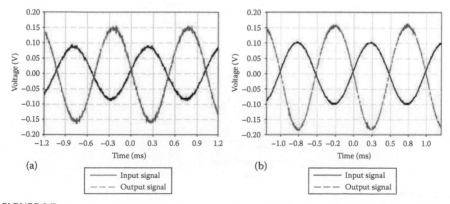

Input signal
Output signal

Input signal
Output signal

FIGURE 9.7
Measured input and output signals from inverting op-amp with SWNTs as reference resistors. (a) Small-signal ac gain measurement from CNTs on two-dimensional electrodes. $R_i = 22$ KΩ; measured output gain is about –1.95. (b) Three-dimensional electrodes with small-signal ac gain measured about –1.98 (multi-finger electrodes). (Copyright IEEE, 2008.)

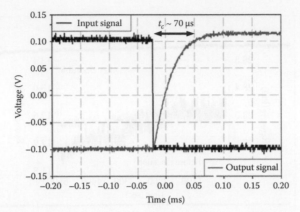

FIGURE 9.8
Frequency response from inverting op-amp with SWNTs as feedback resistors. Cutoff frequency is estimated at 9.5 kHz. (Copyright IEEE, 4/2009.)

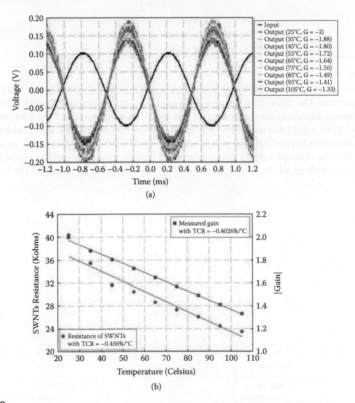

(a)

(b)

FIGURE 9.9
(a) Measured output voltage of CMOS amplifier with integrated SWNT feedback resistor in response to variations in temperature. (b) Measured temperature response of SWNT thermal sensor. The measured temperature coefficient of resistance (TCR) calculated from the operational amplifier output is −0.4%/°C and agrees well with the measured TCR of −0.43%/°C obtained from assembled SWNTs on plain electrodes.

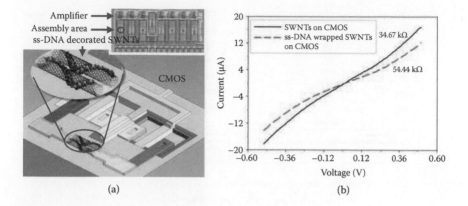

(a)   (b)

FIGURE 9.10
(a) Optical photograph of CMOS chip and schematic of DNA decorated SWNT sensors integrated onto CMOS circuitry. (b) I–V characterization of SWNTs on CMOS circuitry before and after ss-DNA decoration. The decoration was found to increase the resistance while maintaining sufficient conduction.

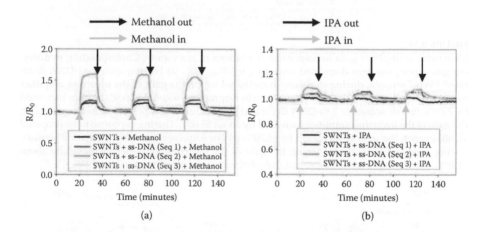

(a)   (b)

FIGURE 9.12
Change in sensor resistance upon chemical vapor exposure. Resistances are normalized to the value when exposed to air. (a) Bare SWNTs respond to methanol vapor (black line) with about $13.41 \pm 1.03\%$ increase in resistance. The SWNTs decorated with ss-DNA show enhanced response to methanol (red, green and yellow lines for sequences 1, 2, and 3, respectively) with resistance increases of about 18.43 ± 0.81, 58.02 ± 3.36, and $24.7 \pm 1.34\%$, respectively. (b) Bare SWNTs respond to isopropanol alcohol vapor (black line) with about $3.23 \pm 0.50\%$ increase in resistance. The same SWNTs decorated with ss-DNA show enhanced response to isopropanol alcohol (red = sequence 1; green = sequence 2; yellow = sequence 3) with resistance increases of about $5.65 \pm 0.30\%$, $11.25 \pm 0.33\%$, and $7.38 \pm 0.49\%$.

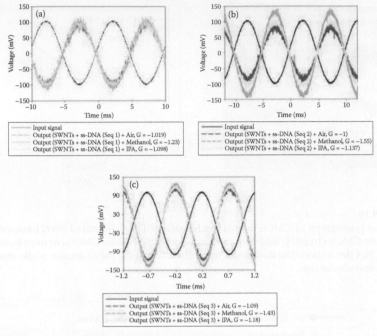

FIGURE 9.14

Measured ac amplifier gain in response to variations in gas vapors. Corresponding to different gas vapors, the gain decreased according to $-R_{SWNT}/R_i$. (a) For sequence 1 ss-DNA decorated SWNT sensors, during exposure to methanol vapor, the gain of the inverting amplifier increased by about 20.71%. During exposure to isopropanol alcohol vapor, the gain increased by 7.75%. (b) For sequence 2 ss-DNA decorated sensors, during exposure to methanol vapor, the gain of the inverting amplifier increased by about 55.00%. During exposure to isopropanol alcohol vapor, the gain is increased by 13.70%. (c) For sequence 3 ss-DNA decorated sensors, during exposure to methanol vapor, the gain of the inverting amplifier increased by about 31.19%. During exposure to isopropanol alcohol vapor, the gain increased by 8.25%.

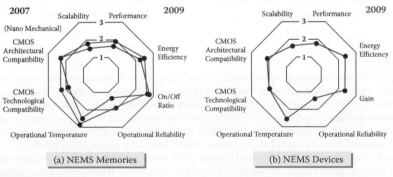

FIGURE 10.2

Technology performance evaluation predicted by the ITRS (2009). (a) NEMS memories; the blue and red lines correspond to the data available for years 2007 and 2009, respectively. (b) NEMS devices. Various properties are indicated with numbers; 1, 2, and 3 correspond to poor, average, and excellent, respectively.

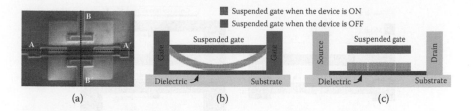

FIGURE 10.3
Suspended gate (SG) transistor structure. (a) SEM of fabricated device. (b) Cross-sectional view along AA′ line when the device is in the on and off states. (c) Cross-sectional view along BB′ line when the device is in the on and off states. (From Abelé, N., Fritschi, R., Boucart, K. et al. 2005. *Proceedings of International Electron Devices Meeting*, pp. 479–481. With permission.)

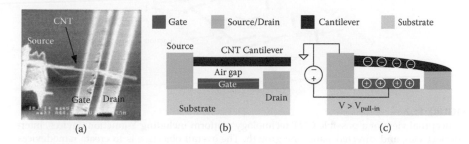

FIGURE 10.4
Vertically actuated cantilever-based NEMS relay. (a) SEM of fabricated device (From Lee S., Lee, D., Morjan, R. et al. 2004. *Nano Lett.*, 4, 2027–2030. With permission.). (b) Schematic of device in off state. (c) Schematic of device in on state.

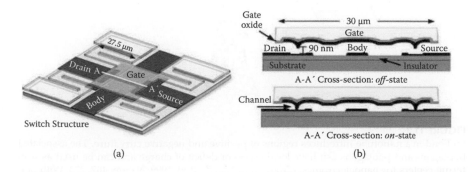

FIGURE 10.12
Schematic of NEMS device. (a) Top view. (b) Cross sections along AA′ in the off state (top) and on state (bottom). (From Chen, F., Spencer, M., Nathanael, R. et al. 2010. *Proceedings of International Solid-State Circuits Conference*, pp. 26–28. With permission.)

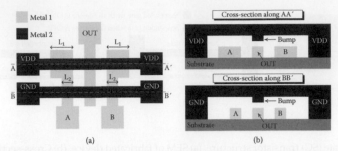

FIGURE 10.14

NEMS-based NAND. (a) Top view diagram of logic gate. (b) Cross sections of structure along its beams [AA′ (top) and BB′ (bottom) lines]. (From Akarvardar et al. 2007. With permission.)

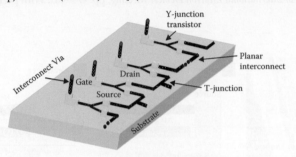

FIGURE 11.2

Conceptual view of a possible CNT technology platform including Y-junction devices, interconnect vias, and directed nanotube growth. The overall objective is to create nanodevices with novel functionalities that go beyond existing technologies.

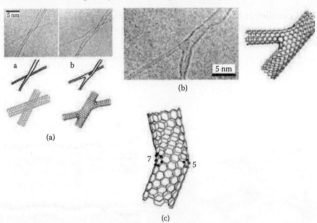

FIGURE 11.3

(a) Bend in a nanotube introduces regions of positive and negative curvature. The associated heptagons and pentagons can have local excess or deficit of charge and can be used as scattering centers for nanoelectronics. (From From Yao, Z. et al. 1999. *Nature*, 402, 273. With permission.) (b) X-shaped and (c) Y-shaped nanotube molecular junctions can be fabricated by irradiating crossed single-walled nanotube junctions with high energy (~1.25 MeV) and beam intensity 10 A/cm²) electron beams. (From Terrones, M. et al. 2002. *Physics Review Letters*, 89, 075505. With permission.)

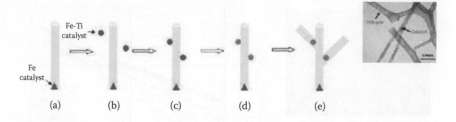

FIGURE 11.4
Postulated growth sequence of Y-junction nanotube. (From Gothard, N. et al. 2004. *Nanoletters*, 4, 213. With permission.) (a) Initial seeding of straight nanotube through conventional catalytic synthesis. (From Teo, K.B.K. et al. 2004. In Nalwa, H.S., Ed., *Encyclopedia of Nanoscience and Nanotechnology*. Stevenson Ranch, CA: American Scientific Publishers. With permission.) (b) Ti-doped Fe catalyst particles (from ferrocene and $C_{10}H_{10}N_4Ti$) attach (c) to sidewalls and nucleate (d) the side branches (e).

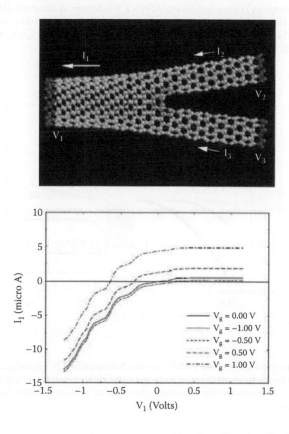

FIGURE 11.6
Asymmetry and rectification-like behavior in current and voltage characteristics of single-walled Y-junction nanotube is indicated through quantum conductivity calculations. (From Andriotis, A.N. et al. 2001. *Physics Review Letters*, 87, 066802. With permission.)

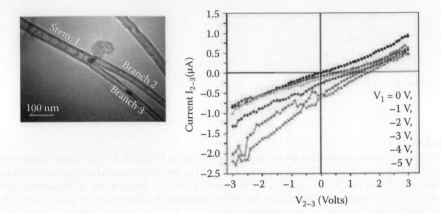

FIGURE 11.9
Current (I) and voltage (V) characteristics of MWNT Y-junction. A constant direct current voltage is applied on the stem, while the I–V behavior across branches 2 and 3 is monitored. The gating action of the stem voltage (V_1) and the asymmetric response are to be noted. (From Bandaru, P.R. et al. 2005. *Nature Materials*, 4, 663. With permission.)

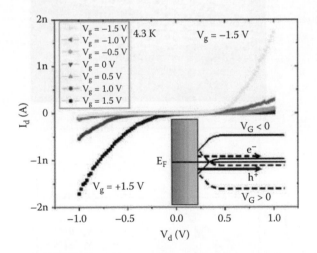

FIGURE 11.10
I–V measurements on single walled Y-CNTs in which a metallic CNT interface with a semiconducting CNT (see band diagram in inset) indicates ambipolar behavior as a function of applied gate voltage on the semiconducting nanotube. (From Kim, D.H. et al. 2006. *Nanoletters*, 6, 2821. With permission.)

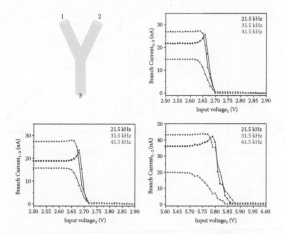

FIGURE 11.11

Abrupt modulation of current through two branches of Y-junction, indicative of electrical switching, revealed by varying voltage on the third branch. The voltage at which the switching action occurs on the two branches (1 and 2) is similar and smaller (~2.7 V; see a and b) compared to the turn-off voltage (~5.8 V) on the stem (3) in c. Such abrupt switching characteristics are seen up to 50 kHz, the upper limit arising from the capacitive response of the Y-junction. (From Bandaru, P.R. et al. 2005. *Nature Materials*, 4, 663. With permission.)

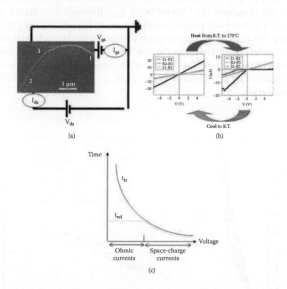

FIGURE 11.13

(a) Scanning electron micrograph showing circuit arrangement used to probe the current (I) and voltage (V) characteristics of Y-CNT. (b) Reversible current blocking behavior induced in CNT Y-junction. The individual segments' electrical transport characteristics exhibit different blocking and linear characteristics and are geometry dependent. (c) Transition from ohmic behavior to space charge behavior is a function of voltage and can be accelerated at higher temperatures. The ratio of the transit time (t_{tr}) to dielectric relaxation time (t_{rel}) determines the dynamics of the carrier transport.

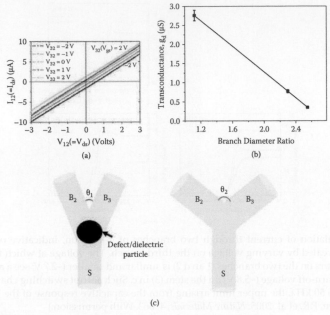

(a)

(b)

(c)

FIGURE 11.14
(a) Current (I) and voltage (V) characteristics of Y-CNT consisting of metallic branches. The inset shows the circuit diagram. (b) Conductance (g_d) through two nanotubes of different diameters in the Y-CNT is inversely proportional to their ratio (= $d_>/d_<$). (c) Proposed Y-junction CNT based switching devices. (i) Smaller angle (θ_1) between the branches (B_2 and B_3) can result in a higher gating efficiency for the stem (S). (ii) Y-CNT with uniform gating and electrical switching characteristics can be fabricated by synthesizing all the constituent nanotubes to be of the same diameter and θ_1 = 120 degrees. (From Park, J. et al. 2006. *Applied Physics Letters*, 88, 243113. With permission.)

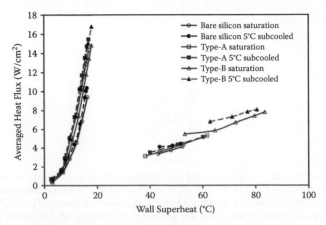

FIGURE 12.1
Pool boiling curve of PF-5060 on silicon and on CNT-coated surfaces. (From Sathyamurthi, V. et al. 2009. *J. Heat Transfer*, 131, 071501. With permission.)

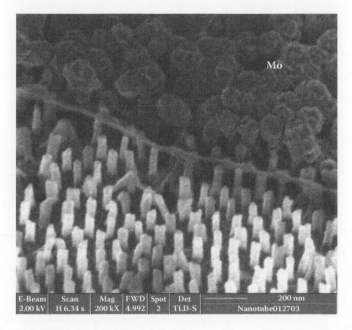

FIGURE 13.1
Scanning electron micrograph of vertical array of carbon nanotubes on silicon substrate. (From Choi, D. et al. 2003. *Proceedings of IEEE Nano 2003 Conference*, Seoul. With permission.)

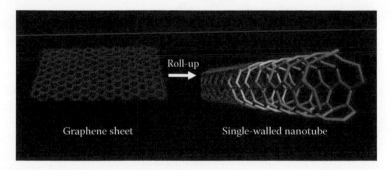

FIGURE 13.3
Conversion of graphene sheet to single-walled nanotube.

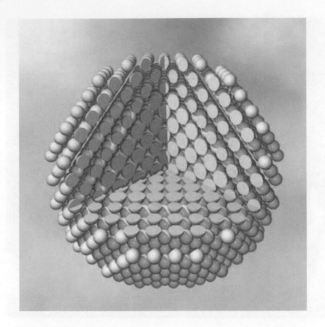

FIGURE 13.6
Gold core–silver shell bimetallic nanoparticle.

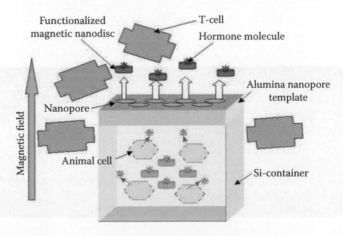

FIGURE 13.7
Drug delivery system based on magnetic nanodiscs manipulated in nanochannels. The animal cells produce hormones or antibodies responding to environmental changes caused by T cells. The functionalized magnetic nanodiscs attached by hormone molecules are transported out of the container via alumina nanopores that selectively apply a magnetic field.

5

Nanoscale Electromechanical Devices Enabled by Nanowire Structures

Philip X.-L. Feng

CONTENTS

ABSTRACT This chapter presents a brief overview of the emerging nanoscale electromechanical devices based on nanowire structures enabled by state-of-the-art nanofabrication techniques. In particular, the overview focuses on nanowire structures that have controllable mechanical degrees of freedom and can be engineered into functional electromechanical devices. Especially interesting are movable nanowires that can vibrate in their mechanical resonant modes. We discuss here some of the recent important advances in materials, nanofabrication techniques, device technologies, and potential applications of nanowire-based electromechanical systems.

Developments of very high frequency nanomechanical resonators with high quality factors employing various nanowires in metal, semiconductor, and polymer materials represent noticeable research milestones and appear to be promising for various sensing technologies. These advances have been made possible by new approaches in realizing suspended nanowires and by novel means of precision detection of nanowire motions. At the frontiers of the area, we face intriguing challenges such as improving device merit and performance (e.g., scaling up both resonance frequency and quality factor of nanowire resonators), interfacing electronics, and monolithic and large-scale integration. While addressing such challenges, the growing research efforts are expected to generate new enabling technologies for sensing, computing, and other novel information processing functions.

5.1 Introduction

The past decade has witnessed significant growth of attention and dramatic increases of research efforts dedicated to nanowire structures. Usually such structures have widths or diameters on the orders of a few nanometers to a few hundred nanometers, and lengths ranging from sub-microns to hundreds of microns and even millimeters. This low-dimensional nature in structure makes such nanowires a generic type of one-dimensional (1D) nanomaterials, often with large aspect ratios (e.g., a 10-μm long, 10-nm thick nanowire readily yields an aspect ratio of 1000).

Such 1D structures lead to interesting properties at the nanoscale, mainly related to the novel transport characteristics of energy and information carriers. For example, phonon and electron transport in nanowires could also be approximately viewed as 1D process in certain cases. In fundamental research, remarkable progress has been made in probing and understanding the properties of nanowires, primarily in the areas of materials science, chemistry, and physics. Great effort is being expended toward engineering nanowire structures into various functional devices for a very wide spectrum of technological applications.

Exploration and understanding of nanowire material properties, principles of synthesis and nanofabrication and the transport characteristics of carriers in nanowires are all very intriguing and important. They help lay a foundation for nanowire-based device engineering and technology development. Based on their attractive and highly promising properties, semiconducting and metallic nanowires have spawned significant research interest and efforts in nanowire device physics and engineering targeting on many potential applications. While we appreciate the exponential growth of research and advances in the field of nanowire devices, the overall scope of nanowire research is too vast to cover and far beyond the scope of this brief book chapter.

Recent excellent review articles have addressed the achievements and challenges of various aspects of this rapidly expanding field. A partial list may include, for examples, focused reviews of chemical synthesis and characterization of nanowires and other 1D materials (Xia and Yang et al., 2003; Law et al., 2004), control of growth details (Fortuna and Li, 2010), growth and electronic properties of silicon nanowires (Schmidt et al., 2009), semiconductor nanowires as potential logic building blocks for future electronics (Lu and Lieber, 2006, 2007), nanowire electronic and optoelectronic devices, lasers, and nanophotonics (Li et al., 2006; Pauzauskie and Yang, 2006; Yan et al., 2009; Zimmler et al., 2010), nanowire plasmonics (Ditlbatcher et al., 2005; Yan et al., 2009), integrating nanowire electronic and photonic devices on flexible substrates (McAlpine et al., 2005), and nanowire biosensors with potentials for medicine and life sciences (Patolsky and Lieber, 2005; Patolsky et al., 2006).

Moreover, nanowires have recently also been very actively explored for energy conversion and potential renewable energy solutions such as thermoelectrics (Hochbaum et al., 2008; Boukai et al., 2008) and photovoltaics (Law et al., 2005; Tian et al., 2007 and 2009; Garnett and Yang, 2008 and 2010; Kelzenberg et al., 2008 and 2010; Boettcher et al., 2010). We note that these represent only limited selections from developing research efforts targeted at exploring and exploiting nanowire structures and devices. Today, nanowires and related research topics continue to be important foci of scientific journals and technical conferences and symposia. Certain highly developed nanowire materials and device technologies are being adopted in relevant industry sectors.

Mechanical motion is another fundamental feature that can be explored and exploited, in addition to all the aforementioned attractions of nanowire structures. Enabling mechanical degrees of freedom in nanowires allows the construction of movable devices, sensors, and transducers that can directly couple to the mechanical domain, and nanowire machines that can both sense and actively actuate other structures or devices.

In particular, the nanomechanical devices that operate in their resonance modes—resonant nanoelectromechanical systems (NEMS)—are of great interest due to the remarkable benefits attained simply by scaling the device dimensions (Roukes, 2000 and 2001; Craighead, 2000). Such benefits include surprisingly high resonance frequency and operating speed; ultrasmall mass, volume, and footprint; and ultralow power consumption, to name a few. In principle these characteristics imply ultrafast or ultrawide bandwidth, ultralow power devices and their capability for very large scale integration (VLSI). These attractions have been well envisioned by researchers in physics where curiosity about various mesoscopic systems and quantum-limited phenomena continue to drive searches for all types of ultrasmall and sensitive devices as probes. The benefits have also been well perceived by engineers from the conventional microelectronics domains where further miniaturization of microelectromechanical systems (MEMS) has been an important impetus.

Nanowires have been highly interesting for NEMS because they offer the unique advantage of allowing NEMS devices at molecular scales with ideally terminated surfaces. Moreover, nanowires offer a variety of materials choices and device platforms, thus enabling possibilities of diverse emerging applications. This chapter focuses on introducing nanowire-based NEMS and resonant devices: the materials, nanofabrication, devices and their performance, technology perspectives, and potential applications.

5.2 Nanowire Materials

In the past two decades of active research, many nanowire materials have been synthesized, a number of generic techniques for producing nanowire structures have been developed, procedures established, and recipes optimized. These activities have led to a growing library of processes and property data along with a versatile toolbox that many researchers have found interesting for engineering nanoscale devices with new functions.

In developing NEMS devices, one would mainly consider the mechanical properties of the nanowire materials that could be exploited and interesting electromechanical coupling effects that could be engineered. The constitutive properties that could be important include (1) elasticity and plasticity, (2) thermal transport capability, (3) electromechanical effects native to the material, (4) crystalline structures and defects, and (5) temperature dependence of mechanical properties. Other factors specific to making NEMS resonators include Young's modulus E_Y, the square root of modulus-to-density ratio $(E_Y/\rho)^{1/2}$ (proportional to the speed of sound in the material), crystalline microstructure (single- or poly-crystal, isotropic or anisotropic, polytype if applicable), surface roughness, and internal friction.

Table 5.1 summarizes a limited set of materials that we believe could be of interest for nanowire NEMS device research. It also lists their basic properties of interest.

5.3 Fabrication of Nanowire NEMS

Growing materials chemically with precise controls at a molecular or nanometer length scale has been an important theme in modern materials and chemistry research. This so-called bottom-up paradigm of nanoscience and nanotechnology led to many advances in new materials, devices, and technologies, and opened up opportunities to exploit electronic and optical processes in innovative nanoscale systems. Important examples of functional nanomaterials developed via the bottom-up approach (Figure 5.1) include nanocrystals, nanotubes, and nanowires (NWs).

Before the advent of nanowires, the existing fabrication techniques in NEMS research were almost completely limited to the top-down method Integrating NWs grown with NEMS from the bottom up will allow exploitation of the superb crystal quality and control at a scale of 10 nm or smaller into a regime previously difficult to achieve in the top-down paradigm.

TABLE 5.1

Selected Nanowire Materials of Interest for NEMS

Nanowire Material	Elastic Modulus E_Y (GPa)	Speed of Sound (m/s)	Crystal Lattice	Growth Orientation	Catalyst or Seed Nano- particles	Synthesis Temperature (°C)	Nanowire Cross Section
Si	~150–200	~2,200	FCC	[111]	Au, Pt, Fe, Ti	~850	Hexagonal
Ge	~100–150	~5,400	FCC	[112]	Au	~880	Hexagonal
SiC	~400–500	~7,520	FCC	[111]	Au	~1200	Hexagonal
AlN	~350	~6,000	Wurtzite	[001]	Au	~900	Triangle, hexagonal
GaN	~250–350	~4,400	Wurtzite	[100]	Fe	~900	Triangle, hexagonal
GaAs	~80–90	~2,500	Zinc Blend	[111]	Au	~890	Triagonal, hexagonal
ZnO	~20–140	~2,800	Zinc Blend	[001]	Au	~900	Hexagonal
Ag	~85	~2,600	FCC, 4H	[111]	Ag	~25 to >100	Pentagonal
Pt	~160	~2,800	FCC	[111]	Pt	~25 to >100	Hexagonal

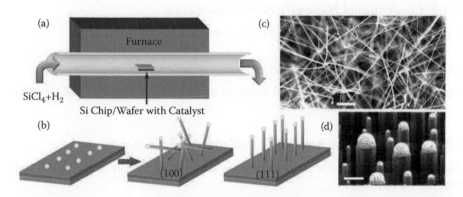

FIGURE 5.1

(See color insert). Typical examples of conventional bottom-up nanowire growth processes. (a) Schematic of a typical chemical vapor deposition (CVD) reactor (hot wall furnace) for growing semiconductor NWs employing the vapor–liquid–solid (VLS) mechanism. While here the specific vapor precursors are for Si NWs growth, the same thermal CVD setup can be used for other NWs by varying precursors. (b) Simplified schematics showing the VLS growth process, using Si NWs growth as an example, on Si (100) and (111) substrates, respectively. The gold color spheres are the catalyst nanoparticles, and the light blue rods highlight the Si NWs. (c) A typical SEM image of Si NWs grown on Si (100) substrate. Scale bar is 1μm. (d) A typical SEM image of vertically aligned Si NWs grown on Si (111) substrate. Scale bar is 300nm. (From Gao, D., et al., 2005, *J. Am. Chem. Soc.* 127, 4574–4575. With permission).

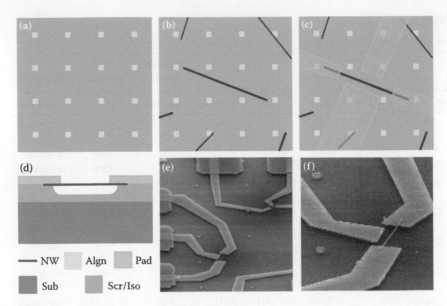

FIGURE 5.2
(See color insert.) Conventional nanowire and other one-dimensional NEMS device fabrication processes. (a) EBL patterning markers. (b) NWs are distributed on the substrate and coordinates identified. (c) EBL patterning of electrodes. (d) Device suspension. (e) and (f) SEM images of devices.

5.3.1 Conventional Nanowire NEMS Fabrication

Figure 5.2 illustrates typical NEMS devices fabrication processes.

5.3.2 Transfer Print Approach

Figure 5.3 describes the typical process of the recently developed transfer print technique.

5.3.3 Hybrid Approach with Controlled Epitaxial Growth

Figure 5.4 illustrates the typical process of a novel hybrid (bottom-up and top-down) approach.

5.4 Nanowire Electromechanical Resonators

We have developed the first silicon nanowire (SiNW) very high frequency (VHF) NEMS devices (Figure 5.5). The VHF NEMS resonators are based upon bottom-up epitaxially grown SiNWs with well terminated surfaces. Metallized SiNW resonators operating near 200 MHz are realized with quality factor $Q \approx 2,000$ to 2,500 (Table 5.2).

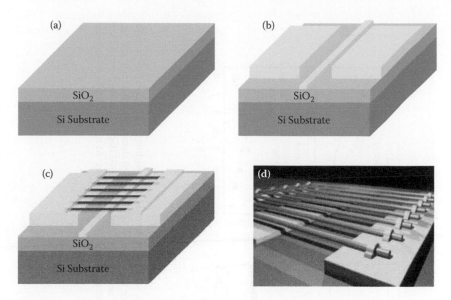

FIGURE 5.3
(See color insert.) Recent development of new transfer print technique for making self-aligned nanowire NEMS devices and arrays.

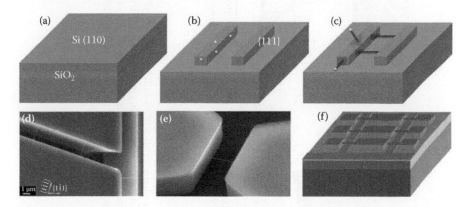

FIGURE 5.4
(See color insert.) Recently developed novel hybrid bottom up and top down approach for making self-aligned suspended nanowire NEMS devices and arrays.

Pristine SiNWs with fundamental resonances as high as 215 MHz are measured using a VHF readout technique that is optimized for these high resistance devices. The pristine resonators provide the highest Qs (as high as ≈13,100) for an 80MHz device. SiNWs excel at mass sensing; characterization of their mass responsivity and frequency stability demonstrates sensitivities approaching 10 zeptograms. We have developed SiNW NEMS-based phase-locking techniques to perform such real-time measurements.

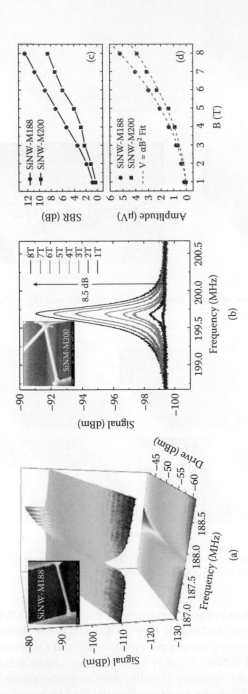

FIGURE 5.5

(See color insert.) Measured resonance characteristics of metallized SiNW VHF resonators. (a) Magnetomotively transduced response [referred-to-input (RTI) of first-stage amplifier] from SiNW-M200 at varying *B* fields (RF drive to device is −46 dBm). (b) Achieved detection efficiency and signal-to-background ratio (SBR) in decibels. (c) Actual voltage signal amplitude of the SiNW resonances displaying the B^2 dependence expected from magnetomotive transduction. (d) Magnetomotive response (RTI) from SiNW-M188 as RF drive to the device sweeps from −61 to −41 dBm (at *B* = 6 tesla).

TABLE 5.2

Representative Examples of Nanowire NEMS Resonators.

Representative Devices	Materials	Nanofabrication Technique	Resonance Signal Transduction	Device Specifications and Performance
	Pt (Husain et al., 2003)	Synthesis of Pt NWs (via electrodeposition) + Electron-beam lithography	Lorentz-force based magnetomotive excitation and detection	$d = 43$ nm, $L = 1.3$ µm, $f_0 = 105.3$ MHz, $Q \approx 8500$, $k_{eff} \sim 2.5$ N/m, $S_F^{1/2} \sim 13$aN/$\sqrt{Hz}$ (at $T \approx 4K$)
	GaN (Nam et al., 2006)	GaN (Wurtzite structure) grown on Si substrate, using Ga_2O_3 and NH_3 in VLS growth with Au/Pd catalyst, NWs assembled on W tips	Counter electrode's electrostatic actuation, transmission electron microscope (TEM), discrete observation	$d \approx 84$ nm, $L \approx 5.5$ µm, $f_0 = 2.2$ MHz, $Q \approx 2800$ *(at room temperature)*
	Si (Feng et al., 2007)	Naturally suspended Si NWs in microtrenches; Hybrid bottom-up VLS growth and top-down patterning	Magnetomotive (Lorentz-force) excitation & detection, specially engineered circuit	$d \approx 81$ nm, $L \approx 1.69$ µm, $f_0 = 215.4$ MHz, $Q \approx 5750$ $f_0 \times Q \approx 1.24 \times 10^{12}$ *(at $T \approx 25K$)*
	Si (He & Feng et al., 2008)	New-generation, thinner suspended Si NWs in micron and sub-micron trenches; Hybrid bottom-up VLS growth and top-down lithography	On-chip integrated transduction: electrostatic actuation & piezoresistive self-detection (at 2ω)	(i) $d \approx 30$ nm, $L \approx 1.8$ µm, $f_0 = 75.13$ MHz, $Q \approx 700$ (ii) $d \approx 40$ nm, $L \approx 1.8$ µm, $f_0 = 96.45$ MHz, $Q \approx 550$ *(at room temperature)*
	GaN (Tanner et al., 2007)	Gas source molecular beam epitaxy (MBE) growth of GaN NW on Si (111), Chips with as-grown NWs	Excitation by shear mode piezoelectric stack, spot-mode SEM detection	$d \approx 100$ nm, $L \approx 5$–20 µm, (i) $f_0 = 1.088$ MHz, $Q \approx 37800$ (ii) $f_0 = 1.151$ MHz, $Q \approx 37300$ *(at room temperature)*
	Si, Rh (Li et al., 2008)	Si: VLS growth with Au nanoparticle catalyst; Rh: electrodeposited; Integration: electric field assisted alignment and assembly of surface functionalized NWs	Excitation by piezoelectric ceramic actuator, laser interferometer detection	Si: $d \approx 330$ nm, $L \approx 11.8$ µm, $f_0 = 1.928$ MHz, $Q \approx 4830$; Rh: $d \approx 280$ nm, $L \approx 5.8$ µm, $f_0 = 7.186$ MHz, $Q \approx 1080$ *(at room temperature)*

(Continued)

Two principal advantages of the suspended SiNW resonators that we have developed are their ease of fabrication and high yields. By pushing the dimensions of the microtrenches downward and simultaneously optimizing the NW growth conditions, we expect that smaller, even molecular-size, suspended SiNWs should be achievable. These will enable scaling fundamental resonance frequencies into the extreme UHF and low microwave ranges.

TABLE 5.2 (CONTINUED)

Representative Examples of Nanowire NEMS Resonators.

	Si (Belov et al., 2008)	Hybrid bottom-up VLS growth and top-down lithographical patterning, suspended Si NWs, density not really controlled	Excitation by piezoelectric disk actuator, laser interferometer detection	$d \approx 40$ nm, $L \approx 5.2$ μm, $f_0 = 1.842$ MHz, $Q \approx 4200$; $d \approx 130$ nm, $L \approx 9.5$ μm, $f_0 = 13.79$ MHz *(at room temperature)*
	Si (Nichol et al., 2008)	Low-density VLS grown Si NWs (some are tapered)	Using fiber optic interferometry to observe resonant thermal displacement	$d \approx 44$ nm, $L \approx 14.4$ μm, $f_0 = 208$ kHz, $Q \approx 4000$; $S_x^{1/2} \sim 0.5$ pm $\sqrt{\text{Hz}}$, $k_{\text{eff}} \sim 28$ μN/m, $S_F^{1/2} \sim 6$ aN $\sqrt{\text{Hz}}$ *(at room temperature)*
	SnO$_2$ (Fung et al., 2009)	Sb-doped SnO$_2$ NWs, grown using a vapor transport process using high-purity Sn and Sb powders and Au nanoparticles as catalyst	Motion excited capacitively by back gate and detected using NW as a mixer	$d \approx 44$ nm, $L \approx 14.4$ μm, $f_0 = 59$ MHz, $Q \approx 2200$; devices with $f_0 \approx 80$–100 MHz *(at room temperature)*
	Si (Gil-Santos et al., 2010)	Naturally suspended Si NWs in microtrenches; Hybrid bottom-up VLS growth and top-down patterning	Optical interferometry of the thermal vibrations of the Si NWs, mode splitting due to NW cross section asymmetry	$d \approx 100$–300 nm, $L \approx 5$–10 μm, $f_0 = 2$–6 MHz, $Q \sim 2000$ *(at room temperature)*
	Pt/Au/Cr/Al/Ti/Nb/Ni, Si (Melosh et al., 2003)	Superlattice nanowire pattern transfer (SNAP); superlattice based on molecular beam epitaxy (MBE)	Lorentz-force based magnetomotive excitation and detection	Si: $w \geq 20$ nm; $L = 1.3$ μm Pt: $w = 20$ nm (thinnest 8 nm); $L = 0.75$ μm (pitch ~ 150 nm) $f_0 \approx 162.5$ MHz, Q appears low *(at $T \approx 4K$)*
	SiC, Si (Feng et al., 2009, 2010)	Surface nanomachining using heterostructured films such as SiC on Si (and, likewise, Si on Insulator, SiN on Si); Electron-beam lithography	(i) Magnetomotive transduction (ii) Capacitive transduction with impedance matching circuit	$w \approx 55$ nm, $t \approx 50$ nm, $L \approx 3$ μm, $f_0 \approx 74$ MHz, $Q \sim 1825$ $w \approx 50$ nm, $t \approx 50$ nm, $L \approx 2.5$ μm, $f_0 \approx 100$ MHz, $Q \sim 1500$ *(at room temperature)*

These SiNW resonators offer significant potential for applications in resonant sensing, quantum electromechanical systems, and high frequency signal processing. We have found that VHF SiNW resonators vibrating ~200 MHz typically have displacement sensitivity of ~5 fm/Hz$^{1/2}$ and force sensitivity of 50 to 250 aN/Hz$^{1/2}$, set by thermomechanical fluctuations. They have critical amplitude ~1 nm and intrinsic dynamic range of 90 to 110 dB.

In the development of compact solid state transducers, ingeniously engineering the strain in miniscule mechanical structures is an important factor and is now becoming a key for resonant NEMS. We have demonstrated that for very thin SiNWs, their time-varying strain can be exploited for self-transducing the device's resonant motion at very high frequencies. The strain that is only second-order in doubly clamped structures enables efficient displacement transducers due to the enhanced piezoresistance effect in these SiNWs.

By combining the piezoresistive self-transducing feature of SiNWs with off-chip piezoelectric-disk actuation and on-chip electrostatic excitation, we demonstrated resonators that operate in the VHF band (Figure 5.6).

For devices with widths ranging from 90 to 30 nm and lengths of 1.8 to 5 μm, the resonators operate at frequencies from 20 to 100MHz, with quality factor Q values in the range of 550 to 1200, at room temperature and non-stringent vacuum conditions (around millitorr range). The measured Qs do not drop until pressure is raised to ~1 torr and are still appreciable at ~100 torr (e.g., ~300 for a 30 nm-thick Si nanowire operating at 75 MHz).

In parallel to the hybrid bottom-up/top-down approach discussed above, we have also been actively pursuing the development of NW NEMS and arrays based entirely upon top-down nanofabrication. Although at present realizing truly molecular scale thin NWs may seem very challenging, the pure top-down approach has tremendous advantages such as allowing rational designs with great complexity and flexibility. In particular, we developed several top-down processes involving high resolution lithography and surface nanomachining techniques for making NW NEMS (e.g., suspended thin wires ~20 nm).

For example, as shown in Figure 5.7, we devised arrays of NWs with (in-plane coupled) controlling electrodes (gates). These 50 nm thin NWs make robust resonators and demonstrate interesting frequency tunability and non-linear behavior. While it was very challenging to read out such a single NW resonator using conventional interferometry optical detection, we have shown that the conventional scheme can work very well for these very thin NWs when they are made into arrays with controlled dimensions and pitches. We have also been developing all-electronic readout techniques for such arrays. Engineering such NW devices may aid the development of novel coupled and tunable NW resonator arrays and arrayed NW NEMS switches. Currently such research efforts are in progress in a number of groups worldwide.

5.5 Energy Dissipation and Quality Factors (Qs)

An interesting observation is that the uncoated, pristine SiNWs exhibit much higher Qs than metalized ones do. We have measured $Q \approx 13,100$ for SiNW-80, $Q \approx 5,750$ for SiNW-215, $Q \approx 2,500$ for SiNW-M188, and $Q \approx 2,000$ for SiNW-M200, all at $T = 25K$ with $B = 6$ Tesla. The SiNW-215 device has resonance frequency and dimensions ranges similar to SiNW-M188 and SiNW-M200 and a Q two to three times higher. This remarkable difference is not solely a result of geometry and dimension effects (such as those associated with aspect ratios and surface-to-volume ratios). This suggests that metallization may have introduced extra energy loss channels at the surface or material interfaces.

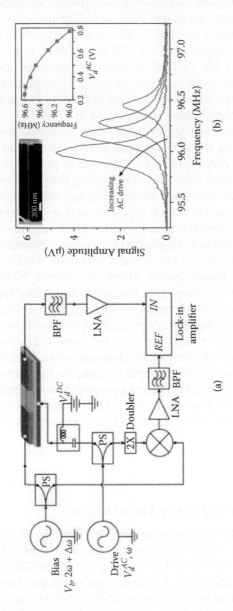

FIGURE 5.6

(See color insert.) Very high frequency (VHF) Si nanowire resonators with fully integrated electrostatic actuation and piezoresistive self-detection. (a) Bias and drive circuitry. (b) Performance of 40 nm-thick, 96MHz nanowire resonator with quality (Q) factor ~550. It is 1.8 μm long and has a dc resistance of 80 kΩ. The AC drive is set at $V_{d,AC}$ = 0.50, 0.63, 0.71, and 0.79 V for the curves, respectively, with DC voltage fixed at $V_{d,DC}$ = 0.2 V. The left inset is an SEM image of the device. The right inset shows the drive dependence of the resonance frequency. (c) Performance of a 30 nm-thick, 75MHz Si nanowire resonator with Q ~700. Its length is 1.8 μm and its dc resistance is 300 kΩ. The curves are taken at different bias voltages with the same drive. The insets show the SEM image and the linear dependence on bias voltage. (d) Data taken at varied drives with the same bias. Inset shows quadratic dependence on drive voltage. (Continued)

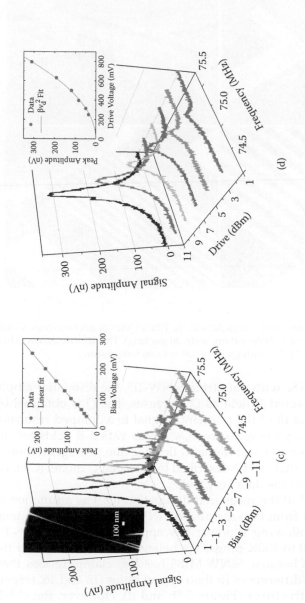

FIGURE 5.6 (CONTINUED)

(See color insert.) Very high frequency (VHF) Si nanowire resonators with fully integrated electrostatic actuation and piezoresistive self-detection. (a) Bias and drive circuitry. (b) Performance of 40 nm-thick, 96MHz nanowire resonator with quality (Q) factor ~550. It is 1.8 μm long and has a dc resistance of 80 kΩ. The AC drive is set at $V_{d,AC}$ = 0.50, 0.63, 0.71, and 0.79 V for the curves, respectively, with DC voltage fixed at $V_{d,DC}$ = 0.2 V. The left inset is an SEM image of the device. The right inset shows the drive dependence of the resonance frequency. (c) Performance of a 30 nm-thick, 75MHz Si nanowire resonator with Q ~700. Its length is 1.8 μm and its dc resistance is 300 kΩ. The curves are taken at different bias voltages with the same drive. The insets show the SEM image and the linear dependence on bias voltage. (d) Data taken at varied drives with the same bias. Inset shows quadratic dependence on drive voltage.

FIGURE 5.7
Top-down NW NEMS resonators and arrays. (a) and (b) Tilted views of device arrays. (c) and
(d) Close-in views of an array of 10 NWs (50 nm wide, 10 μm long). The devices shown exhibited multiple resonances at ~10 MHz with Qs of ~4,000 at room temperature.

As shown in Figure 5.8e with data from SiNW-215, the resonance amplitude decreases with lowered B field as Q increases. The Q is obtained by
optimizing the nonlinear fit of the resonance signal to a damped resonator
model. Figure 5.9 displays the measured Qs versus varied B fields for the
four SiNWs. The decrease in Qs at higher B fields is due to magnetomotive
(eddy-current) damping that can be described by a loaded Q model (Cleland
and Roukes, 1999) that fits the data very well.

As shown in Figure 5.10 the measured dissipation (Q^{-1}) as a function of
temperature, with data from two metalized SiNWs, increases as the temperature is elevated, following a power law approximately from $Q^{-1} \propto T^{0.3}$
to $Q^{-1} \propto T^{0.4}$ in the $T = 20$ to 100K range. We believe the offset between the
two data traces occurs because SiNW-M188 has less clamping loss than
SiNW-M200 due to the differences in their aspect ratios (18 and 16, respectively) and clamping structures (Figure 5.8b and d). However, the $Q^{-1} \propto T^{\beta}$
($\beta = 0.3$ to 0.4) power law is not readily explained by geometric or surface
effects. It may relate to some intrinsic dissipation processes in the single
crystal SiNW and/or metallization materials.

High Qs for SiNW resonators are important both for resonant sensing
and other device applications, and also to enable exploration of dissipation

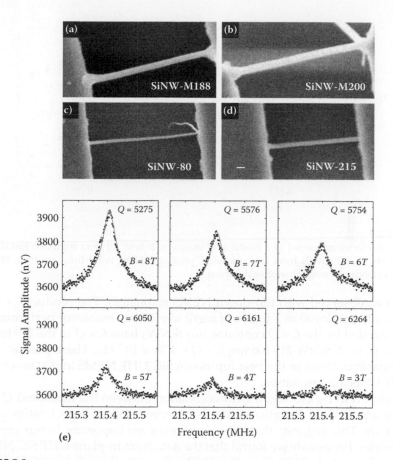

FIGURE 5.8
Scanning electron micrographs and measured resonance and Q of VHF SiNW resonators.
(a) Metallized 188 MHz SiNW. (b) Metalized 200 MHz SiNW. (c) Pristine 80 MHz SiNW. (d)
Pristine 215 MHz SiNW. Qs of SiNW-215 at varied B fields are extracted from optimized least-
square nonlinear fitting (e) of measured resonance signals. Scale bar is 100 nm for all four
micrographs.

sources. Figure 5.11a displays $Q \approx 5{,}750$ of SiNW-215 and $\approx 13{,}100$ of SiNW-80.
The pristine SiNW-215 exhibits a Q of ~2.3 to 3.0 times higher than the Qs of
2,500 and 2,000 from the metallized SiNW-M188 and SiNW-M200, despite
their similar dimensions and resonance frequencies. This difference may
arise from metallization-induced losses such as surface and interface modification and internal friction in the metal layers.

Figure 5.11b presents the measured Qs of the SiNWs and a comparison of
their performance to other types of NEMS resonators such as bottom-up Pt
nanowires (PtNW; Husain et al., 2003), and top-down SiC beams. These are
loaded Qs [measured while embedded in the measurement circuitry in an
applied magnetic field $(B = 6\ T)$]. For both metallized and pristine SiNWs, we
find that Q decreases with increasing resonance frequency f_0. This trend has

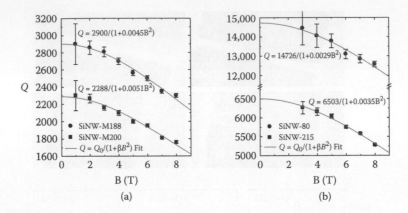

FIGURE 5.9

Quality (Q) factor measured as a function of magnetic B field. (a) Data from two metallized SiNWs. (b) Very high Qs from two uncoated, pristine SiNWs. Solid lines are from fit to a loaded-Q model for magnetomotive damping.

also been observed with top-down NEMS resonators. The product $f_0 \times Q$ is commonly employed as a figure of merit to evaluate resonator performance. As illustrated by the $f_0 \times Q$ contours, our SiNWs have $f_0 \times Q$ in the 10^{11} to 10^{12} Hz range, with SiNW-215 having $f_0 \times Q = 1.24 \times 10^{12}$ Hz. These products are comparable to those of the best top-down SiC UHF NEMS and state-of-the-art MEMS beam resonators.

As shown in Figure 5.8c, a clear correlation between the measured Q and device aspect ratio (length/width) is observed; larger aspect ratios yield higher Qs. This suggests that clamping losses are important in our present geometries. Previously we found that the data from in-plane UHF SiC NEMS beams (Feng et al., 2006) fit well with the theoretical prediction $Q \sim (L/w)^3$. With the thinner SiNWs we have a much wider aspect ratio range in which it appears that likely the Qs of SiNWs do not scale as aggressively as $Q \sim (L/w)^3$.

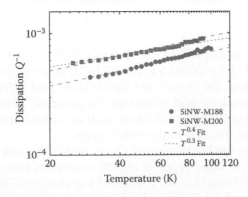

FIGURE 5.10

Measured dissipation (Q^{-1}) in two metallized SiNWs, as function of temperature. A phenomenological dependence of about $Q^{-1} \mu T^{0.3-0.4}$ is observed in the range of $T = 20$ to $100\ K$.

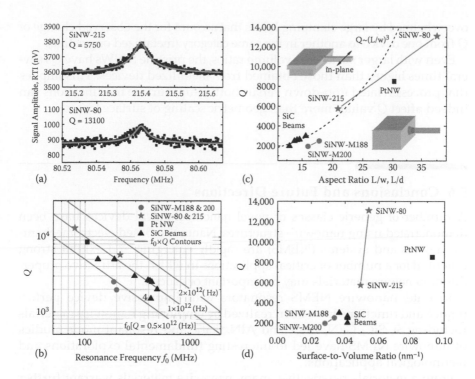

FIGURE 5.11
Quality (Q) factors of pristine and metallized VHF SiNW resonators. (a) Resonance signals (RTIs) with high Qs from pristine SiNWs (B = 6 T, squares are data points, solid lines are fit from the damped simple harmonic oscillator model). Measured Qs as functions of (b) device resonance frequency, (c) aspect ratio, and (d) surface-to-volume ratio, in comparison with previously measured NEMS devices under similar experimental conditions with the same transduction scheme. Insets: differences in clamping structures of in-plane top-down NEMS resonators (upper left) and pristine suspended SiNWs (bottom right).

We suspect that the precise geometry of the NW clamping depicted in the bottom-right inset in Figure 5.11c plays an important role. The support region for an in-plane beam is a plate as thick as the beam but much wider. This may provide much stiffer clamping (more isolation) for in-plane motion than for out-of-plane motion. These considerations lead to clamping-losses scaling as $Q \sim (L/w)^3$ and $\sim (L/w)$, respectively.

The actual points of NW attachment, shown in SEM images (Figures 5.1 through 5.3) have conical structures and sometimes a backward-growing parasitic wire at one end. Presumably the resulting isolation is less than achieved for in-plane beams, but may be comparable to that for out-of-plane beams. Nonetheless, single crystal pristine SiNWs with large aspect ratios (L/d) in the range of 20 to 40 exhibit very high Qs. Figure 5.11d shows that Q for similar devices (metallized or pristine) does not decrease with increasing surface-to-volume ratio. This suggests that possible surface loss phenomena, if present, are

overwhelmed by large clamping losses that account for the observed change of
Q from one device to another in the same category (metallized or pristine).

Even with larger surface-to-volume ratios, the pristine SiNWs have Qs several times higher than those obtained from metallized devices. This implies that process-related (top-down versus bottom-up) surface modifications can indeed affect Q values more than geometric scaling of surface losses.

5.6 Conclusions and Future Directions

A number of generic classes of novel nanomechanical devices have been demonstrated using nanowire structures. Nanowire-based electromechanical devices and systems (NEMS) are rapidly emerging and exhibit strong potential for a number of critical applications in which the attractive properties of nanowire materials may be important or crucial.

To date, nanowire NEMS resonators with impressive device performance and functions have been realized by employing nanowire materials including Si, SiC, GaN, ZnO, and AlN. Static and resonant mode studies of nanowire NEMS have led to interesting fundamental explorations and technological applications.

From a materials perspective, many nanowire materials warrant further investigation, with emphasis on mechanical properties and NEMS device technology. Such materials include ultrathin metallic nanowires, various III–V nanowires, and core-shell heterostructure nanowires, to name a few.

From a devices and systems perspective, the challenges concern engineering for highly efficient electromechanical coupling effects in nanowires for making sensors and transducers, developing reliable techniques for device assembly and large scale integration, and controlling device uniformity and yield. From technology and application perspectives, we are only at the beginning of exploring and utilizing nanowire structures and the interesting devices they have enabled. Tremendous opportunities are emerging for elaborate engineering of nanowire NEMS applications ranging from sensing to computing.

References

Belov, M., Quitoriano, N., Sharma, S. et al. 2008. Mechanical resonance of clamped silicon nanowires measured by optical interferometry. *Journal of Applied Physics*, 103, 074304.

Boettcher, S.W., Spurgeon, J.M., Putnam, M.C. et al. 2010. Energy-conversion properties of vapor–liquid solid-grown silicon wire array photocathodes. *Science*, 327, 185–187.

Boukai, A.I., Bunimovich, Y., Tahir-Kheli, J. et al. 2008. Silicon nanowires as efficient thermoelectric materials. *Nature,* 451, 168–171.

Cleland, A.N. and Roukes, M.L. 1999. External control of dissipation in a nanometer-scale radiofrequency mechanical resonator. *Sensors and Actuators A—Physical,* 72, 256–261.

Craighead, H.G. 2000. Nanoelectromechanical systems. *Science,* 290, 1532–1535.

Ditlbacher, H., Hohenau, A., Wagner, D. et al. 2005. Silver nanowires as surface plasmon resonators. *Physical Review Letters,* 95, 257–403.

Feng, X.L., Zorman, C.A., Mehregany, M. et al. 2006. Dissipation in single-crystal 3C-SiC ultra-high frequency nanomechanical resonators. *Digest of Technical Papers, Solid-State Sensors, Actuators and Microsystems Workshop,* Hilton Head, SC. Transducers Research Foundation. pp. 86–89.

Feng, X.L., He, R.R., Yang, P.D. et al. 2007. Very high frequency silicon nanowire electromechancial resonators. *Nano Letters,* 7, 1953–1959.

Feng, X.L., Matheny, M.H., Karabalin, K.B. et al. 2009. Silicon carbide (SiC) top-down nanowire electromechanical resonators. *Digest of Technical Papers of 15ᵗʰ IEEE International Conference on Solid State Sensors, Actuators and Microsystems.* Denver, pp. 2246–2249.

Feng, X.L., Matheny, M.H., Zorman, C.A. et al. 2010. Low voltage nanoelectromechanical switches based on silicon carbide nanowires. *Nano Letters,* 10, 2891–2896.

Fortuna, S.A. and Li, X.L. 2010. Metal-catalyzed semiconductor nanowires: a review on the control of growth directions. *Semiconductor Science and Technology,* 25, 024005.

Fung, W.Y., Dattoli, E.N., and Lu, W. 2009. Radio frequency nanowire resonators and in situ frequency tuning. *Applied Physics Letters,* 94, 203104.

Gao, D., He, R.R., Carraro, C. et al. 2005. Selective growth of Si nanowire arrays via galvanic displacement processes in water-in-oil microemulsions. *J. Am. Chem. Soc.,* 127, 4574–4575.

Garnett, E.C. and Yang, P.D. 2008. Silicon nanowire radial p–n junction solar cells. *Journal of American Chemical Society,* 130, 9224–9225.

Garnett, E.C. and Yang, P.D. 2010. Light trapping in silicon nanowire solar cells. *Nano Letters,* 10, 1082–1087.

Gil-Santos, E., Ramos, D., Martinez, J. et al. 2010. Nanomechanical mass sensing and stiffness spectrometry based on two-dimensional vibrations of resonant nanowires. *Nature Nanotechnology,* 5, 641–645.

He, R.R., Feng, X.L., Roukes, M.L. et al. 2008. Self-transducing silicon nanowire electromechanical systems at room temperature. *Nano Letters,* 8, 1756–1761.

Hochbaum, A.I., Chen, R.K., Delgado, R.D. et al. 2008. Enhanced thermoelectric performance of rough silicon nanowires. *Nature,* 451, 163–167.

Husain, A., Hone, J., Postma, H.W. et al. 2003. Nanowire-based very high frequency electromechanical resonator. *Applied Physics Letters,* 83, 1240–1242.

Kelzenberg, M.D., Boettcher, S.W., Petykiewicz, J.A. et al. 2010. Enhanced absorption and carrier collection in Si wire arrays for photovoltaic applications. *Nature Materials,* 9, 239–244.

Kelzenberg, M.D., Turner-Evans, D.B., Kayes, B.M. et al. 2008. Photovoltaics measurements in single-nanowire silicon solar cells. *Nano Letters,* 8, 710–714.

Law, M., Goldberger, J., and Yang, P.D. 2004. Semiconductor nanowires and nanotubes. *Annual Review of Materials Research,* 34, 83–122.

Law, M., Greene, L.E., Johnson, J.C. et al. 2005. Nanowire dye-sensitized solar cells. *Nature Materials,* 4, 455–459.

Li, M.W., Rustom, B.B., Morrow, T.J. et al. 2008. Bottom-up assembly of large-area nanowire resonator arrays. *Nature Nanotechnology*, 3, 88–92.

Li, Y., Qian, F., Xiang, J. et al. 2006. Nanowire electronic and optoelectronic devices. *Materials Today*, 18–27.

Lu, W. and Lieber, C.M. 2006. Semiconductor nanowires. *Journal of Physics D*, 39, R387–R406.

Lu, W. and Lieber, C.M. 2007. Nanoelectronics from the bottom up. *Nature Materials*, 6, 841–850.

McAlpine, M.C., Friedman, R.S., and Lieber, C.M. 2005. High-performance nanowire electronics and photonics and nanoscale patterning on flexible plastic substrates. *Proceedings of IEEE*, 93, 1357–1363.

Melosh, N.A., Boukai, A., Diana, F. et al. 2003. Ultra high density nanowire lattices and circuits. *Science*, 300, 112–113.

Nam, C.Y., Papot, J., Tham, D. et al. 2006. Diameter-dependent electromechanical properties of GaN nanowires. *Nano Letters*, 6. 153–158.

Nichol, J.M., Hemesath, E.R., Lauhon, L.J. et al. 2008. Displacement detection of silicon nanowires by polarization-enhanced fiber-optic interferometry. *Applied Physics Letters*, 93, 193110.

Patolsky, F. and Lieber, C.M. 2005. Nanowire nanosensors. *Materials Today*, 20–28.

Patolsky, F., Zheng, G.F., and Lieber, C.M. 2006. Nanowire sensors for medicine and the life sciences. *Nanomedicine*, 1, 51–65.

Pauzauskie, P.J. and Yang, P.D. 2006. Nanowire photonics. *Materials Today*, 36–45.

Roukes, M.L. 2000. Nanoelectromechanical systems, *Digest of Technical Papers of Solid State Sensors, Actuators and Microsystems Workshop*. Hilton Head, SC. Transducers Research Foundation, pp. 367–376.

Roukes, M.L. 2001. Nanoelectromechanical systems face the future. *Physics World*, 14, 25–31.

Schmidt, V., Wittemann, J.V., Senz, S. et al. 2009. Silicon nanowires: a review on aspects of their growth and their electrical properties. *Advanced Materials*, 21, 2681–2702.

Tanner, S.M., Gray, J.M., Rogers, C.T. et al. 2007. High-Q GaN nanowire resonators and oscillators. *Applied Physics Letters*, 91, 203117.

Tian, B.Z., Kempa, T.J. and Lieber C.M. 2009. Single nanowire photovoltaics. *Chemical Society Reviews*, 38, 16–24.

Tian, B.Z., Zheng, X.L., Kempa, T.J. et al. 2007. Coaxial silicon nanowires as solar cells and nanoelectronic power sources. *Nature*, 449, 885–889.

Xia, Y.N., Yang, P.D., Sun, Y.G. et al. 2003. One-dimensional nanostructures: synthesis, characterization, and applications. *Advanced Materials*, 15, 353–389.

Yan, R.X., Gargas, D., and Yang, P.D. 2009. Nanowire photonics. *Nature Photonics*, 3, 569–576.

Yan, R.X., Pausauskie, P., Huang, J.X. et al. 2009. Direct photonic–plasmonic coupling and routing in single nanowires. *Proceedings of National Academy of Sciences of USA*, 106, 21045–21050.

Zimmler, M.A., Capasso, F., Müller, S. et al. 2010. Optically pumped nanowire lasers. *Semiconductor Science and Technology*, 25, 024001.

6

Silicon Etching and Etch Techniques for NEMs and MEMs

M. David Henry and Axel Scherer

CONTENTS

ABSTRACT Fabrication of precision micro- and nanoscale structures in silicon demands exacting control over pattern transfer during etching. The most utilized method to enable this pattern transfer anisotropically is plasma etching with an inductively couple plasma reactive ion etcher. When properly used, these machines can etch high-aspect-ratio structures in silicon that are hundreds of microns in height to tens of nanometers in width. The work presented here will detail how to tune parameters associated with a plasma etcher and manipulate the gas chemistry to achieve and control anisotropic silicon etches at both micro- and nanoscale. Etch masks that ensure a high pattern transfer fidelity and complement the gas chemistries employed here will also be detailed. Finally several etching techniques that use the principles detailed to fabricate extraordinary structures in silicon will be described.

6.1 Introduction

The fabrication of microelectromechanical (MEMs) and nanoelectromechanical (NEMs) devices in silicon has played a critical role in the advancement of science and in our understanding of the natural world. These systems rely upon the fabrication techniques extensively developed by the microelectronics industry (Campbell, 2001), namely lithography, precision control over depositions of layers, and anisotropic etching.

New semiconductor MEMs devices such as accelerometers (Yazdi et al., 1998), movable mirrors, micromechanical switches, and gyroscopes have seen complete fabrication and packaging in silicon substrates (Craighead, 2010). As silicon fabrication in microelectronics pushed into the nanoscale regime to meet Moore's law, new NEMs devices took advantage of the small volumes and high surface-to-volume ratios. Extraordinary sensitivities and resolutions have been realized in planar devices such as optomechanical crystals (Eichenfield et al., 2009) and nanobeam cantilevers (Ekinci et al., 2005). Although these structures are generally planar, next generation structures will require fabrication of more three-dimensional structures and integration of the micro- and nanoscales.

Critical to the new requirements will be the ability to plasma etch structures in semiconductors. Although wet chemical etching presents advantages and should remain a significant tool in the fabrication line (Stern et al., 2007), plasma and ion etching is just as critical. To utilize this tool to its maximum potential, an understanding of how plasma etching is controlled is vital.

This chapter will detail how basic plasma etching parameters can be tuned to optimize a silicon etch, how etch chemistry can be established to generate highly anisotropic etch profiles, and how the proper selection of an etch mask can improve pattern transfer fidelity from the mask to the substrate. A few techniques will also be described to demonstrate how to make unique structures in silicon using these etching concepts. This chapter is intended to introduce silicon etching and etch techniques to scientists and engineers who must understand precision etching.

6.1.1 Evolution of Etching Machines

To explain the critical qualities necessary to achieve anisotropic micro- and nanoscale silicon etches, it is useful to recount a history of silicon etching machines (Coburn, 2000; Ziegler et al., 1985; Sugawara, 1998; Johnson, 1995). This account is not chronological but examination of etching machines in the following manner is useful for developing an understanding of silicon etching.

To begin, basic etching of silicon involves the bombardment of ions or neutrals to remove the silicon atoms. Machines can generate both species using plasmas with different atoms, and accelerate the atoms to the silicon substrate where they can either chemically bond with the silicon or mechanically impinge on the silicon and move the atoms away. Etching machines such as ion mills can generate plasmas using inert gases such as argon and accelerate the ions under a large voltage or electric field. The resulting collisions between ions and silicon can mill trenches and features into the silicon.

The etch structures are defined by using etch masks more resilient to the ions than the silicon substrate. Note that resiliency is not simply a mechanical property, but also depends greatly on thickness and the chemistry between the ions and the mask. With slight modifications, injecting chemically reactive gases such as chlorine can cause the ions to impinge on the substrate, freeing the silicon atoms, and the reactive gases can combine with the freed silicon. This technique, known as chemically assisted ion beam etching (CAIBE), dramatically increases the etching rate and reduces the roughness of the etched surfaces. A popular implementation of this technique uses ionized argon to impinge a gallium arsenide substrate while injecting chlorine gas to chemically assist in the etching.

With the invention of the capacitive coupled plasma reactive ion etcher (CCP RIE) a chemically reactive gas can ionize and create a plasma using radio frequency (RF) voltages (Johnson, 1995). The capacitive connection of the RF power to the table holding the substrate generates a strong negative charge from the build-up of electrons on the table. This creates an electric potential difference between the table and plasma. This induced static electric field, known as the biasing voltage, can then accelerate the ions from the plasma to the silicon substrate where the silicon is etched using a combination of chemical and mechanical milling.

Although ions are dominant, significant contributions to the etch rate are made by neutrals and radicals; these are ions that have recombined with electrons. By reducing the needed accelerating voltage from thousands of volts typical of milling to a few hundred volts typical of RIE, the damage to the silicon substrate is reduced concurrently with improved resolution, quality, and achievable silicon structures. Increased etch rate is achieved by increasing the RF power applied to the plasma. This effect occurs since increasing the RF power generates more collisions between the ionized electrons and the injected gas.

The increased collisions result in generating more of the ionized and chemically active species. However, increasing the power also increases the build-up of negative charge on the plate; subsequently the DC biasing voltage also increases. With a larger electric field and more ions accelerated down to the substrate, a higher etch rate is achieved. Increasing the RF power jointly increases the number of ions in the plasma (plasma density) and the ion energy. Unfortunately, the damage to the substrate is also increased. Hence the need arose to separately control the plasma density and the amount of acceleration the ions receive to bombard the substrate.

From this need came the inductively coupled plasma reactive ion etcher (ICP RIE). Although the physics of the RIE segment remain fundamentally the same, the addition of a multiturn coil added an induced magnetic field across the plasma. Similar to the RIE, this magnetic field is also generated using RF power but is different in that it is electrically isolated from the chamber.

This configuration is analogous to the primary and secondary winding of a transformer; the coil acts as the primary and the plasma acts as the secondary. The magnetic field provides an increase in power to the plasma that creates more ionization events. Hence the plasma density can be increased with little change in the biasing voltage. Fundamentally the addition of the ICP coil permits the chemical etching to increase without requiring the mechanical milling to increase. This subtle change enabled a dramatic modification in the ability to etch silicon. The etch rates may now be increased sufficiently to permit through-wafer etching without reducing the selectivity of the mask. However, the increase in the chemical etching rate also reduced the anisotropic nature of the etches. This anisotropy decrease established the need for passivated silicon etches.

6.1.2 Etch Control Using ICP RIE

ICP RIE systems are controlled by four main parameters: forward power (Fwd), ICP power, chamber pressure, and injected gas flow rate. A brief discussion of how each of the parameters affects the etched silicon substrate is useful. The Fwd power is the same power that generates the plasmas for the RIE. Again RF power is capacitively coupled to the plate that generates the plasma and establishes the DC bias.

Since ICP RIE chambers are typically elongated to incorporate ICP coils, the plates of the plasma are further apart than those seen on RIE chambers. This permits a reduction in the required pressure to strike a plasma. This effect implies that the distances between the plasma and the table are longer for ICP etchers than for RIEs. Since the voltage between the plasma and the table is the integral of the electric field over the distance separating the plasma and plate, for a given DC bias the electric field is typically higher in an RIE than in an ICP RIE and acceleration is reduced for the ICP RIE. However, like the RIE, increasing the Fwd power linearly increases the DC

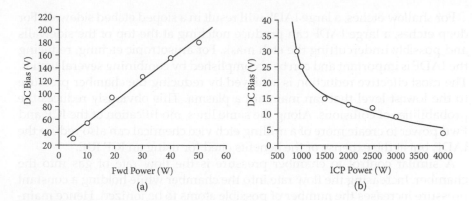

FIGURE 6.1
(a) Effect of Fwd power upon DC bias. (b) Effect of ICP power upon DC bias.

bias. Thus, increasing the Fwd power can be thought of as increasing the milling aspect of the etch (Figure 6.1).

The ICP coil provides an increase in plasma density by increasing the path travelled by an electron through the plasma, thereby creating more collisions (Johnson, 1995). This increase in plasma density creates an increase in both the ions and the neutrals. Thus, the flow of ions to the substrate is controlled by both the ICP power and the Fwd power. Since the forward power remains constant, the product of the DC bias and the ion current also remains constant.

What is typically seen for an increase in ICP power is a decrease in DC bias; this implies that the ion current has increased. Although the increase in ICP increases the milling aspect more significantly, the number of neutrals also increases. This implies an increase in the number of atoms reaching the substrate for chemical reactions. As the neutrals become dominant, typically an increase in ICP power increases the chemical nature of the etch.

Although chamber pressure is critical for establishing a plasma, some of its more subtle effects are more interesting for silicon etching. An increase in pressure can signify an increase in neutral density. A very obvious example is when ICP power is increased in an oxygen plasma; if the pumping rate on the chamber and the oxygen flow rate are held constant, a significant increase in pressure can be observed. This is due to the separation of O_2 molecules into two atoms. An increase in pressure also heralds an increase in undercutting of the etch mask due to the increase in pressure that causes an increase in collisions.

Each collision of an etching atom alters the momentum of the atom's trajectory to the substrate. An addition to the atom's velocity vector parallel to the silicon surface implies that the lateral etch rate will increase. Since these types of collisions are inevitable, quantifying the distribution of velocity vectors as a single function is useful. This function is referred to as the ion angle distribution function (IADF; Jansen et al., 2009).

For shallow etches, a large IADF will result in a sloped etched sidewall. For deep etches, a large IADF can produce notching at the top of the sidewalls and possibly undercutting the etch mask. For anisotropic etching, reducing the IADF is important and can be accomplished by combining several tactics. The most effective reduction is achieved by reducing the chamber pressure to the lowest level that can maintain a plasma. This obviously reduces the probabilities of collisions. Along the same lines, modification of the ICP and Fwd power to create more of a milling etch vice chemical can also reduce the IADF, but at the expense of the benefits cited for using an ICP RIE.

A similar control to chamber pressure is the flow rate of gas into the chamber. Increasing the flow rate into the chamber while holding a constant pressure increases the number of possible atoms to be ionized. Hence maintaining all other parameters at the same values and increasing the flow rate of gas into the chamber will increase the number of ionizations. This increases both the etching rate and the collision rate, so IADF should be expected to broaden with its associated repercussions.

Unfortunately, this continual increase in ionization does not continue indefinitely due to the finite efficiency of ionization and finite reactor size. It is a substantial upgrade for an etch machine to have a higher pump rate that permits a higher gas flow rate. Manufacturers of etching machines typically reflect this fact in purchase costs.

Understanding these adjustments to the ICP RIE can permit a substantial savings in time to a user attempting to control an etch. To assist the reader, a quick summary of tunable controls follows. Increasing the RIE power increases the milling aspect of an etch and hence the etch rate. It also increases the directionality of the etch, thereby reducing the IADF. An increase in the ICP power increases the number of neutrals, or more importantly, the ratio of neutrals to ions. This increase encourages a more chemical etch over milling and also increases the etch rate.

Decreasing the pressure reduces the number of scattering events and in turn reduces the IADF and increases anisotropy. Increasing the injected gas flow rate increases the plasma density and consequently the etch rate.

ICP RIE recipes typically exhibit large ICP powers (1000 to 5000 W), low Fwd powers (5 to 200 W), and low pressures (5 to 20 mT) achievable partly because of the large plate separation and high injected gas flow rates (70 to 300 standard cubic centimeters per minute [SCCM]). These parameters are typical of reactors such as Oxford Instruments PlasmaLab 100 ICP RIE 380 and encourage high etch rates at the cost of a loss of anisotropy when compared to RIE. To control the anisotropy, ingenious gas chemistry is required.

It is prudent to now qualitatively compare RIE machine to ICP RIE systems. Although ICP RIEs have the ability to separate out plasma density from ion energy and achieve very high etch rates, they do not perform as well as an RIE for etching less favored chemical reactions. When the injected gas and the substrate (such as $Si + 4F = SiF4$) interact chemically to produce an exothermic process, ICP RIEs can outperform RIE systems.

However, if the etch process is endothermic ($Pt + 2Cl_2 = PtCl_4$), higher ion energy is needed and the etch is more a milling than a chemical reaction. Furthermore if an etch product is non-volatile, for example, $PtCl_4$, it is sputtered vice pumped away (Coburn, 2000). Non-volatile endothermic etch processes require more milling than chemical etching. As stated previously, the RIE can outperform the ICP RIE in this regime since the generated electric fields and consequent ion energies are higher.

6.2 Passivated Silicon Etches

6.2.1 Controlling Etch Using Gas Chemistry

To control the lateral etch rate, etch chemistry can be improved upon by the addition of a passivation gas. As the selected gas etches the silicon, a second gas is introduced to create a passivation layer removable only by mechanical milling. This shifts an isotropic etch to an anisotropic process. Fundamentally, this is achieved by only passivating the vertical sidewalls. However, this passivation layer is deposited on all the surfaces, both horizontal and vertical.

By increasing the Fwd power, the ions can be accelerated enough to mill through the deposited passivation layers on the horizontal surfaces; equivalently stated, the milling rate of the lateral passivation is higher than the deposition rate. This leaves a protective layer on the sidewalls able to withstand the chemical and ion etching laterally described by the IADF. By adjusting the rate of passivation deposition, one can control the anisotropy of the etched structure.

This chemistry complements the optimized parameters of the ICP RIE for achieving anisotropic etches with high etch rates. In addition to chemistry, temperature can also produce very significant effects. Adjusting the temperature of both the substrate and the etch chamber can alter the passivation deposition rate (Tachi et al., 1988). Typically, oxide passivation occurs at cryogenic temperatures, whereas the polymer-based passivation occurs near room temperature. Substrate temperature and etch chemistry are closely coupled and will be discussed together.

Besides separating the passivation chemistry into oxides and polymers, a second and more general classification scheme can be utilized to describe etch chemistry. Passivation can be accomplished simultaneously as the silicon etches or a cyclic process can be established. A simultaneous process is achieved by injecting the etch and passivation gases together, creating a plasma, and tuning the ICP RIE parameters to achieve the desired anisotropy. This chemistry is referred to as mixed-mode etching (Jansen et al., 2009). Alternatively, the passivation and etching can be temporally separated. First the etch gas is injected, ionized, and sent to etch the silicon substrate. This plasma is then terminated, and a passivation gas injected, ionized, and deposited on the etched silicon. Upon completion of this step, the etching step is then repeated. The sequence is repeated until the desired etch depth is achieved. This chemistry is known as chopping mode etching. Although the sequencing of the etch and passivation gases differs, the same etch chemistries can generate very different results (Isakovic et al., 2008).

Chopping mode etching has the distinct disadvantage that the cyclic nature of the etching leaves scalloping marks along the sidewall. The sizes of the scallops are determined by the duration of the etch step; the duty cycle controls the angle of the etch. A famous example is the Bosch Etch (Rhee et al., 2008).

The advantage this method is the significant improvement in etch rate; typical etch rates using this technique are around 5 to 10 μ/min. The high etch rate makes this system most useful for deep etches such as through-wafer etching in which sidewall roughness is unimportant. The Bosch etch relies on SF_6 gas to generate F ions and neutrals. A favorable chemical reaction can then occur between the silicon and fluorine to create a volatile SiF_4 etch product. The passivation layer is achieved by injecting C_4F_8 (Li et al., 2003). This gas predominantly creates twisted polymer chains of CF_2 (Teflon) that coat and protect the etched silicon sidewalls from damage.

When ionized, this gas can also generate F atoms used for etching. Determination of when the gas transitions from polymerization to etching is delicate, but since ICP etching is typically performed under lower DC bias conditions, the primary effect of the gas is to generate the passivation layer.

6.2.2 Pseudo-Bosch Silicon Etch

Modifying the same etch chemistry to mixed-mode etching establishes a very different silicon etch termed here as pseudo Bosch. By simultaneously injecting SF_6 and C_4F_8, the etch significantly slows down, on the order of a few hundred nanometers per minute and the chopped sidewalls are replaced with atomically smooth sidewalls (Rhee et al., 2009). This etch has become indispensable to nanoscale etching.

This etch is typically performed at room temperature (approximately 20°C) to encourage passivation deposition. To reduce the IADF, pressures are typically as low as possible and highly dependent on the physical geometry of the chamber, around 10 mT. Typical Fwd powers are highly dependent upon the etch mask used. By reducing the Fwd power to lower values, say, 5 to 10 W, the etch masks are milled away at a slower rate.

The ratio of the substrate etch rate to the etch rate of the mask is commonly referred to as selectivity. Increasing the Fwd power permits a higher etch anisotropy, assuming the etch mask does not deteriorate and induce sidewall damage. Increasing the Fwd power also increases the etch rate. Unlike other

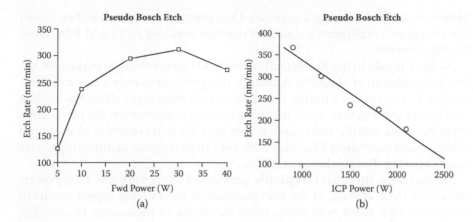

FIGURE 6.2
Pseudo-Bosch etch. (a) Effect of Fwd power upon etch rate. (b) Effect of ICP power upon etch rate.

etches, the Fwd power contributes largely to etch rate. This is not because it generates more etching reactions, but instead because it removes the horizontal passivation faster (Figure 6.2).

To increase the etching reaction, ICP power is increased. However, very counterintuitively, increasing the ICP actually decreases the etch rates. This is because this etch relies on a higher C_4F_8 gas flow than SF_6. Hence increasing the ICP power actually increases the polymerization rate more than the etch rate. Typically, the ICP power is set to approximately 1000 W and then other parameters are used to tune the etch. One such parameter is the ratio of the etch gas SF_6 flow rate to the passivation gas C_4F_8 flow rate. Adjusting the ratio of the two gases permits tuning the ratio of etch rate to passivation rate. The tuning of the two rates permit one to modify the angle of the etch sidewalls. For example, decreasing the C_4F_8 reduces the amount of polymerization and drives the etch angle to more re-entrant. A good starting ratio is approximately 1:2 to 1:3 of $SF_6:C_4F_8$.

The pseudo-Bosch etch provides for a high degree of tuning. Fwd power can control the etch rate and selectivity where the ratio of the gases controls the angle of the etched sidewalls. This etch, when coupled with a good etch mask, can produce extraordinarily small structures, down to 20 nm, with very large aspect ratios, up to 60:1. It is a first choice for top down fabrication in the nanoscale regime.

6.2.3 Cryogenic Silicon Etch

A second mixed mode etch utilizes an oxide for the passivation vice polymer. This approach is called the cryogenic silicon etch because a temperature substrate reduction to below −85°C is required to generate the passivation (Boer et al., 2002; Jansen et al., 2009). Again, SF_6 is utilized as the etch gas.

However, instead of using a polymer, O_2 is injected into the chamber. Under the cryogenic conditions, a chemical reaction bonding Si, O, and F becomes highly favorable.

As the F bonds to the Si sidewall surfaces, O can combine to make a SiO_xF_y oxide (Mellhaoui et al., 2005). Although this polymer is only a few monolayers thick, it can act as a highly protective passivation layer when the milling rate is reduced. Further, upon warming to room temperature, the passivation layer becomes volatile and evaporates away. This etch is capable of very high rates; results exceeding 10 μ/min with very high degrees of anisotropy have been reported (Boufnichel et al., 2002).

Furthermore, this etch is typically performed using very low Fwd powers to reduce IADF etching of the thin passivation layers. This aspect results in very high etch mask selectivity, from hundreds to thousands to one. The low Fwd powers also imply that aspect ratio-dependent etching (ARDE) can become a significant problem. ARDE most commonly manifests when small areas are etched simultaneously with large areas. Although the areas are etched simultaneously, the large areas will be etched more deeply than the smaller areas. Although commonly viewed as a negative aspect, it may be seen as a powerful tool for shaping silicon structures. Unlike the pseudo-Bosch etch, the etching and passivation rates are coupled. This presents another difficulty when tuning this etch for a specific structure.

Control of the etch using the machine parameters is similar to the process described earlier. Again, increasing the ICP power increases the etch rate because the plasma generates more ions and neutrals. Typical ICP power is 1000 W but for faster etch rates up to 3000 W are used. Recall that higher ICP powers generate more collisions, increasing the IADF; this effect is more detrimental for the cryogenic etch since the passivation layer is very thin.

Increasing the Fwd power increases the milling rate but also serves to reduce the IADF. Typical Fwd powers are below 10 W. Although the milling rate increases, it is typically insignificant to the overall etch rate. Hence, modification of the Fwd power is usually to improve the anisotropy or prevent black silicon (unwanted spike or spires in the etched regions of the silicon). They are created when the SiO_xF_y passivation layer is not completely removed by the milling ions and provide a micro-masking effect. If black silicon occurs during the etch, the oxygen flow rate can be reduced or the Fwd power increased by 1 to 2 W (Figure 6.3).

It is typically not advisable to reduce the oxygen flow since this flow also controls the passivation rate. Modification of the oxygen flow can control the angle of the etched sidewalls because it controls the passivation. A reduction in O_2 will reduce the achievable passivation, which will in turn cause the sidewall etch to become re-entrant. Typical $SF_6:O_2$ gas flow rate ratios are 10:1.

A second control over passivation comes from the substrate temperature. Typical etch temperatures are –110 to –140°C, with the cooler temperatures providing increased passivation. Since passivation requires the surface of the silicon to bond with fluorine, it should be expected that a few cycles of

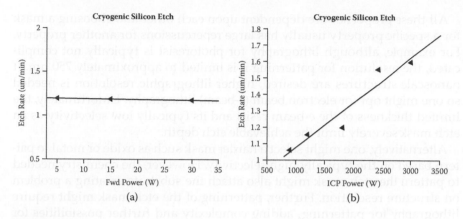

FIGURE 6.3
Cryogenic silicon etch. (a) Effect of Fwd power upon etch rate. (b) Effect of ICP power upon etch rate.

etch and re-passivate will occur due to the presence of impinging ions from IADF. Hence, it is typical to have a lateral etch rate approximately 1% of the vertical etch rate.

The extremely high etch mask selectivity makes the cryogenic method very useful for deep silicon etches or high-aspect-ratio structures. However, this etch is more useful for micron scale etches due to the nature of the passivation. Further, cryogenic etch rates begin around 1 μ/min. This etch rate makes it difficult to precisely control the depth to better than 100 nm; adding a lateral etch rate prevents etching on structures much smaller than 500 nm. This etch is most successful when employed for micron scale structures with hundreds of micron etch depths for smoother sidewalls than can be achieved using the chopping Bosch etch. Further, the highly tunable angle control provides a useful mechanism for undercutting. The advantages of this etch will be demonstrated later in this chapter.

6.3 Etch Masks for Silicon

6.3.1 Choosing Etch Masks

Using an ICP RIE for etching permits a high degree of tuning between chemical etching and milling. Incorporation of passivated silicon etches permits tuning between isotropic etching and highly anisotropic etching. The final component required for precision control over etched structures is the use of an appropriate mask. Critical properties associated with etch masks are their ease of patterning, removal, and selectivity.

All these properties are dependent upon each other and choosing a mask for a specific property usually has large repercussions for another property. For example, although lithography for photoresist is typically not complicated, the resolution for patterning it is limited to approximately 750 nm. If nanoscale structures are desired, higher lithographic resolution is needed so one might opt for electron beam (e-beam) lithography. Unfortunately, the limited thickness of the e-beam resist and its typically low selectivity as an etch mask severely limit the achievable etch depth.

Alternatively, one might select a harder mask such as oxide or metal to pattern based on the typically high selectivity. However, the chemistry needed to pattern the etch mask might also attach the substrate creating a problem on structure resolution. Further, patterning of the etch mask might require lithography for patterning, adding complexity and further possibilities for fabrication failures.

Chemical removal of the etch mask is also very important. For example, choosing a silicon dioxide etch mask for etching the top silicon layer of SOI is not advisable. Although the oxide mask might offer a high selectivity and enable a good etch, removal of the mask upon completion of the etching would etch the buried oxide layer inadvertently. A second example is using silicon nitride as an etch mask for silicon. The chemistry involved with removing the silicon nitride typically requires bonding to the silicon; hence removal of the etch mask also etches the protected structure (Williams et al., 2003). Therefore a certain amount of orthogonality between the etch mask and the substrate is required. Ideally, the etch mask should not react with the plasma chemistry used for etching the substrate and the substrate should not react with the chemistry needed to remove the etch mask.

In MEMs, quite frequently, multiple etches are required to define a structure and lithograph repatterning of the substrate after the first etch is difficult. One technique might be to pattern two different etch masks on the planar substrate. After the first substrate etch is complete, the first mask is removed and the second mask used to further define the structures in the substrate. The complication then arises that removal of the first etch mask must not only be orthogonal to the plasma etch but also to the second etch mask. This technique will be described later.

Finally, selectivity of the etch mask is important. The minimum etch mask thickness is determined from the ratio of the desired etch depth to the selectivity. Hence, for a given etch mask layer, a higher selectivity yields a greater etch depth. This concept is very basic. What typically is overlooked is the fact that the mask thickness should be much greater than the minimum required. If an etch mask is too thin, it will erode slightly faster in certain areas due to the IADF and slight temperature variations. The areas where the mask erodes will then be subjected to etching from the plasma, inducing sidewall damage. We now are back to the problem that thick masks (e.g., silicon dioxide) are too difficult to create or reduce the achievable feature size resolution. Combining all these considerations gives the attributes of the ideal mask: thin construction, high pattern resolution, high selectivity for the chosen plasma etch chemistry, and removal chemistry orthogonal to the substrate.

6.3.2 Resist Etch Masks

Resist masks are the most popular types used and provide a tremendous amount of flexibility. Manufacturers typically provide exact chemical recipes for patterning resist layers on the substrate with precisely controlled varying thicknesses. Patterning of micron scaled features is typically done with photolithography while nanoscale features are patterned using e-beam lithography. Further description of the properties depends on the type of resist and etch chemistry used. For micron scaled features, the cryogenic silicon etch using a Clarion AZ 5214-e photoresist will be described and for the nanoscale features, the pseudo-Bosch silicon etch will be described using polymethyl methacrylate (PMMA).

Photoresist is patterned on silicon by spinning the resist, evaporating off the solvents, exposing the pattern through the mask using ultraviolet light and chemical development. This entire process is very controllable and well understood. Removal of the resist upon completion of the etch is easily achieved using acetone and isopropanol. Note that the removal chemistry does not etch silicon.

Although the photoresist can be patterned to any thickness, the cryogenic etching conditions place a large restriction on this etch mask. Since the mask is patterned at room temperature, cooling the resist down to the cryogenic temperatures contracts the resist and substrate differently enough that the resist film is placed under stress. When the resist is thicker than approximately 1.5 μ, the stress cracks the resist. This places an upper limit on resist thickness.

Fortunately, this etch mask has a very good selectivity with the cryogenic etch; approximately 100:1. This places the maximum achievable etch depth around 120 μ; this includes leaving an adequate amount of masking to prevent sidewall erosion. Note the orthogonal chemistry between removal of the etch mask and the silicon etching chemistry. At the cost of limiting the achievable etch depth to 120 μ, photoresist offers easy patterning and perfect orthogonal chemistry for cryogenic etching—very close to an ideal etch mask (Figure 6.4).

For etching structures close to 100 μ in depth, particular attention should be given to how the etch conditions affect the selectivity. Since the fluorine chemistry does not significantly attack the resist mask, the ICP power can be varied with little repercussions on the selectivity. A large contributor to the high selectivity is from the fact that this etch is performed under very low milling conditions as determined by the Fwd power. Hence one significant factor determining the selectivity is the Fwd power; increasing the Fwd power decreases the mask selectivity.

A second contribution to selectivity is the oxygen that chemically attacks the resist. Although the oxygen flow is typically only a few SCCM, variation

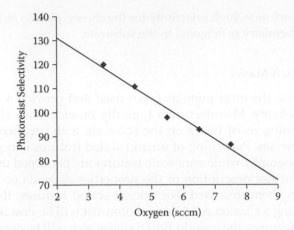

FIGURE 6.4
Loss of photoresist selectivity for cryogenic etch as oxygen is varied. The etch conditions were SF_6 = 70 SCCM, ICP = 900 W, Fwd = 5 W, temperature = –120°C, and pressure = 10 millitorr.

of 5 SCCM can reduce the selectivity from 120:1 to 90:1—a relatively large effect. With selectivity in mind, certain recommendations are given for using the AZ 5214-e resist as the etch mask. Limit the resist thickness on a 3-inch silicon wafer to 1.6 µ. This will ensure thermal stresses do not crack the resist. Limit oxygen to 8 SCCM and Fwd power to 10 W for etch depths no deeper than 100 µ. These few guidelines are typically all that is required for resist usage as a successful cryogenic silicon etch mask (Figure 6.5).

For nanoscale feature etching using pseudo-Bosch, PMMA is a good choice as a resist. However, unlike the cryogenic etch, the pseudo-Bosch has

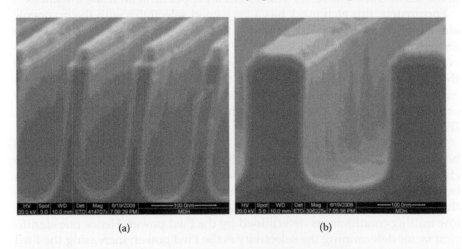

(a)   (b)

FIGURE 6.5
(a) Cross-sectional SEM of 30-nm silicon ridges. (b) Cross-sectional SEM of 100-nm silicon ridges. Both were etched using pseudo-Bosch to 330 nm with PMMA as etch mask.

a very significant proportion of the etch rate due to milling. Further, since nanoscale structures are required, the thickness of the PMMA is limited to the ability to accurately define structures in it. As an example, a photoresist structure defining a 500-nm feature only has a height-to-width aspect ratio of 3 for a 1.6-µ thickness.

For a similar aspect ratio in PMMA, defining a 30-nm structure limits the resist thickness to 100 nm. Hence the combination of higher milling rate and relatively thinner etch mask prohibits deep silicon nanoscale etching using resist etch masks. Typical etch mask selectivity for this etch is approximately 3:1. Although this etch mask exhibits orthogonality for removal with relative ease of patterning, the poor selectivity performance significantly inhibits the use of PMMA as a etch mask. A good alternative is to use the PMMA to pattern an oxide or metal etch mask.

6.3.3 Dielectric Etch Masks

Dielectric etch masks are composed of oxides and nitrides. Nitrides will not be discussed here due to the difficulties mentioned previously. One of the most popular etch masks is silicon dioxide; we will refer to it hereafter as oxide. This material has a distinct advantage in that it is relatively easy to place on silicon using thermal oxidation or chemical vapor deposition (CVD). Oxide is also very resilient both to chemical etching and milling.

Although it etches in fluorinated chemistries, the chemical bonding is strong enough to yield mask selectivity of 200:1 in cryogenic etching. A second significant advantage is that oxide mask removal is easily performed using hydrofluoric acid (HF) which does not significantly etch silicon.

The major limitation is the patterning of the oxide as a mask, limited to two major methods. The first is to pattern the oxide mask by wet etching using HF. Unfortunately, the oxide etches vertically at a similar etch rate to that of lateral etching; the consequence is that feature sizes are limited to approximately the oxide thickness.

A second method of patterning the oxide is to use a harder mask on top of the oxide and pattern the oxide using an anisotropic ICP RIE etch. Although this approach is highly successful, it adds a complication in that the mask for the oxide mask will need to be masked to be patterned. Due to the patterning resolution difficulties, oxide is not recommended for nanoscale features or pseudo-Bosch etching. However, due to high selectivity, oxide is very useful for cryogenically etching MEMS structures (Figure 6.6).

A second dielectric etch mask is aluminum oxide, also known as alumina. This etch mask has been successfully used for both cryogenic and pseudo-Bosch silicon etching (Henry et al., 2009a). The alumina mask is patterned by sputtering alumina over a resist patterned substrate and then lifted off. This leaves the inverse of the resist pattern adhered to the silicon substrate with only the additional processing step of sputtering required.

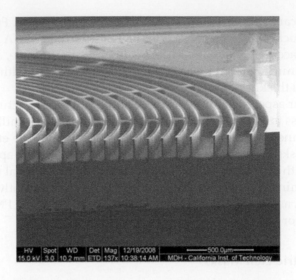

FIGURE 6.6
Cross-sectional SEM of MEMS mechanical resonator cryogenically etched over 150 μ deep using a 5-μ thick silicon dioxide etch mask.

Alumina does not chemically etch in fluorinated environments which leaves only the milling rate as the limiting factor for etch mask thickness. This provides for extraordinary etch selectivity for both the cryogenic and the pseudo-Bosch methods, exceeding 3000:1 and 70:1 respectively.

Since the selectivity is so high, a much thinner etch mask may be used, improving the achievable ultimate resolution. Removal of the etch mask requires an ammonia-based chemistry such as buffered HF, which etches alumina approximately 10 nm per minute, or ammonium-hydroxide solutions such as RCA-1. Note that neither of the solutions etches silicon and the latter has a low etch rate of silicon dioxide.

A minimum mask thickness of 25 nm is required to provide a complete masking layer using sputter deposition. It has been suggested that atomic layer deposition (ALD) of alumina may act as a thinner and better etch mask. The high observed selectivity also enables the cryogenic silicon etch to perform through-wafer etches requiring only modest mask thicknesses. Although it is possible to achieve through-wafer etching using a combination of alumina masks and cryogenic etching, the aspect ratio-dependent etching observed under the lower Fwd powers typically makes this difficult. For this reason, chopping Bosch etching is still more useful for this specific application (Figure 6.7).

6.3.4 Metal Etch Masks

Metal etch masks are commonly used for RIE devices due to the high milling rates. The biggest advantage of metal is that it produces a very durable mask

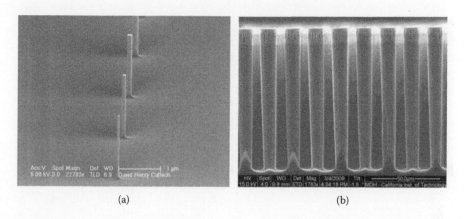

(a)   (b)

FIGURE 6.7
(a) Cross-sectional SEM of 60-, 80-, and 100-nm diameter pillars etched 1.5 μ high using 30 nm of alumina mask. (b) Cross-sectional SEM of 10-μ diameter pillars etched 103 μ tall using 100 nm of alumina mask.

with selectivity at least an order of magnitude higher than resists. Further, patterning of the metal before etching can solve problems involving difficult realignment with the etched silicon structures that develop later.

Although the metals may react with etch chemistries such as Pt with Cl, the combinations are typically nonvolatile and remain on the substrate. When this is the case, the momentum of the impinging ions can inadvertently redistribute a small amount of the metal across the etch structure; this is referred to as re-sputtering. These small islands of re-sputtered metal can then mask the silicon and unintentionally create small structures; this process is referred to as micromasking and is most common for softer metals.

Associated with the mask re-sputtering is contamination of the etch chamber with the metals. It is well known that gold and platinum form deep level electrical traps in silicon that degrade electrical properties (Sze and Ng, 2007). Using these metals as etch masks can contaminate a chamber for several years. Measurements have shown gold-contaminated chamber levels to decrease by only two orders of magnitude in 300 days (Murarka and Mogab, 1979).

Two very common metal masks are chromium and nickel. Chrome is frequently used for masking of silicon dioxide etches and achieves a selectivity better than 30:1. This figure is very impressive since silicon dioxide etches are strongly milling. It is expected to be much higher when chrome is used for cryogenic or pseudo-Bosch etching. It is easily patterned using common metallization techniques.

Standard metallization steps are (1) the resist is lithographically patterned, (2) the metal is evaporated or sputtered, and (3) the metal not attached to the substrate along with the resist is lifted off in acetone. Upon completion of etching, the mask is removed using chrome etchant which does not etch silicon.

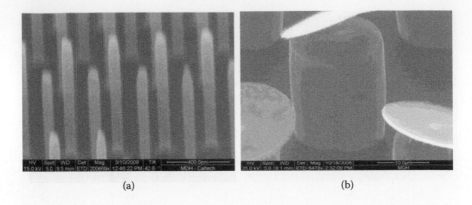

(a)   (b)

FIGURE 6.8
(a) SEM of nickel masked silicon nanopillars 75 nm in diameter and 1 μ tall. Note the notching around the mask at the top. (b) SEM of 20-μm diameter silicon pillar with chrome etch mask. Note that although the pillar is vertical the mask has been completely undercut.

Nickel is also frequently used with similar results. Like chrome, it achieves the high selectivity and is also patterned using standard metallization techniques. The nickel mask can be removed using piranha etching (a hot mixture of sulfuric acid and hydrogen peroxide). Piranha is commonly used to clean silicon and it chemically oxides the silicon only a few nanometers. Although these masks seem ideal by meeting the selectivity, ease of patterning, and etch orthogonality requirements, under the low milling powers seen in the ICP RIE, they present an unusual problem (Figure 6.8).

Metal masks can encourage mask undercutting that appears to be an electrostatic problem in which ions are more attracted to the region very close to the mask. It has been hypothesized that the metal provides a reservoir of electrons for the incoming ions, encouraging localized etching. Since the passivation is very thin for the pseudo-Bosch and extremely thin for cryogenic etching, the increased etch rate of the passivation can exceed the deposition rate on a localized scale.

After the etching passes through the passivation layer, thin layers of silicon have been observed to etch effectively and chop off the etch mask. Although chrome and nickel appear to work as perfect etch masks, under the new low milling–high chemical etch regime, they tend to create more problems. It should be noted that although the pseudo-Bosch is less susceptible than the cryogenic etch, the effect is still noticeable. For this reason, the recommendation is to avoid using metal etch masks unless they solve other problems in a fabrication sequence.

The implanted gallium etch mask (Henry et al., 2010) is a particularly useful type. Using a focused ion beam (FIB), Ga ions can be implanted into silicon. Although some of the ions bond with the silicon, most of the ions destroy the silicon matrix and remain as interstitial atoms. The thickness of this amorphous silicon–Ga mixture is predictably increased by using higher implantation voltages. Typical thicknesses range from 10 to 30 nm for implantation voltages of 5 to 30 kV.

When using this implanted Ga layer as an etch mask for fluorinated silicon etches, extremely high selectivity is seen for both the cryogenic and the pseudo-Bosch methods. The cryogenic etch has been observed to have a selectivity better than 3000:1; it is 60:1 for the pseudo-Bosch etch. The minimum areal dose required to enable useful masking begins at 2×10^{16} cm^{-2}. Increasing the areal dose increases the achievable etch depth with the upper limit of the areal dose limited to where the Ga begins to segregate, approximately 2×10^{17} cm^{-2}. Although the exact masking mechanism is unclear, it seems that Ga bonds to either F or O which remains non-volatile.

A significant advantage of this technique is the ease of patterning. Patterning is performed by rastering a highly focused Ga$^+$ beam in the FIB. A typical beam width is approximately 10 to 20 nm. Establishing the implanted mask is similar to how patterning is performed using electron beam lithography but is different in that resist is not required. This gives the implanted Ga$^+$ etch mask its most significant advantage: it is a completely dry lithography that does not require a planar substrate. Furthermore, with a precisely focused beam, feature sizes down to 30 nm have been achieved.

Upon completion of etching, two methods are common for removing the etch mask and repairing the substrate. To repair the silicon, thermal anneals are employed. The anneal also activates the Ga, in which case the Ga bonds with the silicon and acts as a P-type dopant. If dosing is heavy, sequestered Ga moves to the surface and can be removed using RCA-1 cleans. Note that neither annealing nor cleaning completely removes the Ga mask; this is the Ga mask's most distinctive disadvantage.

Further, this technique alters N-type silicon to P-type silicon. Using this etch mask with the cryogenic etch also demonstrates undercutting similar to that seen with chrome. Although still in early development, the implanted Ga etch mask offers an intriguing method that should not be limited to silicon or Ga alone. It is conceivable that an ion beam implanted in a substrate could offer similar results if the idea of etch orthogonality is upheld (Figure 6.9).

6.4 Etching Techniques

Thus far, three major topics have been discussed: the fundamentals of ICP etching, two useful silicon etches, and several etch masks. We also explain when a tuning parameter, etch, or mask is applicable or appropriate. This section will demonstrate the use of these parameters to achieve structures in silicon that would otherwise be difficult to fabricate.

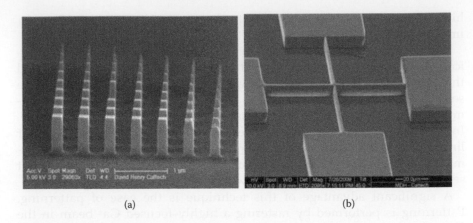

(a)　　　　　　　　　　　　　　　　　　　(b)

FIGURE 6.9

(a) Ga masked, square silicon pillars starting at 200-nm widths and decreasing down to 40 nm. The pseudo-Bosch etched the pillars to a height of 710 nm. (b) Ga masked micron scaled silicon structure. The cryogenic etched this to approximately 15 μ.

6.4.1 Multiple Orthogonal Etch Masks

A common need in experimental semiconductor fabrication is the ability to pattern and etch multiple levels. For example, a large system might be created with different height structures on top of or next to a larger component. After etching has defined the first structure, patterning the second layer becomes difficult using resist, mainly because the silicon is no longer a planar surface and spinning resist becomes more complicated. To achieve structures on top of one another, patterning of both the larger and smaller structures can be done prior to etching so long as the requirement of orthogonality is met.

Orthogonality between the etch chemistry, masks, and substrate was one of the critical ideas explained in selecting an etch mask. The chemistry for removing the etch mask should not etch the substrate, and the plasma etch chemistry should not chemically etch the mask. If multiple etch masks are needed, the orthogonality requirement becomes more complex. The second etch mask should not be removed by either the plasma etch chemistry or the removal chemistry of the first etch mask. If a third etch mask is needed, the orthogonality requirement becomes even more complicated.

One combination of etch masks uses alumina and photoresist to create a transition between nanoscale and micron scaled structures. For this sequence, the alumina is used to mask the silicon to create nanoscale structures using the pseudo-Bosch etch chemistry. A photoresist etch mask is used to mask micron scaled structures for cryogenic silicon etch chemistries. Both etch masks should have high selectivity values for the given plasma etch chemistries and the orthogonality required for removal etch chemistries, RCA-1 and acetone, respectively. The combination of the two masks

(a)   (b)

FIGURE 6.10

(a) SEM of 100-nm diameter pillars masked using alumina and etched 1 μ tall. The pillars sit on the micron-sized pillar in (b). (b) SEM of 20-μ diameter pillar masked using photoresist and etched 40 μ high with array of nanopillars on top.

requires that the alumina mask is not removed in acetone; this requirement is met. With the orthogonality met, the following fabrication sequence can be implemented.

First the nanoscale pattern is defined using the alumina etch mask; 100-nm dots are used as examples. Second, the micron scale pattern is then defined using photoresist. For this mask, 20-μ micron diameter dots were placed over the 100-nm alumina dots. Third, the cryogenic etch was performed to etch the 20-μ diameter structures to a height of 40 μ. Acetone was then used to remove the photoresist, leaving alumina dots on the micropillars. The fourth step was a pseudo-Bosch etch of the 100-nm dots to a height of 1 μ. The alumina mask was then removed using a RCA-1 clean.

The end structure consisted of an array of 100-nm diameter nanopillars etched 1 μ tall, sitting on a 40-μ tall 20-μ diameter micropillar. In this way, a nanostructure can be integrated with a microstructure when the etch masks are defined orthogonally (Figure 6.10).

Another possible combination of etch masks uses implanted gallium and photoresist. Again the required orthogonality between the etch masks and etch chemistry is evident. If the resist is used to mask a silicon or cryogenic etch first, it can then be removed to expose a gallium etch mask. The Ga+ mask can then define the second structure. Admittedly, since the Ga+ mask can be implanted after the first etch is performed, this sequence only helps convey the idea of using multiple etch masks. Finally, a combination of all three of the masks (implanted Ga+, alumina, and resist) could be potentially utilized for three etch layers.

6.4.2 Metallization with Etch Mask

One of the most significant advantages of the cryogenic etch is the extraordinary selectivity observed when using photoresist and highly controllable sidewalls. These attributes can be utilized to accomplish metallization in silicon that would otherwise be very difficult with photoresist alone.

Metallization is the process of depositing patterned metal on silicon, as was described earlier. Standard metallization relies on a layer of photoresist patterned on silicon that meets several stringent requirements. First, the photoresist must be significantly thicker than the thickness of metal required—typically two to three times as thick. Second, the profile of the resist must be very vertical to slightly re-entrant to ensure a break in the metal between the horizontal surfaces on the top of the resist and in the defined trench. This break permits acetone to dissolve the resist and lift off the metal deposited on the resist surface. After these requirements are met, the surface on which the silicon is to be deposited must be cleaned. Adhesion is known to be difficult if any residue or oxide remains between the metal and the silicon (Figure 6.11).

A combination of a photoresist etch mask and a cryogenic silicon etch can achieve higher degrees of control and repeatability (Henry et al., 2009). First, a photoresist pattern is lithographically defined on a substrate, then a cryogenic etch is performed to a depth two times deeper than the desired metal thickness. The metal is then sputtered or evaporated into the trench, and liftoff in acetone completes the sequence. This sequence achieves all the requirements for metallization noted above.

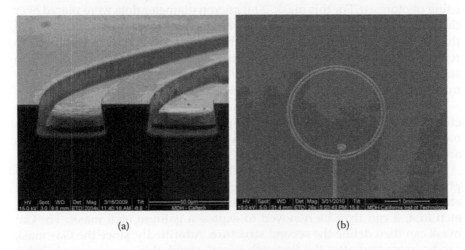

(a)   (b)

FIGURE 6.11

(a) Embedded copper in silicon using photoresist etch mask and cryogenic silicon etch. (b) Planar copper microcoil using metallization technique described in text. The metal is insulated by oxide that also tunes the coil's resonant frequency.

By controlling the etching time, the needed step height between the surface of the resist and the bottom of the trench is set. The angle of the etched sidewalls is highly repeatable and can set the needed angle for evaporation. Finally, since the surface at the bottom of the trench was recently etched, any native oxide or polymer was removed. After the etch process is established, the recipe can be used for any metal thickness or pattern.

This technique can be used for sputtered layers other than metals as well. As an example, this technique was modified to create current-carrying conductors embedded in silicon for improved thermal dissipation to the substrate. Upon completion of the etching step, oxide is deposited. Conformal oxides can be deposited under temperatures lower than the resist cross-linking temperature by either sputtering the oxide or ICP CVD. A thin oxide layer will prevent current from leaving the metal conductor to the substrate. The metallization deposition is then performed followed by liftoff.

As a demonstration of this fabrication sequence, 10-µ thick planar copper microcoils were fabricated in silicon. By controlling the area between the metal wires and the silicon substrate and oxide thickness, the parasitic capacitance can tune the microcoil to a desired resonant frequency. Microcoils such as these provide an integrated solution for achieving integrated inductance or magnetic fields needed for spectroscopy or testing platforms.

6.4.3 ARDE and Angle Control for Cryogenic Etching

Although it can hinder etching patterns, ARDE can be utilized to achieve a high degree of control over an etch when properly characterized. Formally, ARDE is the reduction of the silicon etch rate as the aspect ratio of the etched trench is increased. This reduction is problematic when etching different geometries on the same substrate. ARDE typically occurs when milling is reduced and the plasma etch becomes more chemical in nature in ICP RIE.

For the etch chemistries described here, ARDE is only relevant for the cryogenic etch. Although the exact reason for the phenomenon is not well understood, it is clear that the Knudsen transport of ions or neutrals is likely responsible, with some effects notable from ion shadowing and charging of the substrate (Gottscho, 1992). New theories such as ion-neutral synergy (Blauw et al., 2003) attempt to describe the phenomenon but are complex and require multiple variables to be understood and quantified. Alternatively, one might instead describe the effect using etching rate equations (Henry et al., 2009b). The advantage this method offers is that only two coefficients need be determined: the ideal etch rate (E_o) and a linear reduction to the etch rate (b) due to the aspect ratio (AR). Solving this differential equation is trivial and yields a predictive equation with only two fitting coefficients. By performing a single etch with multiple widths, the two coefficients can be determined and further etch depths (Ds) predicted based on etch time (t) and desired trench width (w) (Figure 6.12).

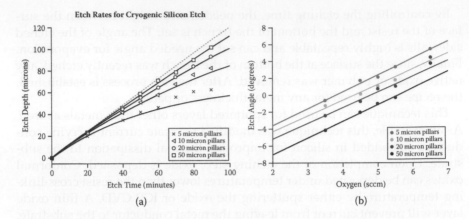

FIGURE 6.12
(See color insert.) (a) Graph of theoretical and experimental results of cryogenically etching silicon that illustrates the aspect ratio-dependent etching behavior. (b) Graph of experimental data showing etch angle dependence on oxygen flow rate for cryogenic silicon etching.

$$\frac{dD}{dt} = E_o - b \cdot AR \Rightarrow D(t,w) = \frac{W \cdot E_o}{b}\left(1 - \exp\left(\frac{-bt}{w}\right)\right) \quad (6.1)$$

The power of treating ARDE in this fashion is that a predictive model can be rapidly established, thus allowing the same recipe to be used for multiple etch projects. Since the etching rate seems to be dependent on the AR, it stands to reason that the passivation rate would similarly be affected. In fact it is.

Previously we showed that oxygen flow and temperature control the passivation rate, with oxygen having a faster response time than temperature. By controlling the oxygen flow, the angle of the etch can be varied over a range of approximately 15 degrees. As the oxygen flow is reduced, the passivation decreases and the etch becomes re-entrant; the lower limit is determined by the resolution of the oxygen mass flow controller.

Similarly, as the oxygen flow is increased, the angle becomes more positive; the upper limit on flow is determined by the onset of black silicon. As suggested, the angle dependence on aspect ratio is significant. For a set temperature and given oxygen flow, the angle becomes more positive as the aspect ratio is increased.

As an example, for a 5-SCCM oxygen flow rate, a 50-μ trench width was measured to be 2 degrees re-entrant while a 5-μ trench was measured to be 2 degrees positive.

For shallow etches, this may not be problematic, but if deep etches are required, the re-entrant angle can topple a structure. Furthermore, this example suggests that as the etch depth increases the aspect ratio, an etch can change from re-entrant to positive. To describe this effect, the rate

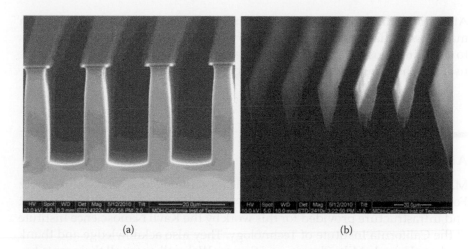

(a)   (b)

FIGURE 6.13
(a) SEM of 30-μ deep silicon ridge with oxygen flow rate set to 4 SCCM. These conditions produced a slightly positive etch angle. (b) SEM of 30-μ deep silicon ridge with oxygen flow rate set to 8 SCCM. Increasing the oxygen flow rate increased the angle of the anisotropic etch.

differential equation is coupled with a width differential equation where β and ε are again coefficients to be determined via a test etch (Figure 6.13).

$$\frac{dw}{dt} = 2 \cdot Tan(\beta + [O_2] \cdot \varepsilon) \cdot (E_o - b \cdot AR) \tag{6.2}$$

It should be emphasized that angle control is not a transition from anisotropic to isotropic. The cryogenic etch remains anisotropic, but with a high degree of angle tenability.

6.5 Conclusions

Etching silicon remains one of the most critical aspects of MEMS and NEMS development. Without precision etches, high-aspect-ratio structures and high-fidelity pattern transfers into silicon substrates would not be possible. This chapter has detailed how to manipulate ICP RIE devices to achieve desired etch structures from a silicon substrate.

Two specific etch chemistries for silicon were also described. One chemistry detailed was the mixed mode passivating etch utilizing the gentle nature of ICP etching to create highly anisotropic silicon etches. The cryogenic silicon etch was demonstrated as a useful micron scale etch and the pseudo-Bosch silicon etch was demonstrated to achieve nanoscale silicon structures.

After the etching machineries and chemistries were explained, several etch masks were discussed. The etch masks were given descriptive quantifiers to assist in choosing masks appropriate for desired structures and severable useful techniques for etching were discussed.

Acknowledgments

M. David Henry gratefully acknowledges the Fannie and John Hertz Foundation for support. Both authors gratefully acknowledge the support and infrastructure provided for this work by the Kavli Nanoscience Institute at the California Institute of Technology. They also acknowledge and thank Andrew Homyk, Mike Shearn, and Sameer Walavalkar for all their contributions. This work was funded in part by grants from DARPA under both EPIC (HR0011-04-1-0054) and NACHOS (W911NF-07-0277).

References

Blauw, M.A., Drift, E. van der, Marcos, G. et al. 2003. Modeling of fluorine-based high-density plasma etching of anisotropic silicon trenches with oxygen side-wall passivation. *Journal of Applied Physics*, 94, 6311.

Boer, M.J.D., Gardeniers, J.G., Jansen, H.V. et al. 2002. Guidelines for etching silicon MEMS structures using fluorine high-density plasmas at cryogenic temperatures. *Journal of Microelectromechanical Systems*, 2, 385–401.

Boufnichel, M., Aachboun, S., Grangeon, F. et al. 2002. Profile control of high-aspect-ratio trenches of silicon. I: Effect of process parameters on local bowing. *Journal of Vacuum Science & Technology B*, 20, 1508–1513.

Campbell, S.A. 2001. *The Science and Engineering of Microelectronic Fabrication*, 2nd ed. New York; Oxford University Press.

Coburn, J.W. 2000. Some fundamental aspects of plasma-assisted etching. In Shul, R.J. and Pearson, S.J., Eds., *Handbook of Advanced Plasma Processing Techniques*. Berlin: Springer, pp. 1–32.

Craighead, H.G. 2000. Nanoelectromechanical systems. *Science*, 290(5496), 1532–1535.

Eichenfield, M., Chan, J., Camacho, R.M. et al. 2009. Optomechanical crystals. *Nature*, 461 (7269), 78–82.

Ekinci, K.L. and Roukes, M.L. 2005. Nanoelectromechanical systems. *Review of Scientific Instruments*, 76(6), 1–12.

Gottscho, R. 1992. Microscopic uniformity in plasma etching. *Journal of Vacuum Science & Technology B*, 10, 2133.

Henry, M.D, Shearn, M.J., Chim, B. et al. 2010. Ga+ beam lithography for nanoscale silicon reactive ion etching. *Nanotechnology*, 21, 245–303.

Henry, M.D, Walavalkar, S., Homyk, A. et al. 2009a. Alumina etch masks for fabrication of high-aspect-ratio silicon micropillars and nanopillars. *Nanotechnology*, 20, 255–305.

Henry, M.D., Welch, C., and Scherer, A. 2009b. Techniques of cryogenic reactive ion etching in silicon for fabrication of sensors. *Journal of Vacuum Science & Technology A*, 27, 1211.

Isakovic, A.F., Evans-Lutterodt, K., Elliott, D. et al. 2008. Cyclic, cryogenic, highly anisotropic plasma etching of silicon using SF_6/O_2. *Journal of Vacuum Science & Technology A*, 26, 1182.

Jansen, H.V., Boer, M.J. de, Unnikrishnan, S. et al. 2009. Black silicon method X: a review on high speed and selective plasma etching of silicon with profile control. *Journal of Micromechanics and Microengineering*, 19, 1–41.

Johnson, W.L. 1995. Electrostatically shielded RF plasma sources. In Popov, O.A., Ed., *High Density Plasma Sources: Design, Physics, and Performance*. New Jersey: Noyes, pp. 100–148.

Li, X., Ling, L., Hua, X. et al. 2003. Characteristics of C_4F_8 plasmas with Ar, Ne, and He additives for SiO_2 etching in an inductively coupled plasma (ICP) reactor. *Journal of Vacuum Science & Technology A*, 21, 1955.

Mellhaoui, X., Dussart, R., Tillocher, T. et al. 2005. SiO_xF_y passivation layer in silicon cryoetching. *Journal of Applied Physics*, 98, 1–10. doi: 10.1063/1.2133896.

Murarka, S.P. and Mogab, C.J. 1979. Contamination of silicon and oxidized silicon wafers during plasma etching. *Journal of Electronic Materials*, 8, 763–779.

Rhee, H., Kwon, H. and Kim, C.K. 2008. Comparison of deep silicon etching using SF_6/C_4F_8 and SF_6/C_4F_6 plasmas in the Bosch process. *Journal of Vacuum Science & Technology B*, 26, 576–581.

Rhee, H., Lee, H.M., Namkoung, Y.M. et al. 2009. Dependence of etch rates of silicon substrates on the use of C_4F_8 and C_4F_6 plasmas in the deposition step of the Bosch process. *Journal of Vacuum Science & Technology B*, 27, 33.

Stern, E., Wagner, R., Sigworth, F.J. et al. 2007. Importance of Debye screening length on nanowire field effect transistor sensors. *Nano*, 1, 1–4.

Sugawara, M. 1998. *Plasma Etching: Fundamentals and Applications*. New York: Oxford University Press.

Sze, S.M. and Ng, K.K. 2007. *Physics of Semiconductor Devices Engineering* 3rd ed. Hoboken: John Wiley & Sons.

Tachi, S., Tsujimoto, K., and Okudaira, S. 1988. Low temperature reactive ion etching and microwave plasma etching of silicon. *Applied Physics Letters*, 52, 616-618.

Williams, K.R., Gupta, K., and Wasilik, M. 2003. Etch rates for micromachining processing II. *Journal of Microelectromechanical Systems*, 12, 761–778.

Yazdi, N., Ayazi, F., and Najafi, K. 1998. Micromachined inertial sensors, *Proceedings of IEEE*, 86(8), 1640–1659.

Ziegler, J.F., Biersack, J.P., and Littmark, U. 1985. *The Stopping and Range of Ions in Solids*. New York: Pergamon.

7

Learning from Biology: Viral-Templated Materials and Devices

Elaine D. Haberer

CONTENTS

ABSTRACT With the rapid advance of nanotechnology, the demand for precision nanoscale assembly techniques has also expanded. We have much to learn about nanoscale assembly from biology. This chapter provides an overview of viral-templated material and device assembly that uses the cowpea mosaic virus, the tobacco mosaic virus, or the M13 bacteriophage as a scaffold. Applications in which viral assembly is advantageous to device performance are emphasized. Specifically, reports of devices that benefit from the long range order of a viral protein, the geometry of a virus, or the location of modifiable binding sites on a viral surface are discussed. Among the devices discussed are conductive nanowires, nanostructured photovoltaics, battery electrodes, humidity sensors, non-volatile memories, and photocatalytic water oxidation systems.

7.1 Introduction

Driven by scaling requirements and the pursuit of novel material properties, nanotechnology has advanced rapidly. Nanostructures in which properties are controlled by size, shape, localized composition, and crystalline structure have been synthesized from countless materials. Innovative nanoscale materials, machines, and devices that utilize the unique characteristics of these tiny building blocks have been envisioned. Unfortunately, versatile, well developed techniques capable of assembling the components are limited. Arrangement of molecular and nanoscale objects with nanometer precision is still rather challenging with man-made tools.

In contrast, Nature has been assembling complex, highly organized nanostructures for millennia. The natural world uses biomolecules such as peptides and proteins that exhibit nanoscale characteristic lengths, considerable chemical diversity, and molecular recognition capabilities to expertly direct the assemblies of inorganic materials. The sizes, shapes, morphologies, topological organizations, and crystal structures of inorganic materials can all be dictated by biomolecules during in vivo assembly (Addadi and Weiner, 1992; Mann, 1993).

The tools and strategies of Nature have evolved to create optimized materials that are application specific. For example, calcium carbonate can be biomineralized to form both sea urchin spicules and mollusk shells, each with a distinctive composition and crystalline structure (Falini et al., 1996; Beniash et al., 1997; Wilt, 1999; Meldrum, 2003; Politi et al., 2008). The organic–inorganic interface controls assembly on multiple length scales ranging from nano- to macroscale, depending on the required function. Sophisticated composite materials are created with unique properties superior to those of the individual components.

Indeed, biomolecules possess a specificity and versatility invaluable in nanoscale assembly. By understanding and harnessing the capabilities of Nature, this extraordinary nanoscale precision can be used to build technological materials and devices that cannot be assembled with conventional synthetic approaches. Moreover, Nature's manufacturing conditions are generally mild, occurring under ambient conditions and in aqueous solutions; therefore the use of biomolecules may lead to environmentally friendly, low cost assembly techniques.

Unbound peptides and proteins harvested from natural organisms or produced synthetically possess the specificity and chemical diversity necessary for local precision assembly. However larger scale organization is also often desired (Sarikaya et al., 2004; Baneyx and Schwartz, 2007). In the future, with greater understanding of Nature's assembly processes, perhaps it will be possible to design a molecule from fundamental principles, with a particular hierarchical structure and the capability of binding to a specific technological material. However, with our present knowledge, such a feat is not readily attainable. We must rely on the longer range order provided by Nature's templates.

A virus is one such template. In general, a virus consists of one or more structural proteins that surround and protect its genetic material [deoxyribonucleic acid (DNA) or ribonucleic acid (RNA)]. The viral capsid or protein coat displays a high density of well ordered biomolecules that make it particularly useful for in vitro biological-based assembly in which hierarchical structure is critical. Viruses cannot survive independently; reproduction requires host organisms. Viruses exist in a variety of shapes and sizes ranging from nearly spherical to high-aspect-ratio nanowires and from tens to hundreds of nanometers in dimension (Douglas and Young, 1999).

These organic nanoparticles (NPs) exhibit exceptional uniformity in physical dimensions and surface chemistry from particle to particle, particularly in comparison to synthetic inorganic NPs. Binding sites for inorganic materials can be integrated into the structural proteins of a virus. These sites may be used to directly template inorganic materials or to covalently bond surface-modifying molecules that serve as linkers for specific inorganic materials.

The locations, quantities, and functionalities of the structural proteins and the binding sites are encoded in the genetic material. Through genetic engineering, the properties and functionalities of these binding sites can be changed, making the virus a programmable scaffold for inorganic material synthesis. Consequently, error-free, large scale production of templates can be easily achieved using a host organism. Viruses are excellent nanoscale building blocks, simultaneously providing the flexibility and uniformity required to reliably construct nanoscale materials and devices.

This chapter focuses on devices made from viral-templated materials. Although reports of viral-templated materials have been numerous, demonstrations of functional devices have been somewhat limited to date. Following a brief survey of material synthesis and assembly efforts, viral-templated functional devices will be examined in depth. Particular attention will be paid to applications in which viral assembly is advantageous to performance. Specifically, devices in which operation depends on the long range order of a viral protein coat, geometry, and location of modifiable binding sites will be reviewed.

7.2 Viral Templates

While almost any virus can be used for templating purposes, the viruses selected for use in assembly are generally well characterized, easily replicated and isolated from the host, and innocuous to humans. The choice of virus for the design of a particular nanomaterial or device is highly application dependent and is usually a function of size, shape, and location of useful surface binding sites. The handful of viruses used as templates in the materials and devices are described in detail in this chapter.

The cowpea mosaic virus (CPMV) and the tobacco mosaic virus (TMV) are both plant viruses frequently used to guide material and device assembly. The CPMV is an icosahedral virus with an average spherical diameter of approximately 30 nm. As shown in Figure 7.1, the virus is composed of 60 asymmetric units, each with L (large) and S (small) subunits, which encapsulate its RNA genome (Lomonossoff and Johnson, 1991; Lin et al., 1999; Wang et al., 2002). This robust virus that infects cowpea plants is stable up to 70°C and remains viable from pH 3 to 10 (Blum et al., 2007). The wild-type CPMV displays addressable residues both on the outer and inner surface of the viral protein coat.

The TMV is named after its host, the tobacco plant. It is a hollow, rod-shaped virus 300 nm in length with 18 nm outer and 4 nm inner diameters, respectively. As shown in Figure 7.1, the viral protein coat is constructed from 2130 identical subunits that surround a single strand of genomic RNA in a helical arrangement (Caspar, 1963; Namba et al., 1989). Like the CPMV, the TMV is a hardy virus that can withstand temperatures up to 90°C and pH 3.5 to 9 environments (Namba et al., 1989; Gorzny et al., 2008). The

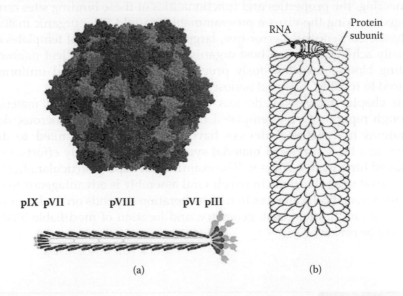

(a) (b)

FIGURE 7.1
(See color insert.) (a) CPMV structure is shown at top. The CPMV is composed of 60 copies of an asymmetric unit made from a small subunit (blue) and a two-part large subunit (green and red). The M13 bacteriophage structure is shown at bottom. The five structural proteins of the M13 virus are labeled. (b) TMV structure. The 2130 identical structural proteins are arranged around the single-stranded RNA that holds its genetic information. (From Caspar, D.L.D. 1963. *Advances in Protein Chemistry*, 18, 37–121; Lin, T.W., Chen, Z.G., Usha, R. et al. 1999. *Virology*, 265, 20–34; Flynn, C.E., Lee, S.W., Peelle, B.R. et al. 2003. *Acta Materialia*, 51, 5867–5880. With permission.)

amino acids displayed on the outer and inner surfaces of the native TMV are distinct from one another and include several genetically modifiable binding sites.

The M13, a filamentous virus that infects male Escherichia coli bacteria has also been used often as a viral template. It is rod-shaped like the TMV, but has a much higher aspect ratio. The native M13 bacteriophage is 6.5 nm in diameter and approximately 890 nm in length (Sidhu, 2005). The viral coat, depicted in Figure 7.1 is composed of five structural proteins: pIII, pVI, pVIII, pIX, and pVII (Flynn et al., 2003a). Approximately three to five copies of each pIII and pVI minor coat protein are located at the proximal or infectious end of the virus. Likewise, about three to five copies of each pIX and pVII minor coat protein are located at the distal end of the virus. Twenty-seven hundred copies of the pVIII major coat protein run the length of the virus in a right-handed α helix with fivefold rotational symmetry and a twofold screw axis (Sidhu, 2005).

The M13 is a combinatorial phage display peptide library "workhorse" with pIII and pVIII virion proteins used most frequently for this purpose (Barbas et al., 2001; Clackson and Lowman, 2004; Sidhu, 2005). Both pIII and pVIII virion proteins are capable of displaying high copy number random linear or constrained peptide fusions and act as adaptable binding sites (Cabilly, 1998; Barbas et al., 2001).

7.3 Viral-Templated Material Synthesis and Assembly

Within the past few decades, a vast range of inorganic materials have been constructed using viruses. Among them are various metals (Au, Ag, Pd, Co, Ni, Co-Pt, Fe-Pt), metal oxides (IrO_2, $FePO_4$, SiO_2, Co_3O_4), and II–VI semiconductors (ZnS, CdS, and PbS) (Manocchi et al., 2010; Shenton et al., 1999; Fowler et al., 2001; Dujardin et al., 2003; Flynn et al., 2003b; Knez et al., 2003; Mao et al., 2003; Knez et al., 2004; Klem et al., 2005; Lee et al., 2005; Slocik et al., 2005; Lee et al., 2006; Nam et al., 2006; Balci et al., 2007; Tsukamoto et al., 2007; Bromley et al., 2008; Lee, L.A. et al., 2009; Lee, Y.J. et al., 2009; Steinmetz et al., 2009; and Nam et al., 2010).

The morphology of the nucleated materials ranges from isolated, nearly spherical Au NPs to thicker, continuous films of SiO_2. Crystallinities vary from amorphous $FePO_4$ to Ag single crystals (Shenton et al., 1999; Fowler et al., 2001; Huang et al., 2005; Slocik et al., 2005; Balci et al., 2007; Bromley et al., 2008; Nam et al., 2008a; Lee, L.A. et al., 2009; Lee, Y.J. et al. 2009; Steinmetz et al., 2009).

Many characteristics of these templated materials have been controlled by binding moiety, nucleation and growth conditions, and viral-template selection. Reported binding mechanisms have varied in specificity. Combinatorial phage display techniques have been used to identify high affinity peptides for target materials. These peptides were shown to be extremely selective, exhibiting a preference for one crystallographic structure or direction over another, defects over pristine materials, and subtle variations in composition (Whaley et al., 2000; Flynn et al., 2003b; Mao et al., 2003; Sinensky and Belcher, 2006). Other binding mechanisms were more promiscuous relying on simple metal-binding residues such as cysteine, histidine, methionine, arginine, aspartate, and glutamate or electrostatic interactions (Knez et al., 2004; Nam et al., 2006; Royston et al., 2008; Lee, L.A. et al., 2009; Lee, Y.J. et al., 2009; Liu et al., 2009).

Some material–virus combinations require an activation agent such as Pd or Pt to initiate biomineralization (Knez et al., 2003; Lee et al., 2005; Royston et al., 2008). With multiple available binding sites on a viral surface, spatially selective binding has also been possible. For example, Co and Ni were bound to either the external surface or inner channel of the TMV, depending on the nucleation conditions, and a metal oxide was nucleated on the major coat protein of the M13 while the pIII minor coat protein bound the carbon nanotubes (CNTs; (Knez et al., 2004; Nam et al., 2010).

Moreover, multiple protein copies within a single viral coat permitted co-binding of materials. ZnS/CdS, Au/Co_3O_4, and $Au/FePO_4$ material pairs were concurrently bound to the M13 major coat protein (Flynn et al., 2003b; Mao et al., 2003; Nam et al., 2006; Lee, Y.J. et al., 2009). The nanoscale shape and crystallographic structure have been controlled by the binding molecule.

Depending on the peptide fusion, spherical, zinc–blend, or elongated, wurtzite ZnS NPs were mineralized on the M13 virus (Flynn et al., 2003b; Mao et al., 2003). The use of viral-templates led to significantly reduced polydispersity as observed with Co_3O_4 NPs templated with the M13 virus (Nam et al., 2006). Viral-templated assembly uses mild conditions to create materials of similar or higher quality than those synthesized under harsher conditions without a template. For instance, the M13 was used to nucleate Co_3O_4 and $FePO_4$ materials with properties similar to materials created at 500 to 600°C (Lee et al., 2006; Nam et al., 2006).

These biomolecular structures are equally as capable of nanoscale building block assembly as biomineralization. A variety of pre-made nanoscale components have been organized on viral surfaces. Gold NPs were immobilized on CPMV, TMV, and M13 mutants using the thiol groups in cysteine residues, exposed RNA strands, and gold-binding peptides found by combinatorial phage display, respectively (Huang et al., 2005; Balci et al., 2007; Blum et al., 2007).

Linkers or bifunctional molecules have been used to extend the range of attachable materials to include water-soluble colloidal quantum dots. For example, a streptavidin-binding peptide was used to attach a streptavidin-conjugated CdSe NPs to the pIII minor coat protein of the M13 bacteriophage and 1-ethyl-3-[3-dimethylaminopropyl]carbodiimide (EDC) chemistry was used to covalently link ligand molecules on the surfaces of CdSe/ZnS core/shell quantum dots to a viral surface residue on the CPMV (Huang et al.,

2005; Portney et al., 2007; Portney et al., 2008). Multi-functional linkers were also used to connect CNTs to TMV, creating a parallel alignment of the two nanotubes (Holder and Francis, 2007).

7.4 Viral-Templated Devices

7.4.1 Nanowires

A conductive nanowire allows electrical current to flow from one location to another. Although simple, these connections are critical to the performance of nanoscale electrical circuits. The geometries of anisotropic viruses such as M13 and TMV are particularly suitable for nanowire templates and are used frequently for this purpose.

Rod-shaped viruses allow nanowires to be fabricated at more mild conditions than most conventional chemical synthesis techniques. In addition to shape, properties such as connectivity on the template, type of material, and crystallinity affect electrical conductivity and ultimately the functionality of viral-templated nanowires. For these reasons, viruses that biomineralize rather than assemble pre-made building blocks, display high copy numbers of the binding moiety, and preferentially bind metals are favored as templates.

Electrically conductive, viral-templated nanowires have been realized in various shapes, sizes, and materials. Gorzny and colleagues (2008) reported individual viral-templated Pt nanowires. The TMV was used for electroless deposition of a layer of Pt approximately 6 nm on the external protein coat. Figure 7.2 is a scanning electron microscope (SEM) image of the Pt nanowire.

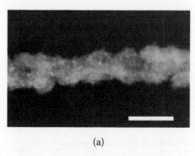

(a)

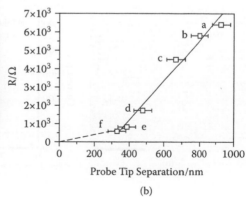

(b)

FIGURE 7.2
(a) SEM image of a TMV-templated Pt nanowire. Scale bar is 50 nm. (b) Resistance versus probe tip separation showing variations in resistance along length of nanowire. (From Gorzny, M.L., Walton, A.S., Wnek, M. et al. 2008. *Nanotechnology*, 19, 165704–165709. With permission.)

Four-point probe measurements were used to evaluate the electrical properties of individual nanowires. Resistances on the order of 10^3 Ω were measured, but as shown in the figure, variations were observed along the length of the nanowire. The source of the variation was likely change in the continuity of the Pt coating.

Given the uncertainty in nanowire resistance, calculated electrical resistivities ranged from 1294 to 235 $\mu\Omega$-cm. Both values are significantly higher than the resistivity of bulk Pt (10.5 $\mu\Omega$-cm). Gold nanowires were fabricated using a genetically modified M13 virus by Huang and coworkers (Huang et al., 2005). Gold NPs, 5 nm in diameter, were bound to the major coat protein with a gold-binding peptide fusion. Subsequently, electroless deposition was used to grow the NPs into a continuous metallic nanowire approximately 40 nm in diameter. An individual nanowire resistance of approximately 588 Ω was measured, resulting in a resistivity of approximately 180 $\mu\Omega$-cm or about 100 times greater than that of bulk gold.

The comparatively low conductivity was attributed to variations in diameter, surface scattering, and grain boundaries. Much larger Ag nanowires were made from TMV virus bundles with long-range alignment by Kuncicky and co-workers (2006). Films of TMV fibers were created by pulling the meniscus of a viral suspension over a glass substrate in a controlled manner. Gold NPs, 3.5 nm in diameter, were bound to the bundles of TMV and used as seeds for the electroless deposition of Ag. An optical image of the nanowire bundles and current voltage data are shown in Figure 7.3. The measured parallel and orthogonal resistances of the oriented, fiber films were 0.06 and 0.59 Ω, respectively. Due to the uncertainty of the fiber cross-sectional area, electrical resistivity values were not calculated.

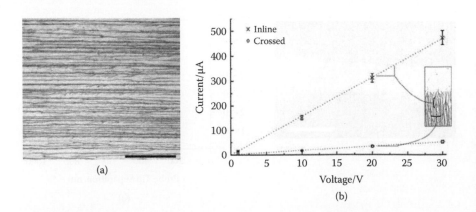

(a)

(b)

FIGURE 7.3
(a) Bright field optical image showing aligned Ag nanowires. Scale bar is 500 μm. (b) Current-voltage plot for Ag wires taken parallel to and normal to viral bundle alignment. (From Kuncicky, D.M., Naik, R.R., and Velev, O.D. 2006. *Small*, 2, 1462–1466. With permission.)

7.4.2 Nanostructured Photovoltaics

Nanostructured photovoltaics use quantum-confined materials to convert sunlight into electrical energy. Nanoscale materials exhibit size-dependent effects that include the ability to tune the material absorption wavelength to increase overlap with the solar spectrum—a rather alluring property for photovoltaics. Unfortunately, the typically low conductivity of nanoscale materials can severely limit photovoltaic efficiency by preventing the extraction of photogenerated carriers.

One approach to improving carrier collection in quantum-confined materials is to form hybrid materials using zero-dimensional semiconductor nanocrystalline materials to decorate one-dimensional nanostructures such as nanowires and nanotubes made from a more conductive material. The intimate contact of the nanocrystals with the conductive nanowire or nanotube may allow carriers to be separated and collected more readily along the length of the nanowire or nanotube, improving quasi-1D transport.

Viral-templated, hybrid devices using the M13 bacteriophage were reported by Haberer and colleagues (2009). The hybrid gold/CdSe material was assembled by chemical bath deposition to decorate one-dimensional viral-templated gold nanowires with nanocrystalline CdSe (zero-dimensional component). Viral-templating was chosen for gold nanowire assembly because of its capacity to modify the conductivity of an ensemble of gold nanowires over many orders of magnitude.

The virus was genetically modified to uniformly display a peptide fusion known for its affinity to gold on the major coat protein. Gold NPs, 5 nm in diameter, were selectively bound to the phage and used as seed crystals for the electroless deposition of gold. The additional gold increased and controlled the size, connectivity, and electrical conductivity of the gold nanowires. Subsequently, approximately 400 nm of CdSe nanocrystalline material was selectively deposited on the viral-templated gold nanowires. Figure 7.4 is an SEM image of the hybrid Au/CdSe nanowire film.

Haberer et al. (2009) made test devices from films of the hybrid Au nanowire and nanocrystalline CdSe. Current-voltage measurements of the gold nanowire films were linear, confirming ohmic behavior. The conductivity of the templated gold nanowires was controlled by the duration of the electroless deposition. Absorption and photocurrent action spectra of the hybrid materials confirmed that the films were photosensitive and that the carriers collected by the photocurrent measurements were predominantly generated in the CdSe nanocrystalline material.

Haberer and co-workers found that the photocurrent for the hybrid gold/CdSe material was approximately 29 times greater than for the CdSe nanocrystalline material used as a control. The incorporation of one-dimensional gold nanowires within the zero-dimensional CdSe film significantly increased the photocurrent collection of the hybrid material system. As shown in Figure 7.4, experiments also revealed that in a hybrid gold/CdSe

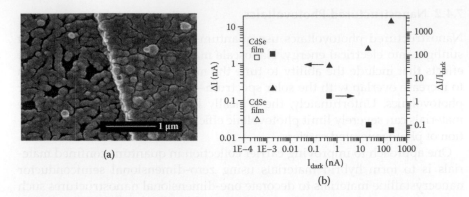

(a)

(b)

FIGURE 7.4
(a) SEM image of gold/CdSe nanowire film. Right side shows an electrical contact pad.) (b) Photocurrent and photocurrent-to-dark-current ratios for several hybrid gold/CdSe devices plotted along with corresponding values from a CdSe only device. (From Haberer, E.D., Joo, J.H., Hodelin, J.F. et al. 2009. *Nanotechnology*, 20, 415206–415213. With permission.)

viral-templated device, the broad spectrum photocurrent may be increased as much as 350-fold over devices made from nanostructured CdSe alone by further increasing the conductivity of the gold nanowires.

7.4.3 Battery Electrodes

Batteries are used to convert chemical energy to electrical energy. They are composed of two electrodes and an ionic conducting electrolyte. The negative electrode or anode reduces and the positive electrode or cathode oxidizes. Throughout history, many battery technologies have been developed including alkaline, lead–acid, nickel–cadmium, nickel metal hydride, and lithium ion. Yet, the demand for an environmentally friendly battery with greater capacity, higher energy density, faster charging, and increased battery life continues. Scientists are working to develop and test new electrode materials to meet consumer demands. The quest for improved battery technology has included numerous designer nanostructured materials that may improve electrochemical behavior. Specifically, viral-templated materials reduce aggregation of incorporated NPs and increase surface area, enhancing electrical and ionic transport. Furthermore, viral templating can create high quality electrode materials using milder conditions than other methods.

Nam and colleagues (2006) used an M13 viral template to construct nanostructured Co_3O_4-based anode materials for Li ion batteries. Two M13 templates were created and assessed for this purpose. In one template, the genome was modified to express a tetraglutamate (EEEE–) peptide fusion on the major coat protein with 100% display. Glutamic acid has a carboxyl side group that aids in binding positive ions using ion exchange. The tetraglutamate motif was used to synthesize Co_3O_4 NPs along the length of the M13 virus.

The genetically modified phages were incubated with a cobalt chloride precursor, reduced, and oxidized in aqueous solution. The highly crystalline Co_3O_4 NP morphology was varied from monodispersed 2 to 3 nm diameter particles to larger branched particles by controlling the peptide-cobalt ion interactions. The quality of the Co_3O_4 materials assembled at room temperature with the viral template was comparable to the qualities of particles synthesized with other much higher temperature techniques. In the other template, the tetraglutamate mutant was used with a phagemid system to incorporate a low copy number of gold-binding peptides into the pVIII protein such that the gold-binding peptide and the tetraglutamate peptide fusion were simultaneously displayed. Hybrid composite Au/Co_3O_4 anode materials were assembled with this template. The Au/tetraglutamate M13 mutant was incubated with pre-formed 5 nm gold NPs that selectively attached to the gold-binding peptides. The template was subsequently mineralized with Co_3O_4 NPs using the same procedure as for the tetraglutamate mutant. Notably, control samples of Co_3O_4 NPs synthesized without a phage template or with a non-specific binding phage resulted in large or irregular precipitates.

Electrode materials assembled with the two viral templates were separately evaluated by these authors using simple battery cells (Nam et al., 2006). Devices that incorporated viral-templated Co_3O_4 within the electrode material exhibited measured charge and discharge capacities between 600 and 750 mAh/g that stabilized at approximately 600 mAh/g over 20 cycles. These values are approximately two times higher than current carbon-based electrodes. Moreover, the test cells fabricated with Au/Co_3O_4 electrode materials showed an additional 30% increase in specific capacity. Nam and colleagues hypothesized that the gold NPs improved device performance through increased electrical conductivity or electrochemical catalytic activity. The virus was found to be electrochemically inactive and stable over a range of measurement conditions.

As a further demonstration of the technique's utility, functional microbatteries were fabricated (Nam et al., 2008b). The net charge and high-aspect-ratio of the tetraglutamate-modified phage was used to form highly ordered, liquid crystal virus films on multi-layered polymer surfaces. Through a stamping process, the film was subsequently transferred to another substrate and mineralized with Co_3O_4. The storage capacity of each 8 μm diameter electrode was estimated to be 625 to 766 pAh.

Another demonstration of viral-templated materials used for battery technology was reported by Y.J. Lee and colleagues (2009). Amorphous $FePO_4$/single-walled carbon nanotube (SWNT) hybrid cathode materials were assembled using a two-gene system in which both the major and minor coat proteins of the M13 bacteriophage were modified. As in the work by Nam at al., a tetraglutamate chain (EEEE–) was fused to the N terminus of each copy of the pVIII protein (Nam et al., 2006). However, in this case, the pIII was also modified to display a CNT-binding peptide fusion selected using a 12-mer combinatorial phage display.

The additional negative charge of the tetraglutamate peptide fusion enabled positively charged ions to bind electrostatically. Using the affinity of the tetraglutamate peptide fusion for positively charged metal ions, the pVIII was pre-loaded with silver NPs, and amorphous (a-) $FePO_4$ NPs were nucleated along the length of the virus. The a-$FePO_4$ NPs were 10 to 20 nm in diameter and anhydrous. These authors suggested that the a-$FePO_4$ was dehydrated due to the formation of AgCl during growth.

Subsequent incubation of the two-gene phage with SWNTs resulted in selective binding to the pIII. The SWNTs were bound to the pIII in 4 to 5 nm diameter bundles of six to eight tubes each. The two-gene system created nanostructured materials in which the SWNTs were more uniformly dispersed and aggregation reduced compared to materials assembled with the one-gene system in which only the tetraglutamate peptide was fused to the major coat protein and the minor coat protein binding was non-specific. The authors hypothesized that the presence of dispersed SWNTs would increase the connectivity and long range electrical conductivity of the cathode material.

Test cells were made using nanostructured a-$FePO_4$ and a-$FePO_4$/SWNT hybrid materials in the cathode (Lee, Y.J. et al., 2009). The performance of electrodes fabricated with room temperature, viral-templated a-$FePO_4$ was comparable to that of electrodes synthesized at high temperatures. The low discharge rate (C/10) specific capacity was 165 mAh/g and high discharge rate (1C) specific capacity was 110 mAh/g. C/n is defined as the rate required to charge or discharge to the material's theoretical capacity in n hours. For anhydrous a-$FePO_4$, C is 178 mA/g (Lee, Y.J. et al., 2009). The authors speculated that the enhanced electrical conductivity of the a-$FePO_4$ from the Ag NPs and the nanostructuring of the viral template resulted in improved performance.

Batteries with a-$FePO_4$/SWNTs hybrid cathodes assembled with the two-gene system had a specific capacity of 170 mAh/g at the low discharge rate of C/10, whereas control devices assembled using the one-gene system had a specific capacity of 143 mAh/g at the same rate. Moreover, at a high discharge rate of approximately 10C and a specific power of 4000 W/kg, the energy density of devices using the two-gene system was three times higher than the densities of control devices. The battery was cycled 50 times without loss in capacity. Thus, Lee and co-workers concluded that the improved SWNT dispersion associated with the two-gene system is critical to battery performance. The percolation network of uniformly distributed SWNTs within the cathode coupled with the viral-templated a-$FePO_4$ resulted in specific capacities similar to those of leading edge c-$LiFePO_4$ electrodes.

Royston and co-workers (2008) used genetically modified TMV to assemble cobalt oxide and nickel oxide battery electrodes. A cysteine residue was fused to the amino terminus of each external coat protein at the third amino acid position. Cysteine has a high affinity for several metals. To fabricate electrodes, the TMVs were incubated with a gold-coated substrate, activated with Pd catalyst, and plated with nickel or cobalt using electroless deposition. The TMV was vertically oriented, normal to the substrate.

The cysteine residue is located within a groove that runs the length of the virus. The slightly recessed position may prevent the cysteine residues from binding to the gold-coated substrate. In contrast, the cysteine residues near the end of the virus are a bit more exposed, allowing stronger interactions with the gold substrate. The TMVs were shown to align end to end, allowing for rod lengths that are multiples of a single viral length. The metal coatings are continuous and uniform along the viral surface with a 20 to 40 nm thickness, as shown in Figure 7.5. The surface area of the viral-templated electrodes is increased by a factor of 6 to 13 (depending on the concentration of viruses) compared to electrodes without viruses.

To make test cells, the nickel viral-templated electrodes were oxidized in air at room temperature for 72 hours. Control electrodes were made without viruses (Royston et al., 2008). The measured capacity for the viral-templated electrode devices was two times those without viruses. Using a calculated material density of 10^{-4} g/cm^2, the specific capacity of the electrodes was said to be on the order of 10^5 mAh/g or 1.2 mAh/cm^2 after 30 cycles, as shown in Figure 7.5.

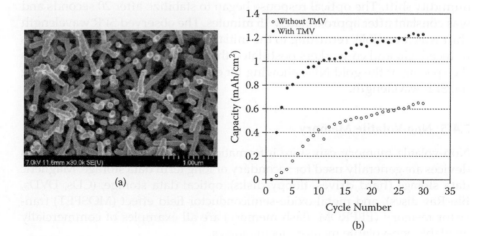

(a)

(b)

FIGURE 7.5
(a) SEM image of viral-templated Ni-based electrode material. (b) Discharge capacity versus cycle number for battery fabricated with high surface area viral-templated electrode material. (From Royston, E., Ghosh, A., Kofinas, P. et al. 2008. *Langmuir*, 24, 906–912. With permission.)

7.4.4 Humidity Sensors

Humidity sensors are used to assess the water vapor content in air or other environmental gases for a number of applications including heating, ventilating, and air conditioning (HVAC) systems, process control, meteorology stations, and consumer goods. Relative humidity is commonly measured using sensors based on changes in capacitance, resistance, or thermal conductivity. Modern humidity sensors are sensitive, highly accurate, resistant to contamination, and relatively inexpensive; nonetheless no single sensor is effective for all applications. Niche applications may require new humidity sensor designs.

Liu and colleagues (2009) used layer-by-layer assembly to construct a composite Au NP-M13 optical humidity sensor. A tetraarginine peptide (RRRR–) was fused to the N terminus of each copy of the M13 major coat protein. Arginine has a cationic side chain that modified the slightly negative surface charge of the virus, allowing electrostatic assembly of 8 nm anionic gold NPs.

Layer-by-layer assembly is used to create hybrid composite films. A poly (vinylsulfate)/poly (diallyldimethylammonium chloride) (PVS/PDDA)-coated quartz slide was repeatedly dipped, alternating between Au NP and M13 solutions. The electrostatic interactions between positively charged virus and negatively charged particles formed a uniform composite film. No film was formed with wild-type phage. Similarly, the tetraarginine mutant could not assemble cationic Au NPs.

The surface plasmon resonance (SPR) of the gold NPs within the layered films was humidity sensitive. The SPR wavelength is red shifted with decreasing relative humidity. The sensor was effective in a 20 to 100% relative humidity shift. The optical response began to stabilize after 20 seconds and was constant after approximately 15 minutes. The observed SPR wavelength shift was reversible, returning to the initial resonance when the humidity is reduced. Liu and co-workers postulated that a change in humidity modifies the spacing of the gold NPs, allowing for enhanced interaction and shifting the SPR wavelength.

7.4.5 Non-Volatile Memory

Non-volatile memory can store information without a power source. Such devices are generally used for secondary or long term data storage. Magnetic data storage (hard drives, floppy disks), optical data storage, (CDs, DVDs, Blu-Ray discs), and metal-oxide-semiconductor field effect (MOSFET) transistor memory (EEPROM, flash memory) are all examples of commercially available non-volatile memory technologies.

Critical performance parameters include speed, retention time, endurance time, and data storage density. MOSFET-based non-volatile memory technology is currently the electrically addressable memory standard because it is compatible with silicon-based technologies, can be easily integrated onto a computer chip, and is scalable in the same manner as other silicon-based technologies. Unfortunately, conventional chip manufacturing techniques are rapidly approaching scaling limits. Alternative approaches to electrically addressed, non-volatile memory are required.

Recently, viral-templated non-volatile memory devices were realized. Tseng et al. (2006) and Portney et al. (2007 and 2008) reported functional memory devices assembled using metal NP decorated TMV viruses and functionalized colloidal semiconductor quantum dots bound to CPMV, respectively. The bi-stable storage devices can be electrically addressed using two terminals. Operation is based on the presence of trapped electrical charge in the NPs or quantum dots that cause the device to switch between low and high conductivity conditions producing logical 0 and 1 states. Write, read, and erase processes are controlled by applied bias.

Tseng and colleagues (2006) used electroless deposition to form Pt NPs 7 to 15 nm in diameter on the external protein coat of the TMV. An average of 16 NPs was bound to each TMV electrostatically or via complex formation with specific functional groups on the virion surface. The composite nanowires were dispersed in polyvinyl alcohol (PVA) and placed between two aluminum electrodes with one or two viruses per device.

Electrical bi-stability was observed in these test devices with on/off ratios in the 100 to 1000 range. The high on/off ratios persisted for roughly 400 write, read, and erase cycles. Retention times on the order of 1 to 2 hours were measured near room temperature. A greater number of cycles led to reduced device performance as the device no longer readily returned to the off state. Upper working voltages were on the order of 2 to 4 V (positive for write and negative for erase); therefore the authors suggested that Joule heating of the organic virus material during device operation was the source of the degradation. No bi-stability was observed in devices made with TMV or Pt NPs alone.

The function of the viral-templated non-volatile memory was described in detail by Tseng and co-workers (2006). The TMV protein capsid provided structural organization for the metal NPs, while the aromatic rings served as charge donors. Initially, the Pt NPs that functioned as traps for mobile carriers were essentially empty. Only a small thermionic emission current flowed at low bias, and the device was in the off or low conductivity state.

The authors found that with enough applied bias, charge tunneled from the aromatic residues, through the energy barrier of the capsid, and was trapped in the Pt NPs. The device was programmed by increasing the bias until the number of NPs occupied by an electron reached a threshold value and a large non-linear increase in current flow arose from charge tunneling through the NPs.

Without sufficient kinetic energy to overcome the energy barrier of the capsid, the electrons remained trapped. The device remained on or in the high conductivity state until a sufficient number of traps was emptied by reducing the energy barrier with a negative applied bias through the erase process. The logic state of the device was read by biasing between the on and off states.

The CPMV-T184C mutant was used by Portney and colleagues (2007 and 2008) to concurrently assemble two sizes of CdSe/ZnS core/shell colloidal quantum dots (QDs). Sixty cysteine residues were arranged periodically on the surface of this mutant, in addition to the five lysines per asymmetric repeat unit regularly found on wild-type CPMV. Using linker molecules, red- and green-emitting quantum dots were covalently bonded to the virus at the cysteine and lysine residues, respectively.

An atomic force microscope (AFM) image (Figure 7.6) shows a sample virus with one red and one green quantum dot bound to the surface. Memory devices were made from the viral hybrid materials using a conductive atomic force microscope (CAFM) probe tip as one of the two terminals. As shown in the figure, on/off ratios of 10/100 were achieved for at least six cycles in which the devices were exposed to write, read, and erase biases

A 6 minute retention time was recorded. Portney et al. noted that the limitation in performance for these devices may relate to thermal drift of the CAFM probe tip. This suggestion is supported by measurements of devices in which the composite materials were dispersed in PVA and placed between two metal contacts. These devices sustained approximately 100 cycles without degradation. Control devices made from CPMV or green CdSe/ZnS alone showed no data storage potential.

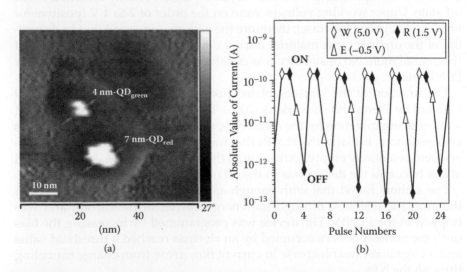

FIGURE 7.6
(See color insert.) (a) AFM phase image of CPMV with one red and one green quantum dot bound to the CPMV surface. (b) Current measurements for several write-read-erase cycles applied to the CPMV-QD device. (From Portney, N.G., Martinez-Morales, A.A., and Ozkan, M. 2008. *ACS Nano*, 2, 191–196. With permission.)

The observed electrical bi-stability, as described by Portney and collaborators (2007 and 2008), was caused by a change in the conduction mechanism with applied bias. In the off state, the conductivity was controlled by thermionic emission from the electrode to the viral hybrid and was therefore low. In the on state, the larger applied bias decreased the energy barrier associated with the ZnS shell, permitting tunneling into the CdSe core and resulting in a large increase in conductivity through Poole–Frenkel emission.

When the applied bias was removed, the charges injected into the QD cores during the write bias cycle remained trapped until the energy barrier of the QD shell was sufficiently reduced through a negatively biased erase cycle. As in the Pt TMV devices, the device was read by biasing between the low and high conductivity states.

7.4.6 Photocatalytic Water Oxidation Systems

Photocatalytic water splitting uses light from the sun to produce hydrogen and oxygen, allowing solar energy to be stored chemically. Solar energy stored in the form of hydrogen can be used as a fuel that produces only water when burned. For this reason, interest in a hydrogen-based economy has grown steadily. Two half-reactions are required to split water: oxidation and reduction.

A water oxidizing system generally consists of a photosensitizer and a catalyst. The photosensitizer absorbs sunlight, creating an electron-hole pair. The electron is quickly transferred to an electron acceptor, spatially separating the electron and hole and allowing the hole to be transferred to the catalyst. The holes in the catalyst stimulate water oxidation.

The kinetic activity between the photosensitizer and catalyst is critical to water oxidation. Smaller than optimal spacing can result in catalyst degradation and reduced oxidation. Viral templating offers spatial precision and nanoscale control and thus is an attractive method of assembly for water oxidation systems.

A genetically modified M13 bacteriophage was used to construct hybrid nanostructures for photocatalytic water splitting (Nam et al., 2010). The viral template was used to spatially arrange the photosensitizer, Zn(II) deuteroporphyrin IX 2,4 *bis*-ethylene glycol (ZnDPEG), and metal oxide catalyst, IrO_2, of the light harvesting system. An IrO_2-binding peptide was chosen using phage display technology and fused to the pVIII protein coat with 100% display. Using a carbodiimide reaction, ZnDPEG was conjugated to approximately half of the 5400 primary amines on viral surfaces. Spectral changes were observed in porphyrin absorption after conjugation, indicating exciton delocalization through energy transfer. IrO_2 nanoclusters were precipitated onto the virus varying the IrO_2:ZnDPEG molar ratio.

The water oxidation activity of the multi-component photocatalytic system was measured under UV illumination (Nam et al., 2010). Three control samples were used to aid in analysis: (1) IrO_2 hydrosol clusters with unconjugated ZnDPEG, (2) viral-templated IrO_2 nanowires with unconjugated ZnDPEG, and (3) IrO_2 hydrosol clusters with viral-conjugated ZnDPEG.

The IrO_2-ZnDPEG nanowire system exhibited turnover numbers and turnover rates significantly greater than the three control samples. Nam and colleagues concluded that the enhanced oxygen production was a result of the cooperative effect of the energy transfer between photosensitizer molecules and the proximity of the photosensitizers to the catalyst. The IrO_2:ZnDPEG molar ratio also had a large effect on oxygen evolution. As the molar ratio decreased, the oxidation activity increased.

The presence of excess catalyst may extend the chemical stability of the photosensitizer. The quantum yield measured at 0.86 was also higher for the IrO_2-ZnDPEG nanowire system. In comparison, the quantum yield of the IrO_2 hydrosol clusters with unconjugated ZnDPEG was 0.47. Nam's group also demonstrated that encapsulating IrO_2-ZnDPEG nanowires in porous microgels prevented aggregation and allowed the materials to be recycled. The oxygen production of the enveloped water-splitting system decreased only by 6% after the first cycle; however, after the fourth cycle, a 35% decrease was observed. The performance likely declined after several successive cycles due to photosensitizer degradation.

7.5 Summary and Outlook

As nanotechnology matures, schemes for material and device assembly with nanoscale precision and control are indispensable. Nature possesses a comprehensive toolkit for nanoscale assembly that has been refined by thousands of years of evolution. Using peptides and proteins, biology can manipulate the structures of inorganic materials from the molecular level to the macroscale for a variety of applications. Much can be learned from biology and the templates it provides for nanoassembly.

Within recent decades, scores of nanoscale materials have been mineralized or assembled using virus scaffolds. Control of material properties such as crystallographic structure and orientation, particle size, morphology, and composition has proven possible. As discussed, viral-templated assembly is also responsible for enhancing the performance of several devices. Despite these exciting accomplishments, much more remains to be learned. Binding mechanisms, origins of specificity, and positional control are just a few subjects that have yet to be fully explored. Greater knowledge and understanding of biological templates such as viruses will fundamentally advance nanoassembly methods, potentially enabling the rational design of future human-made templates.

Acknowledgments

The author would like to acknowledge the efforts of and convey gratitude to those involved in the nanostructured viral-templated photovoltaic development efforts discussed herein, including current and past collaborators: J. H. Joo, J. F. Hodelin, C. E. Flynn, J. C. Hsieh, A. J. Shi, A. M. Belcher, and E. L. Hu.

References

Addadi, L. and Weiner, S. 1992. Control and design principles in biological mineralization. *Angewandte Chemie International (English Edition)*, 31, 153–169.

Balci, S., Noda, K., Bittner, A.M. et al. 2007. Self-assembly of metal–virus nano dumbbells. *Angewandte Chemie International*, 46, 3149–3151.

Baneyx, F. and Schwartz, D.T. 2007. Selection and analysis of solid-binding peptides. *Current Opinion in Biotechnology*, 18, 312–317.

Barbas, C.F., Burton, D.R., Scott, J.K. et al. 2001. *Phage Display: A Laboratory Manual*. New York: Cold Spring Harbor Press.

Beniash, E., Aizenberg, J., Addadi, L. et al. 1997. Amorphous calcium carbonate transforms into calcite during sea urchin larval spicule growth. *Proceedings of the Royal Society of London Series B*, 264, 461–465.

Blum, A.S., Soto, C.M., Wilson, C.D. et al. 2007. Electronic properties of molecular memory circuits on a nanoscale scaffold. *IEEE Transactions on Nanobioscience*, 6, 270–274.

Bromley, K.M., Patil, A.J., Perriman, A.W. et al. 2008. Preparation of high quality nanowires by tobacco mosaic virus templating of gold nanoparticles. *Journal of Materials Chemistry*, 18, 4796–4801.

Cabilly, S., Ed. 1998. *Combinatorial Peptide Library Protocols*. Totowa, NJ: Humana Press.

Caspar, D.L.D. 1963. Assembly and stability of the tobacco mosaic virus particle. *Advances in Protein Chemistry*, 18, 37–121.

Clackson, T. and Lowman, H.B., Eds. 2004. *Phage Display: A Practical Approach*. New York: Oxford University Press.

Douglas, T. and Young, M. 1999. Virus particles as templates for materials synthesis. *Advanced Materials*, 11, 679–681.

Dujardin, E., Peet, C., Stubbs, G. et al. 2003. Organization of metallic nanoparticles using tobacco mosaic virus templates. *Nano Letters*, 3, 413–417.

Falini, G., Albeck, S., Weiner, S. et al. 1996. Control of aragonite or calcite polymorphism by mollusk shell macromolecules. *Science*, 271, 67–69.

Flynn, C.E., Lee, S.W., Peelle, B.R. et al. 2003a. Viruses as vehicles for growth, organization and assembly of materials. *Acta Materialia*, 51, 5867–5880.

Flynn, C.E., Mao, C.B., Hayhurst, A. et al. 2003b. Synthesis and organization of nanoscale II–VI semiconductor materials using evolved peptide specificity and viral capsid assembly. *Journal of Materials Chemistry*, 13, 2414–2421.

Fowler, C.E., Shenton, W., Stubbs, G. et al. 2001. Tobacco mosaic virus liquid crystals as templates for the interior design of silica mesophases and nanoparticles. *Advanced Materials,* 13, 1266–1269.

Gorzny, M.L., Walton, A.S., Wnek, M. et al. 2008. Four-probe electrical characterization of Pt-coated TMV-based nanostructures. *Nanotechnology,* 19, 165704–165709.

Haberer, E.D., Joo, J.H., Hodelin, J.F. et al. 2009. Enhanced photogenerated carrier collection in hybrid films of bio-templated gold nanowires and nanocrystalline CdSe. *Nanotechnology,* 20, 415206–415213.

Holder, P.G. and Francis, M.B. 2007. Integration of a self-assembling protein scaffold with water-soluble single-walled carbon nanotubes. *Angewandte Chemie International* 46, 4370–4373.

Huang, Y., Chiang, C.Y., Lee, S.K. et al. 2005. Programmable assembly of nanoarchitectures using genetically engineered viruses. *Nano Letters,* 5, 1429–1434.

Klem, M.T., Willits, D., Solis, D.J. et al. 2005. Bio-inspired synthesis of protein-encapsulated CoPt nanoparticles. *Advanced Functional Materials,* 15, 1489–1494.

Knez, M., Bittner, A.M., Boes, F. et al. 2003. Biotemplate synthesis of 3-nm nickel and cobalt nanowires. *Nano Letters,* 3, 1079–1082.

Knez, M., Sumser, M., Bittner, A.M. et al. 2004. Spatially selective nucleation of metal clusters on the tobacco mosaic virus. *Advanced Functional Materials,* 14, 116–124.

Kuncicky, D.M., Naik, R.R., and Velev, O.D. 2006. Rapid deposition and long-range alignment of nanocoatings and arrays of electrically conductive wires from tobacco mosaic virus. *Small,* 2, 1462–1466.

Lee, L.A., Niu, Z.W., and Wang, Q. 2009. Viruses and virus-like protein assemblies: chemically programmable nanoscale building blocks. *Nano Research,* 2, 349–364.

Lee, S.K., Yun, D.S., and Belcher, A.M. 2006. Cobalt ion mediated self-assembly of genetically engineered bacteriophage for biomimetic Co-Pt hybrid material. *Biomacromolecules,* 7, 14–17.

Lee, S.Y., Royston, E., Culver, J.N. et al. 2005. Improved metal cluster deposition on a genetically engineered tobacco mosaic virus template. *Nanotechnology,* 16, S435–S441.

Lee, Y.J., Yi, H., Kim, W.J. et al. 2009. Fabricating genetically engineered high-power lithium ion batteries using multiple virus genes. *Science,* 324, 1051–1055.

Lin, T.W., Chen, Z.G., Usha, R. et al. 1999. The refined crystal structure of cowpea mosaic virus at 2.8 angstrom resolution. *Virology,* 265, 20–34.

Liu, A.H., Abbineni, G., and Moo, C.B. 2009. Nanocomposite films assembled from genetically engineered filamentous viruses and gold nanoparticles: nanoarchitecture and humidity-tunable surface plasmon resonance spectra. *Advanced Materials,* 21, 1001–1005.

Lomonossoff, G.P. and Johnson, J.E. 1991. The synthesis and structure of comovirus capsids. *Progress in Biophysics and Molecular Biology,* 55, 107–137.

Mann, S. 1993. Molecular tectonics in biomineralization and biomimetic materials chemistry. *Nature,* 365, 499–505.

Manocchi, A.K., Horelik, N.E., Lee, B. et al. 2010. Simple, readily controllable palladium nanoparticle formation on surface-assembled viral nanotemplates. *Langmuir,* 26, 3670–3677.

Mao, C.B., Flynn, C.E., Hayhurst, A. et al. 2003. Viral assembly of oriented quantum dot nanowires. *Proceedings of the National Academy of Sciences of USA,* 100, 6946–6951.

Meldrum, F.C. 2003. Calcium carbonate in biomineralisation and biomimetic chemistry. *International Materials Reviews,* 48, 187–224.

Nam, K.T., Kim, D.W., Yoo, P.J. et al. 2006. Virus-enabled synthesis and assembly of nanowires for lithium ion battery electrodes. *Science*, 312, 885–888.

Nam, K.T., Lee, Y.J., Krauland, E.M. et al. 2008a. Peptide-mediated reduction of silver ions on engineered biological scaffolds. *ACS Nano*, 2, 1480–1486.

Nam, K.T., Wartena, R., Yoo, P.J. et al. 2008b. Stamped microbattery electrodes based on self-assembled M13 viruses. *Proceedings of the National Academy of Sciences of USA*, 105, 17227–17231.

Nam, Y.S., Magyar, A.P., Lee, D. et al. 2010. Biologically templated photocatalytic nanostructures for sustained light-driven water oxidation. *Nature Nanotechnology*, 5, 340–344.

Namba, K., Pattanayek, R., and Stubbs, G. 1989. Visualization of protein–nucleic acid interactions in a virus-refined structure of intact tobacco mosaic virus at 2.9-Å resolution by x-ray fiber diffraction. *Journal of Molecular Biology*, 208, 307–325.

Politi, Y., Metzler, R.A., Abrecht, M. et al. 2008. Transformation mechanism of amorphous calcium carbonate into calcite in the sea urchin larval spicule. *Proceedings of the National Academy of Sciences of USA*, 105, 17362–17366.

Portney, N.G., Martinez-Morales, A.A., and Ozkan, M. 2008. Nanoscale memory characterization of virus-templated semiconducting quantum dots. *ACS Nano*, 2, 191–196.

Portney, N.G., Tseng, R.J., Destito, G. et al. 2007. Microscale memory characteristics of virus–quantum dot hybrids. *Applied Physics Letters*, 90, 214104.

Royston, E., Ghosh, A., Kofinas, P. et al. 2008. Self-assembly of virus-structured high surface area nanomaterials and their application as battery electrodes. *Langmuir*, 24, 906912.

Sarikaya, M., Tamerler, C., Schwartz, D.T. et al. 2004. Materials assembly and formation using engineered polypeptides. *Annual Review of Materials Research*, 34, 373–408.

Shenton, W., Douglas, T., Young, M. et al. 1999. Inorganic–organic nanotube composites from template mineralization of tobacco mosaic virus. *Advanced Materials*, 11, 253–256.

Sidhu, S.S., Ed. 2005. *Phage Display in Biotechnology and Drug Discovery* Boca Raton: CRC Press/Taylor & Francis.

Sinensky, A.K. and Belcher, A.M. 2006. Biomolecular recognition of crystal defects: a diffuse selection approach. *Advanced Materials*, 18, 991–996.

Slocik, J.M., Naik, R.R., Stone, M.O. et al. 2005. Viral templates for gold nanoparticle synthesis. *Journal of Materials Chemistry*, 15, 749–753.

Steinmetz, N.F., Shah, S.N., Barclay, J.E. et al. 2009. Virus-templated silica nanoparticles. *Small*, 5, 813–816.

Tseng, R.J., Tsai, C.L., Ma, L.P. et al. 2006. Digital memory device based on tobacco mosaic virus conjugated with nanoparticles. *Nature Nanotechnology*, 1, 72–77.

Tsukamoto, R., Muraoka, M., Seki, M. et al. 2007. Synthesis of CoPt and FePt3 nanowires using the central channel of tobacco mosaic virus as a biotemplate. *Chemistry of Materials*, 19, 2389–2391.

Wang, Q., Kaltgrad, E., Lin, T.W. et al. 2002. Natural supramolecular building blocks: wild-type cowpea mosaic virus. *Chemistry and Biology*, 9, 805–811.

Whaley, S.R., English, D.S., Hu, E.L. et al. 2000. Selection of peptides with semiconductor binding specificity for directed nanocrystal assembly. *Nature*, 405, 665–668.

Wilt, F.H. 1999. Matrix and mineral in the sea urchin larval skeleton. *Journal of Structural Biology*, 126, 216–226.

8

Principles and Methods for Integration of Carbon Nanotubes in Miniaturized Systems

A. John Hart, Sei Jin Park, Michael F.L. de Volder,
Sameh H. Tawfick, and Eric R. Meshot

CONTENTS

ABSTRACT The outstanding properties of carbon nanotubes (CNTs) make them compelling materials for use in applications including nanoscale circuits, flexible conductors, battery electrodes, and reinforced composites; and their combination of attractive mechanical, thermal, and electrical properties makes them truly unique. However, as with any new material, practical use requires efficient and scalable processing methods, and commercial development requires these methods to be compatible with existing manufacturing platforms such as semiconductor processing. This chapter presents a thorough review of methods for integrating CNTs into micro- and nanodevices and systems, focusing on challenges of controlling CNT size and structure; precisely placing and packing CNTs; and achieving device-compatible growth and processing methods. Examples are selected across the spectrum from early-stage research to present commercialization. Overall, while we are far from meeting the aforementioned challenges, near-term uses of CNTs will include interconnects, non-volatile memory, inertial sensors, and thermal interfaces.

8.1 Promises and Challenges of Carbon Nanotubes

As seamless cylinders of carbon atoms arranged in a hexagonal lattice, CNTs represent the third allotropic form of carbon in addition to diamond and graphite (Dresselhaus and Avouris, 2001). The nature of CNTs as long continuous molecules imparts exceptional material properties, including several times the strength of steel piano wire at one-fifth the density, at least five times the thermal conductivity of copper, and high electrical conductivity and current-carrying capacity (Dresselhaus et al., 2001; Rotkin and Subramoney, 2005).

While research and development in science and applications of CNTs have boomed during the past 15 years, filamentous carbon has been studied and used for over a century. Many researchers unknowingly worked with CNTs well before the atomic structure of a CNT was first observed by Sumio Iijima of NEC Research Laboratories in 1991, using high-resolution transmission electron microscopy (TEM) (Iijima, 1991; Iijima and Ichihashi, 1993). Iijima's publication identified the possibility for unique properties depending on the chirality of CNTs, thereby igniting a broad base of research.

The exceptional properties of CNTs, along with application-oriented characteristics (Baughman et al., 2002; Endo et al., 2004; Endo et al., 2006) such as high surface area, diverse capabilities for chemical modification and functionalization, and strong interactions with polymers and composite host materials, create many possible opportunities for their use in next-generation micro- and nanodevices and systems (N/MEMS devices). However, as with any new material or process, many challenges exist in making CNTs compatible with the existing microfabrication processes common in the microelectronics and MEMS industries. To frame the content of this chapter, we identify five major challenges for the integration of CNTs in N/MEMS devices:

1. Controlling the chirality, wall number, and defect density of individual CNTs during synthesis or via post-synthesis processing
2. Achieving precise and directed placement of individual CNTs or CNT bundles in a repeatable fashion over large areas
3. Controlling the orientation and packing of CNTs into dense assemblies such as in vertical pillars or lateral films
4. Establishing low-resistance (e.g., electrical, thermal) contact to CNTs and strong interconnections of CNTs and supporting materials.
5. Achieving compatibility of growth and/or deposition of CNTs with device fabrication processes (typically lithography)

This chapter will address: (1) the structures and properties of CNTs; (2) methods of CNT synthesis by chemical vapor deposition; (3) methods of integrating CNTs into N/MEMS devices; and (4) selected examples of devices that utilize CNTs for electrical, mechanical, electromechanical, and thermal functions.

8.2 Structures and Properties of CNTs

8.2.1 CNT Structures

CNTs are distinguished by their number of concentric layers (with spacing 0.34 nm), and their chirality. The most distinct structures based on numbers of walls are single-wall CNTs (SWNTs), double-wall CNTs (DWNTs), and multiwall CNTs (MWNTs) as illustrated in Figure 8.1. Typically, SWNTs are 0.4 to 5 nm in diameter and MWNTs are up to 100 nm in diameter. Further, many interesting supramolecular constructs have been demonstrated, such CNTs packed with fullerenes (C_{60}); these are called "peapods" (Smith et al., 1998).

If we visualize the formation of a CNT as the action of rolling a graphene sheet into a seamless cylinder, the chirality of a CNT is established by the orientation of the graphene lattice with respect to the axis of rolling. The chirality is denoted by (n,m) indices (Figure 8.2) according to a convention introduced by Dresselhaus et al. (1995 and 2001). The chirality determines the band gap, and therefore the electronic properties of a CNT. For example, "armchair" SWNTs that have straight edges of hexagonal lattices perpendicular to the tube axes, have small band gaps and are typically metallic.

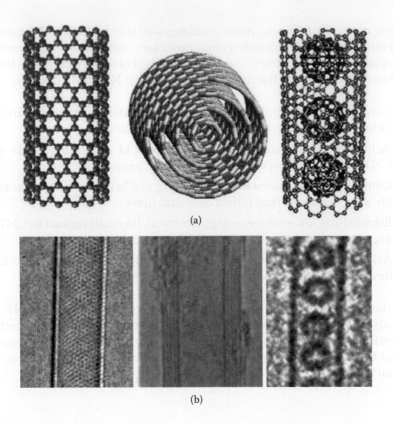

(a)

(b)

FIGURE 8.1
(a) CNT structures. Left to right: single-wall CNT (SWNT), multi-wall CNT (MWNT), double-wall CNT (DWNT), and SWNT peapod. (Adapted from Dresselhaus, M.S., Lin, Y.M., Rabin, O. et al. 2003. *Materials Science and Engineering C*, 23, 129–140. With permission.) (b) TEMs. Left to right: SWNT, MWNT, and SWNT peapod (Adapted from Iijima, S. 1991. *Nature*, 354, 56–58; Zhong, G., Hofmann, S., Yan, F. et al. 2009. *Journal of Physical Chemistry C*, 113, 17321–17325; Koshino, M., Niimi, Y., Nakamura, E. et al. 2010. *Nature Chemistry*, 2, 117–124. With permission.)

Conversely, "zigzag" SWNTs that have straight edges of the hexagonal lattices parallel to the tube axes are metallic in one-third of cases and semiconducting in the remaining two-thirds.

The band gap of a CNT is inversely proportional to the tube diameter. All CNTs larger than approximately 3 nm in diameter have band gap energies less than the thermal energy at room temperature and therefore exhibit metallic behavior. Hence, while SWNTs may be semiconducting or metallic, MWNTs are almost always metallic. Overall, hundreds of possible CNT structures present different diameter and chirality pairings. Geometric and energetic calculations suggest that the pentagon or hexagon arrangement of the cap uniquely determines the chirality of a SWNT (Reich et al., 2005).

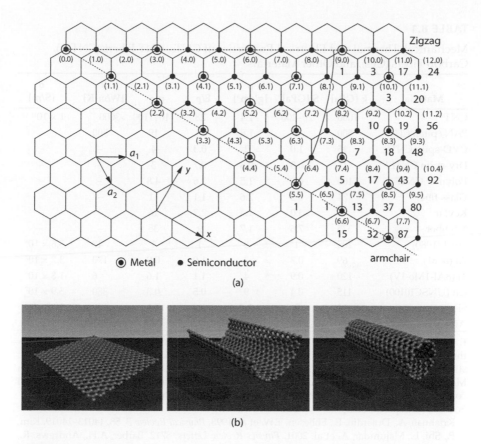

FIGURE 8.2
(See color insert.) (a) Chiral orientations of CNTs represented by alignment of CNT axis with particular lattice points of graphene sheet that determine n,m index of CNT. (From Dresselhaus, M.S., Dresselhaus, G., and Saito, R. 1995. *Carbon*, 33, 883–891. With permission.) (b) Visualization of SWNT structure by wrapping of graphene sheet.

8.2.2 CNT Properties

Table 8.1 summarizes the electrical, mechanical, and thermal properties of CNTs and various other materials including gold and silicon that are widely used in N/MEMS. In the following subsections, we discuss the electrical, mechanical, thermal, and fluidic properties of CNTs.

8.2.2.1 Electrical Properties

Interest in the electrical characteristics of CNTs arose because the semiconducting character of SWNTs makes them useful in transistors and other circuit elements and because CNTs can exhibit very high electrical conductivity due to their small dimensions and high structural quality. In particular, room temperature ballistic electron transport has been demonstrated over

TABLE 8.1

Mechanical, Thermal and Electrical Properties of CNTs, CNT Fibers (CNTFs), Carbon Fibers (CFs), and Other Materials

Material	E [GPa]	S [GPa]	ρ [g/cm³]	E/ρ	σ/ρ	k [W/m-K]	σ [S/m]
CNT[a]	1000	20–200	1.4	28	110–1100	>3000	1×10^{11}
PAN/pitch CF[b]	500	4	1.8	11	19	500	–
CVD-spun CNTF[c]	15	1.0	2	0.3	4	–	–
Dry-spun CNTF[d]	15	0.3	0.8	0.7	3	–	–
Polymer CNTF	100	0.9	1.5	2.6	4.8	–	–
Glass fiber	74	3.5	2.6	1.1	11	–	–
Kevlar	87	3.0	1.4	2.3	16	–	–
M5 fiber	375	7.6	1.7	9	35	–	–
Steel (1080)	205	1	8	1	1	48	5.6×10^6
Al (6061)	69	0.3	3	1.0	0.9	170	3.7×10^7
Ti (8Al-1Mo-1V)	120	0.9	4	1.1	1.6	6	1.8×10^6
Cu (UNSC10100)	115	0.4	9	0.5	0.3	380	5.9×10^7
Si	112	7[e]	2.3	1.9	24	149	10^{-3}
Au	77	0.1	19.3	0.2	0.05	301	4.5×10^8

E = stiffness. S = tensile Strength. ρ = density. E/ρ = specific stiffness. σ/ρ = specific strength. k = thermal conductivity. σ = electrical conductivity. Listed values of specific strength and specific stiffness are normalized relative to those of steel. Properties of metals were obtained from Matweb (Website). Properties of carbon and synthetic fibers are as listed in NAS (2005). Thermal conductivities of polymer and CNT fibers are generally lower than for carbon fibers.

Sources (with permission):

[a] Krishnan, A., Dujardin, E., Ebbesen, T.W. et al. 1998. *Physical Review B*, 58, 14013–14019. Kim, P., Shi, L., Majumdar, A. et al. 2001. *Physics Review Letters*, 8712. Barber, A.H., Andrews, R., Schadler, L.S. et al. 2005. *Applied Physics Letters*, 87, 203106. Li, H.J., Lu, W.G., Li, J.J. et al. 2005. *Physics Review Letters*, 95, 086601.

[b] Li, Y.L., Kinloch, I.A., and Windle, A.H. 2004. *Science*, 304, 276–278.Motta, M., Li, Y.L., Kinloch, I. et al. 2005. *Nano Letters*, 5, 1529–1533.

[c] Zhang, M., Atkinson, K.R., and Baughman, R.H. 2004. *Science*, 306, 1358–1361.

[d] Dalton, A.B., Collins, S., Razal, J. et al. 2004. *Journal of Material Chemistry*, 14, 1–3. Erickson, L.M., Fan, H., Peng, H.Q. et al. 2004. *Science*, 305, 1447–1450.

[e] Madou, M.J. 2002. *Fundamentals of Microfabrication: The Science of Miniaturization*. CRC Press.

micrometer-scale distances in SWNTs (Javey et al., 2003; Wind et al., 2003) and multichannel MWNTs (Li et al., 2005). Under these conditions, the electrical conductivity of CNTs exceeds that of copper.

These attributes—chirality-dependent band gap, and low-resistance transport—make SWNTs more frequently studied for use in next-generation transistor devices (Tans et al., 1998; Javey et al., 2003; Chen et al., 2006), and MWNTs are sought for their high conductivity due to their metallic character. Further, while individual CNTs have been studied widely for their electronic properties as active device elements, a larger scale and nearer-term opportunity exists for assemblies of CNTs to replace copper (Cu) in vertical and horizontal electronic interconnects, at both local and global chip-scale dimensions (Peercy, 2000).

At emerging linewidths (~20 to 40nm and smaller), Cu will be unsuitable for interconnects due to electromigration and boundary scattering; however, CNTs are resistant to electromigration, have micrometer-scale electron mean free paths (Javey et al., 2003; Li et al., 2005), and withstand higher current densities (10^9 A/cm^2 versus 10^7 A/cm^2) than Cu (Naeemi et al., 2005; Naeemi and Meindl, 2007). Calculations also suggest CNTs can decrease switching energy consumption; gigahertz operation of individual and bundled CNTs as a horizontal interconnects has been demonstrated (Close et al., 2008; Tselev et al., 2008).

Importantly, assemblies of CNTs exhibit much lower values of electrical conductivity than predicted by linear scaling of the properties of individual CNTs, as shown in Figure 8.3. Such shortfalls are due to structural defects inherent in as-grown CNTs and non-uniform and non-ideal metal contacts that do not necessarily address all walls of every CNT. Further, in isotropic networks such as tangled CNT films, CNT–CNT junction resistances can far exceed the bulk resistance of CNTs over short distances. In this regard, continuous aligned CNTs compose an ideal network morphology that does not suffer from such CNT–CNT resistances.

Making low-resistance electrical contact to CNTs has been a practical challenge and relies on intimate engagement between the metal and the conduction path through the CNT. Ohmic contact to metallic CNTs was first achieved by deposition and subsequent annealing of gold electrodes (Soh et al., 1999), and later it was found that Pd-CNT interfaces do not exhibit Schottky barriers and thereby enable observation of ballistic transport in semiconducting tubes (Javey et al., 2003). Integration of parallel electrically contacted CNTs may be improved by plasma etching to open their ends (Zhu et al., 2006), as demonstrated through room temperature superconductivity measurements of MWNTs having ends rooted in Au (Takesue et al., 2006).

8.2.2.2 Mechanical Properties

The density-normalized mechanical stiffness and strength of CNTs exceed those of all known natural and synthetic bulk materials (Table 8.1). Because a CNT is in essence a rolled sheet of graphene, its tensile elastic modulus is closely approximated by the in-plane modulus of single crystal graphite ($\approx$1000 GPa). Theoretically, this value is independent of CNT diameter, except for a slight decrease due to curvature effects in very narrow SWNTs (Dresselhaus et al., 2003) and this has been confirmed by a wide body of experimental work.

Theory predicts that the ultimate tensile strength of a material is approximately one-tenth of its elastic modulus, or $\approx$100 GPa for small-diameter CNTs; however, the inevitable presence of defects (vacancies, Stone-Wales transformations) prevents this limit from realization. Testing of CVD-grown MWNTs gave a tensile strength distribution (by Weibull statistics)

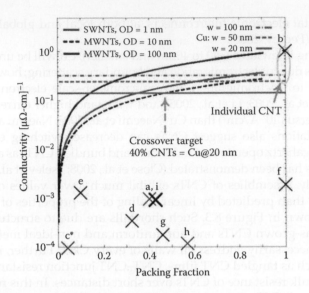

FIGURE 8.3
(See color insert.) Previously published calculations of electrical conductivity along continuous CNTs as related to CNT packing fraction compared to previously measured electrical data for individual CNTs and CNT networks. The packing fraction is defined as the ratio of the CNT areal density (number of CNTs per square centimeter) to the areal density of hexagonally packed CNTs (modeled as cylinders) having the same outer diameter. Solid and dashed x marks are MWNTs and SWNTs, respectively. We assume that the conductivity of an aligned CNT network is linearly proportional to the number of CNTs per unit cross-sectional area and that the constant of proportionality is the conductivity of an individual CNT. (From (a): Fischer, J.E., Zhou, W., Vavro, J. et al. 2003. *Journal of Applied Physics*, 93, 2157–2163. (b): Li, Y.L., Kinloch, I.A., and Windle, A.H. 2004. *Science*, 304, 276–278. (c): Kang, S.J., Kocabas, C., Ozel, T. et al. 2007. *Nature Nanotechnology*, 2, 230–236. (d) Atkinson, K.R., Hawkins, S.C., Huynh, C. et al. 2007. *Physica B*, 394, 339–343. (e): Yokoyama, D., Iwasaki, T., Ishimaru, K. et al. 2008. *Japan Journal of Applied Physics*, 47, 1985–1990. (f): Close, G.F. and Wong, H.S.P. 2007. *Proceedings of IEEE International Electronic Devices Meeting*, pp. 203–206. (g): Wang, D., Song, P.C., Liu, C.H. et al. 2008. *Nanotechnology*, 19. (h): Hayamizu, Y., Yamada, T., Mizuno, K. et al. 2008. *Nature Nanotechnology*, 3, 289–294. i: Tawfick, S., O'Brien, K., and Hart, A.J. 2009. *Small*, 5, 2467–2473. With permission.)

in which 25% of the CNTs fail at or below 55 GPa, and 75% fail at or below 148 GPa (Barber et al., 2005). Further experiments have strained SWNTs to ≈6% elongation before breakage (Walters et al., 1999). Superplastic elongation to strains exceeding 250% has been demonstrated by pulling a SWNT in a transmission electron microscope (TEM) while resistively heating it above 2000°C (Huang et al., 2006).

Beyond individual CNT devices, the prospects for scaling the mechanical properties of CNTs into larger assemblies require effective management of the interconnections between CNTs and engineering of composite materials that bind CNTs to matrix materials. Composite microstructures containing CNTs could significantly enhance micromechanical material properties that could be tuned by the packing fraction of CNTs within the matrix (Ashrafi et al., 2006). Encouragingly, pull-out tests measured CNT–polymer adhesion strength exceeding 100 MPa (Cooper et al., 2002; Barber et al., 2003), suggesting that micrometer-length contacts between CNTs and polymers can effectively impart the native strength and stiffness of CNTs to a composite microstructure.

8.2.2.3 Thermal Properties

Along with their outstanding electrical and mechanical characteristics, high-quality CNTs have thermal conductivity exceeding the in-plane thermal conductivity of graphite (Hone, 2001). The room-temperature thermal conductivity of an individual suspended SWNT has been measured at 3500 W/m-K (Pop et al., 2006), and ballistic phonon transport through MWNTs has been experimentally verified (Chiu et al., 2005).

As a result, CNTs are promising choices among additives for high thermal conductivity fluids and thermal interface materials (TIMs) for thermal management applications. Exceptional improvement of 150% in effective thermal conductivity of oil was achieved with only 1 vol% of CNTs suspended in oil (Choi et al., 2001). A 125% improvement in thermal conductivity of epoxy was also achieved with 1 wt% of CNTs added (Biercuk et al., 2002).

Even higher thermal conductivity may be achieved using organized (e.g., aligned) CNT assemblies; however, as observed for electrical properties, assemblies of CNTs typically have lower thermal conductivities than a single CNT. For example, a magnetically aligned array of SWNTs was measured to have thermal conductivity of 200 W/m-K (Hone et al., 2000). Packing fraction plays a big role in determining the final thermal conductivity of an assembly, as is the case with electrical and mechanical properties. Also, the orientation of the tube within the assembly, intertube junctions, and defects on individual tubes all drive down the bulk thermal conductivity of a given assembly.

For 100-nm diameter MWNTs, an assembly of randomly oriented CNTs has thermal conductivity of 0.17 W/m-K (Prasher et al., 2009), whereas an assembly of VA-CNTs has thermal conductivity of 15 W/m-K (Yang et al., 2002)—almost a 100-fold difference. This discrepancy is due to phonon scattering at tube–tube contacts, again highlighting the importance of organizing and interconnecting CNTs to achieve favorable scaling of individual CNT properties.

8.2.2.4 Fluidic Properties

A fourth and final attractive attribute of CNTs is their ability to confine materials and transport fluids in their inner cavities. Since initial TEM observation of capillary uptake of liquid metals into CNTs (Ajayan and Iijima, 1993), it has been known that CNTs may be used as nanofluidic "pipes." CNT membranes were made by infiltrating a CNT forest with a polymer or ceramic matrix, and opening the CNT ends to facilitate flow through the CNTs (Hinds et al., 2004; Holt et al., 2004).

Experiment and theory indicate that gases and liquids flow through CNTs with very high slip; for example, owing to the smoothness and hydrophobicity of CNT walls, water and organic solvents flow through multi-wall CNTs (7 nm ID) at rates 10^4 to 10^5 times continuum predictions (Majumder et al., 2005).

Double-wall CNTs (1.5 nm ID) exhibit mass-dependent selectivity of gas flows (Holt et al., 2006) and functional groups on the ends of the tubes facilitate ion exclusion from liquids at concentrations at which Debye length is larger than CNT diameter (Fornasiero et al., 2008). Further, CNTs respond electrically to chemistry, rate, and direction of flow by mechanisms of potential ratcheting (Ghosh et al., 2003; Sood and Ghosh, 2004), collision-induced resistivity (Romero et al., 2005), and charge donation (Snow et al., 2006).

8.3 CNT Synthesis by Chemical Vapor Deposition

Since the mid 1990s, catalytic chemical vapor deposition (CCVD or CVD) emerged as the most versatile and scalable method of CNT synthesis due to its low reaction temperatures (relative to arc, laser, and flame methods), high yield, and versatility for gas-phase (floating catalyst) or substrate-bound (fixed catalyst) growth (Dai, 2001; Teo et al., 2004; Terranova et al., 2006). Iijima found CNTs in the deposit on the carbon electrode of a DC electric arc (Iijima, 1991); this apparatus was very similar to that reported for synthesis of large quantities of C60 fullerenes (Kratschmer et al., 1990).

CNTs can also be produced in flames (Howard et al., 1991; Goel et al., 2002; Height et al., 2004), by laser ablation of carbon (Thess et al., 1996), by direct conversion of carbon using microwave energy (Yoon et al., 2006), and by many occasionally used procedures that are too numerous to mention. In concert with the focus of integration with micro- and nanofabrication, this section addresses CVD methods in detail.

In the CVD process, CNTs form by organization of carbon on a nanoscale metal catalyst particle in a high-temperature carbon-containing atmosphere (Figure 8.4a). The carbon dissociates from the source compound and adds to the CNT through surface and/or bulk diffusion at the catalyst. The catalyst particle can remain rooted on the substrate during CNT growth (base growth) or can lift from the substrate and remain at the tip of the advancing CNT (tip growth). In both cases, carbon is added at the catalyst site. The growth kinetics, catalyst–substrate surface interactions, and forces acting on the catalyst particle (e.g., forces induced by an electric field in plasma-enhanced CVD (Merkulov et al., 2001) determine whether base growth or tip growth occurs (Melechko et al., 2002).

A highly simplified picture of CNT synthesis by CVD can be gained via an analogy to the vapor–liquid–solid (VLS) crystal growth first studied during the 1960s for silicon "whiskers" (Wagner and Ellis, 1965). In the VLS model, a vapor-phase precursor dissolves into a liquid growth site (catalyst), and

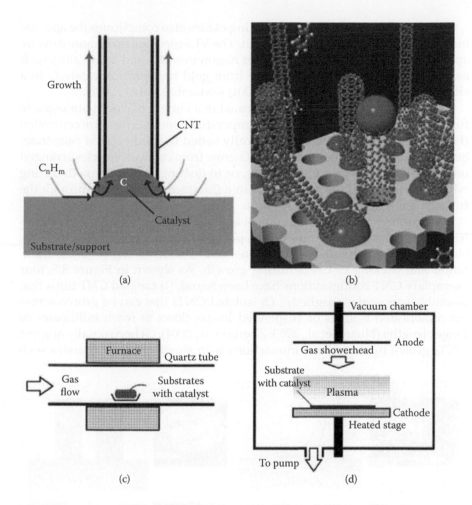

FIGURE 8.4
(See color insert.) Individual substrate-bound CNTs growing by CVD. (a) Base growth of CNT. A gaseous carbon source diffuses at a metal catalyst particle that remains attached to the substrate and a CNT grows upward from the surface of the particle. (b) Base and tip growth of CNTs rooted in nanoporous (zeolite) substrate. (*Source:* Hayashi, T., Kim, Y.A., Matoba, T. et al. 2003. *Nano Letters*, 3, 887–889. With permission). Classical furnace designs for CVD synthesis of CNTs and like nanostructures. (From Teo, K.B.K., Singh, C., Chhowalla, M. et al. 2004. *Encyclopedia of Nanoscience and Nanotechnology.* With permission.) (c) Horizontal tube furnace with fixed catalyst. (d) Low pressure plasma-enhanced (PECVD) chamber with heated stage.

a solid crystal precipitates from the growth site. CNT growth from metal can be described by the VLS model as a high-level example; however, under certain circumstances (e.g., high temperature with liquefied catalyst), carbon may diffuse through the bulk of the catalyst. Under other conditions (e.g., lower temperature) surface diffusion may dominate (Snoeck et al., 1997; Ding et al., 2005; Hofmann et al., 2005).

Growth from many catalysts including oxides also complicates the applicability of the VLS model to CNT growth. The VLS mechanism is more directly applicable to growth of semiconductor nanowires (Wu and Yang, 2001), such as silicon nanowires that precipitate from gold nanoparticle catalysts in a silane (SiH_4) atmosphere (Cui et al., 2001; Kodambaka et al., 2006).

CVD growth of CNTs can be performed in a variety of closed-atmosphere furnace devices in which pressure, temperature, and gas flow are controlled (Figure 8.4b). These systems are typically suited to fixed-catalyst (substrate-bound) CNT growth in which the CNTs grow from catalyst particles arranged on a substrate such as a silicon wafer, or to floating-catalyst growth during which the catalyst particles are held in a fluidized bed or pass through the furnace continuously in the gas phase.

In substrate-bound growth, control of the density, placement of the catalyst particles on the substrate, and application of external forces (gas flows, electric fields, mechanical obstructions) can be used to engineer the orientation and packing of CNTs during growth. As shown in Figure 8.5, four exemplary CNT configurations have been found: (1) tangled CNT films that resemble nanoscale "spaghetti"; (2) isolated CNTs that can be grown across microfabricated bridges or suspended in gas flows to reach millimeter or longer lengths (Huang et al., 2003; Zheng et al., 2004); (3) horizontally aligned CNTs grown by directional interactions with single crystal substrates such

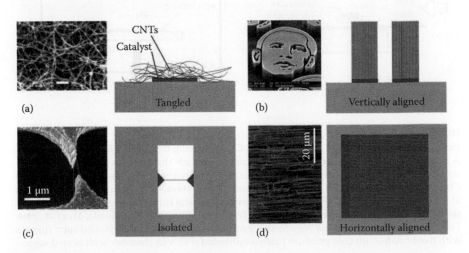

FIGURE 8.5

(See color insert.) Classical morphologies of CNTs grown on substrates. (a) Tangled film that terminates around micrometer thickness due to steric hindrance among CNTs. (From Hart et al., 2006. *Carbon*, 44, 348–359. With permission). (b) Vertically aligned film that can grow uniformly to millimeter thickness. (www.nanobama.com). (c) Single CNT suspended over microfabricated channels. (From Jungen, A., Durrer, L., Stampfer, C. et al. 2007. *Physics Status Solidi B*, 244, 4323–4326. With permission.) (d) Horizontally aligned CNTs on substrate. (From Kocabas et al., 2005. *Small*, 1, 1110–1116. With permission.)

as quartz and sapphire; and (4) vertically aligned CNT "forests" that grow perpendicular to the substrate.

While isolated SWNTs can be grown to millimeter or centimeter lengths when suspended during growth by a gas flow, the density of catalytic sites must be very low to prevent entanglement among CNTs because the CNTs will flutter in the gas flow. At a high catalyst density and CNT growth rate, the CNT forest growth mode is typical. The CNTs self-orient perpendicular to the substrate surface due to initial crowding and continue to grow upward in this direction (Terrones et al., 1997; Fan et al., 1999).

CNT growth by CVD involves many process parameters, and therefore a multi-dimensional parameter space must be explored to develop an empirical model of a particular growth scheme and relate the CNT characteristics to the process conditions. CNT characteristics of interest include diameter, wall structure and chirality, defect density, and length along with the many physical properties that result from these basic characteristics.

8.4 Methods of CNT Integration in N/MEMS

This section reviews emerging methods to integrate CNTs in N/MEMS. We aim to demonstrate how CNTs can be interfaced with device substrates and in functional configurations so that they may be connected mechanically and/or addressed with electrical contacts.

8.4.1 Device-Compatible CNT Growth Methods

Perhaps the most straightforward approach for integrating CNTs into small devices is to synthesize them directly on chips. This offers advantages in scalability for industrial processes since it relies mainly on photolithography and other standard microfabrication methods for defining the catalyst. However, it is challenging to synthesize high-quality CNTs on electrically conductive substrates at temperatures compatible with complementary metal oxide semiconductor (CMOS) processing that are typically limited to 450°C (Awano et al., 2006; Nessim et al., 2009).

Further, repeatable and uniform growth needs to be achieved over 200- or 300-mm wafers for the processes to be relevant for industry. While CNTs must be connected to conductive substrates for active sensing and actuation applications, they can also be connected merely to insulative substrates such as oxides if only passive functionalities are required (such as for springs to support suspended microstructures or passive absorbents for detection or analysis). In the latter case, growth in device configurations is easier because oxides typically tolerate higher processing temperatures than metals.

Nevertheless, the temperature limitation remains an obstacle for achieving high-quality on-chip CNTs, especially ensembles of vertically aligned CNT forests that require high density numbers (10^9 to 10^{12} cm^{-2}) to self-orient during synthesis. Successful CNT growth depends on both effective decomposition and rearrangement of the hydrocarbon precursor and formation of stable catalyst nanoparticles that facilitate deposition of carbon and incorporation into the existing CNT lattice structure.

Low-temperature growth is especially challenging because the decomposition temperatures of most hydrocarbons used for CNT growth well exceed 500°C. However, high temperatures often lead to alloying of catalysts and metallic underlayers that often hinders CNT growth (Nessim et al., 2010). To meet this need, plasma-enhanced CVD has been used widely to achieve low-temperature CNT formation, in which the plasma source provides the energy needed to generate active hydrocarbon species for CNT growth (Chhowalla et al., 2001; Zhong et al., 2005). However, plasma-enhanced methods can also generate undesirable defects in the CNT structure, presumably due to etching by ionized species or radicals (Ren et al., 1998; Lee et al., 2009).

Recent progress in chemically engineering CNT growth has enabled synthesis temperatures below 400°C by an oxidative dehydrogenation method (Magrez et al., 2010); however, more investigation is required to reveal the utility and suitability of this process for small devices.

Alternatively, thermal treatment of the feedstock gases prior to reaching the catalyst substrate (preheating) has been shown to enhance CNT growth for samples with catalysts on both electrically insulating substrates (Lee et al., 2001; Jeong et al., 2002; Mora et al., 2008; Meshot et al., 2009) and conducting (metallic) substrates (Awano et al., 2006; Nessim et al., 2009).

Decoupling gas pretreatment from the catalyst temperature, as shown in Figure 8.6a, allows synthesis at low substrate temperatures without sacrificing control of the gas chemistry. Gas analyses have shown that diverse hydrocarbons including volatile organic compounds (VOCs) and polycyclic aromatic hydrocarbons (PAHs) are formed by thermal treatment of typical feedstock gases (Meshot et al., 2009; Plata et al., 2009). This understanding of the growth ambient has been used to engineer the synthesis process to tune both diameter and quality of the CNTs and also their growth kinetics for rapid (>10 μm/second) growth to millimeter lengths (Meshot et al., 2009). Further, the decoupled CVD method has enabled tunable transformations from fiber-like carbon structures to crystalline CNTs (Figure 8.6b) that have far more desirable electrical properties (Nessim et al., 2011).

8.4.2 Methods of Integrating Individual CNTs in Devices

In this section, we will look into methods to integrate a single, isolated SWCNT or MWCNT in device-relevant configurations. Four exemplary approaches have been pursued. The first three use CNTs grown separately

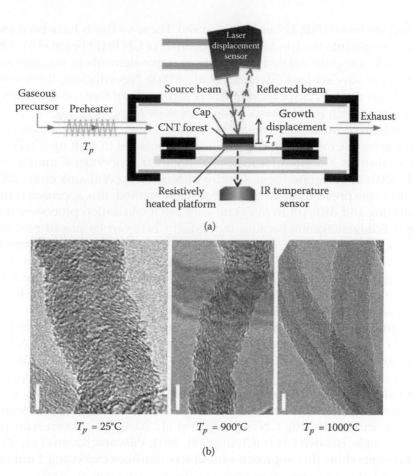

FIGURE 8.6
(a) Custom-built atmospheric pressure CVD apparatus (SabreTube, Absolute Nano) that features local Joule heating of Si platform to independently control catalyst temperature Ts from preheated feedstock gases (at Tp). A laser displacement sensor mounts on the outside of the growth chamber and measures real-time kinetics of the growth process. (From Meshot, E.R. and Hart, A.J. 2008. *Applied Physics Letters*, 92, 113107. With permission.) (b) TEM images of carbon structures synthesized at Ts = 525°C for various Tp values. We observe a distinct evolution from amorphous structures to CNTs with crystalline walls as Tp increases. (From Nessim et al. *Carbon*, 49, 804–810, 2011. With permission.)

from devices and the fourth approach places isolated CNTs by direct growth in a device.

Alignment of microsystems around randomly dispersed CNTs—One of the first methods developed to integrate isolated CNTs in MEMS was depositing them from a solution, resulting in randomly arranged CNTs on a substrate. After the deposition, e-beam lithography was used to define microstructures on or around the individual CNTs (Stampfer et al., 2006a). Alternatively, electrodes can be directly deposited on the CNTs using a focused ion beam (FIB; Ebbesen et al., 1996). These methods have been used to gain insight into the fundamental properties of CNTs (Ebbesen et al., 1996) and also to integrate isolated nanotubes as active elements in microsystems such as pressure sensors (Stampfer et al., 2006a). Nevertheless, these methods rely on random dispersion of CNTs and are therefore limited to serial processing, which is not conducive to large-scale manufacturing.

Direct mechanical placement of CNTs—Using nanomanipulators or atomic force microscopy (AFM) probes, it is possible to pick up CNTs and deposit them at a desired location to study their properties (Cumings and Zettl, 2000) or integrate them directly in N/MEMS (Williams et al., 2002). Similar to the previous dispersion–deposition method, this approach is time consuming and difficult to integrate with batch fabrication processes; however, it is advantageous because individual CNTs can be placed precisely. The strengths of interactions (electrostatic, van der Waals) is important for manipulating individual CNTs. This problem was addressed by Kim (2006) who developed a technique whereby individual CNTs are integrated in a micropellet. This facilitates the handling of the CNTs and also enables the use of parallel self-assembly techniques to increase throughput.

Controlled CNT deposition from solutions—Dielectrophoretic deposition of CNTs from a solution is a common method for integrating CNTs on large numbers of microfabricated electrodes in parallel. CNTs are first dispersed in a solvent (usually with a stabilizing surfactant) and subsequently directed to a predefined location on a chip using an electric field. While this method was first developed to deposit CNT films and simultaneously separate semiconducting CNTs (Krupke et al., 2004), it has been refined to deposit single, isolated CNTs (Chung et al., 2004; Vijayaraghavan et al., 2007). Experiments show this approach can achieve densities exceeding 1 million/cm² of individually contacting SWNTs (Vijayaraghavan et al., 2007).

Direct growth of freestanding CNTs—Using plasma-enhanced CVD and small catalyst island determined by e-beam lithography, free-standing vertically aligned carbon nanofibers (CNFs) can be fabricated. Free-standing CNFs have been used as field emission elements (Guillorn et al., 2001) and electrochemical probes (Guillorn et al., 2002). Alternatively, Hierold and colleagues (2007) demonstrated that CVD growth can also be employed to grow CNTs horizontally to form bridges between microelectrodes in the suspended configuration discussed in Section 8.3.

8.4.3 Transformation and Densification of CNT Forests

When assemblies of CNTs are desired, one limitation to their bulk properties is their low as-grown density. To improve the properties of bulk CNT forests, methods have been developed to increase the CNT density after growth. Two types of methods have been reported. The first approach consists of mechanically pushing CNTs together; the second aggregates CNTs using surface tension to form dense vertical structures and other three-dimensional (3D) geometries.

Mechanical rolling has the ability to simultaneously transform as-grown, low-density, vertically aligned carbon nanotube (VA) CNT forests into higher density, horizontally aligned (HA) CNT networks. The main challenges of the process are maintaining CNT alignment, minimizing defects in the CNTs due to stresses developed during rolling, and controlling the adhesion of the CNTs to the receiving substrate.

One rolling method (Wang et al., 2008) uses a microporous membrane between a large (centimeter) diameter roller and a CNT forest grown from a non-patterned catalyst substrate. By aligning the membrane over the forest before rolling, the CNTs in the forest stick to the membrane after rolling due to van der Waals forces. The membrane with the HA CNTs is then aligned on top of the receiving substrate (e.g., glass) such that the CNTs face the glass, then a solvent such as ethanol is poured from the top of the membrane. While the solvent is evaporated, the CNT film transfers from the membrane to the receiving substrate and is further densified via capillary action.

In a second approach, researchers at Rice University (Pint et al., 2008) used photolithography to pattern the catalyst into 5-μm-wide gratings separated by tens of microns. The idea was to grow the VA CNTs and roll them mechanically so that they slightly overlap to form continuous networks over the substrate area. The group also used an intermediate foil (metal or polymer) between the roller and the CNTs to overcome adhesion problems to the roller. They indicate "shearing" the film along the CNTs before rolling in the same direction.

Finally, guided by the Hertzian contact stress theory, researchers at the University of Michigan (Tawfick et al., 2009) designed a force controlled rolling machine to transform CNTs, as shown in Figure 8.7. The roller size and material are selected to ensure that the CNTs remain on the growth substrate without the need of intermediate foils or membranes. For successful transformation that preserves the CNT alignment without introducing destructive shear forces, the roller diameter must be smaller than the VA CNT height and the line width of the catalyst must be significantly smaller than the roller diameter. These conditions avoid the formation of defects in the HA CNT networks due to shear stresses caused by the rolling action.

These results are confirmed by comparing Raman spectra of the CNT networks before and after rolling. The HA CNT bundles are made of continuous CNTs with lengths up to 1 mm; the thickness of the sheets can be tuned from hundreds of nanometers to several microns. The HA CNT bundles can be transferred to polymer substrates such as PDMS using kinetically controlled peeling (Meitl et al., 2006).

While the rolling process is performed after growth, it is also possible to influence the shape and density of CNT forests during growth. For instance, CNT forests can be grown to conform to the shapes of microfabricated templates (Hart and Slocum, 2006). This process is analogous to micromolding

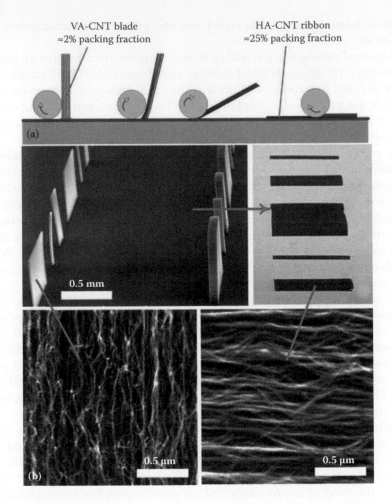

FIGURE 8.7
Transformation of low-density VA CNT forests into HA CNT forests by mechanical rolling. (Adapted from Tawfick, S., O'Brien, K., and Hart, A.J. 2009. *Small*, 5, 2467–2473. With permission.) (a) Process schematic. (b) SEM images of VA CNT blades having initial thickness of 20 μm and 100 μm before and after rolling.

but employs the force generated by the growth process of the nanotubes to fill the templates.

It is also possible to aggregate CNTs using capillary forces. CNT forests are typically submerged in a solvent such as acetone or IPA that aggregates the CNTs as it evaporates because the surface tension of the moving meniscus draws the CNTs together. This approach is simple and versatile; it has been used to segregate CNT forests into cellular foams (Chakrapani et al., 2004; Correa-Duarte et al., 2004; Liu et al., 2004), achieve isotropic contraction of CNT micropillars (Futaba et al., 2006), and perform unidirectional toppling (Hayamizu et al., 2008) of CNT "blades" to create lateral films. However, a

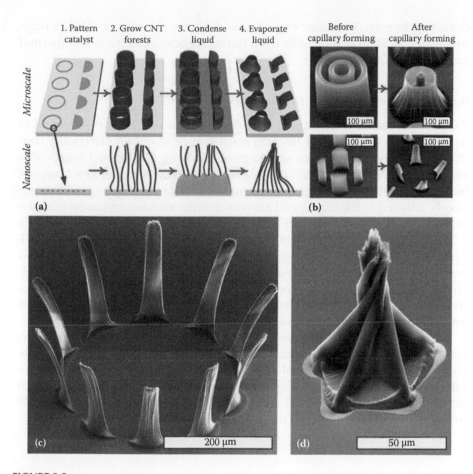

FIGURE 8.8
(See color insert.) Diverse 3D microarchitectures made by capillary forming of carbon nano-tubes. (a) Capillary forming process showing micro- and nanoscale detail. (b) Concentric wells and bending structures before and after capillary forming. (c) Blooming flower of bending structures. (d) Micro-helix made by combining contraction and bending operations. (From De Volder, M., Tawfick, S, Park, S.J. et al. 2010a. *Advanced Materials,* 22, 4384–4389. With permission.)

limitation of the immersion process is that the liquid meniscus can connect adjacent CNT structures and connect CNT structures to the substrate. This can damage small CNT forests that are typically fragile.

To overcome this limitation, recent work shows that by condensing acetone on the structure instead of immersing it, it is possible to densify delicate CNT microstructures, and by carefully designing the initial shape of the forest, it is possible to fabricate complex 3D microstructures including sloped microwells, bent cantilevers, pyramidal microtrusses, and helical micropillars (De Volder et al., 2010). Figure 8.8 shows examples of these structures. De Volder et al. also showed that the condensation densification method increases the Young's modulus of CNT forests from 54 MPa to 5 GPa for densified forests. The latter exceeds the stiffness of common MEMS polymers such as SU-8 and benefits from the outstanding electrical and thermal properties of CNTs.

8.4.4 Composite CNT Microstructures

Extensive research has focused on the dispersion of CNTs in polymers to increase their stiffness and fracture resistance or to obtain electrically conductive polymers. At larger scales, most work on CNT-based composites related to dispersion of SWNTs or MWNTs in polymer matrices; however, bulk CNTs embedded in a polymeric matrix tend to form aggregates that are poorly adhered to the matrix and also concentrate stresses, compromising the reinforcement effects of the CNTs (Thostenson et al., 2001; Garcia et al., 2007).

Instead, Garcia et al. (2007) proposed to infiltrate vertically aligned CNT forests with polymers. Due to the alignment of the CNTs, this method resulted in an increase in Young's modulus of 200% after polymer infiltration. Combining this approach with the capillary densification process described above, Young's moduli of 18 and 25 GPa were achieved using SU-8 and PMMA, respectively, as matrix polymers (De Volder et al., 2010).

Vertically aligned CNT forests have also been infiltrated with ceramics; for example, a CNT framework is filled with polysilicon and silicon nitride using LPCVD (Hutchison et al., 2009). The latter material was employed to fabricate high-aspect-ratio microdevices including electrostatic and thermal actuators. This process could complement other high-aspect-ratio MEMS processes such as DRIE (deep reactive ion etching) and SU-8 processing.

8.5 Examples of CNT Integration in N/MEMS

8.5.1 Electrical Interconnects

Due to their high electrical conductivity and current-carrying capacities, metallic CNTs have been investigated as new materials for vertical interconnect vias. As microelectronics progresses to finer resolution, smaller diameter interconnects are required and they must handle higher current densities. Fujitsu projects that CNTs will be necessary for 32 nm line width processing and beyond, where the required current density will exceed 10^6 A/cm^2; crossover between CNT and Cu performance is expected between before 2015.

Calculations predict that the resistance of a 70-nm diameter vertical via filled with close packed MWCNT of 4 nm diameter having six walls can be as low as that of a Cu via. Fujitsu (Nihei et al., 2004; Horibe et al., 2005; Nihei et al., 2005), Infineon (Kreupl et al., 2002), and IMEC (Chiodarelli et al., 2010)

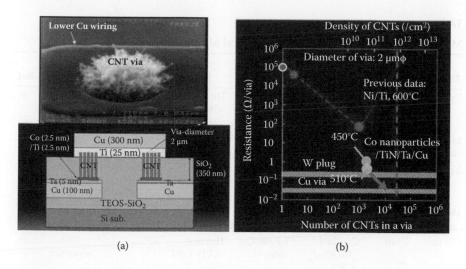

FIGURE 8.9
(See color insert.) Integration and performance of vertical CNTs as microelectronic interconnects by Fujitsu.

reported progress in process technology to achieve this high performance, including direct growth of the CNTs within vias at CMOS-compatible temperatures followed by deposition of a top electrical contact. Figure 8.9 shows this architecture, and Table 8.2 shows the most recently reported results from these groups.

Despite this encouraging progress, several challenges must be met before CNTs can actually replace Cu interconnect vias. The main challenges to overcome for vertical vias are:

1. Improved control over CNT diameter, number of walls, and quality
2. Achieving high CNT areal density, e.g., at least 5×10^{12} CNTs/cm² for MWNTs
3. Achieving ohmic contact to the ends of the CNTs

To achieve high density growth at low temperatures, researchers are decomposing the feedstock gas required for CNT growth using a multi-mode (RF plasma, DC plasma, hot filament, and thermal) CVD chamber to scale up the process to obtain uniform growth on full 300-mm wafers. The catalyst nanoparticles are prepared by dewetting a thin catalyst layer (Co or Ni) through thermal annealing or by direct injection of catalyst nanoparticles from the gas phase. Both variations aim for the highest density of active nanoparticles selectively at the bottoms of the via holes.

More recently, Fujitsu reports a plasma-enhanced formation of small-diameter, closely packed nanoparticles from a deposited layer of Co. The key to the method is to control the nanoparticle size and density and stop

TABLE 8.2

Major Milestones in Development of Vertical MWNT Vias

Via Diameter	CNT Diameter	Stackup Sequence	Catalyst	Recipe	Density [CNTs/cm^2]	Resistance	Ref.
2 μm	10 nm	Cu/Ta/Ti/ Co/CNT/ Ti/Cu	2.5 nm Co layer	Thermal CVD 450°C	10^{10}	5 Ω	a
140–300 nm	15 nm	Cu/Al/ Ni/ CNT/ AuPd	3 nm Ni layer	Thermal CVD 520°C	5×10^{10}	20 Ω	b
2 μm	10 nm	Cu/Ta/ TiN/ Co/ CNT/ Ti/ Cu	4 nm Co nano- particles	Thermal CVD 510°C	9×10^{11} and 10^{11}	0.59 Ω	c
150–300 nm	8–12 nm	TiN/SiC/ PSG/SiC	1.3 nm Ni layer	CVD 400–470°C	2×10^{11} – 7×10^{10}	7.9 kΩ	d
120 nm	7 nm	Cu/TaN/ Tin/Co/ CNT	1.7 nm Co film	Thermal CVD 450°C	10^{12}	Not reported	e

Sources (all with permission):

a Nihei, M., Kawabata, A., Kondo, D. et al. 2005. *Japan Journal of Applied Physics*, 44, 1626–1628.

b Coiffic, J.C., Fayolle, M., Le Poche, H. et al. 2008. *Proceedings of IEEE International Interconnect Technology Conference*, pp. 153–155.

c Sato, S., Nihei, M., Mimura, A. et al. 2006. *Proceedings of IEEE International Interconnect Technology Conference*, pp. 230–232.

d Chiodarelli, N., Li, Y., Cott, D.J. et al. 2010. *Microelectronic Engineering*, 88, 837–843.

e Yamazaki, Y., Katagiri, M., Sakuma, N. et al. 2010. *Applied Physics Express*, 3, 055002.

the aggregation of particles before growth starts. This is achieved using a low temperature (<260°C) and low-power plasma (<0.5 W/cm^2) for annealing. The CNT density reaches 10^{12} CNTs/cm^2 representing a bulk volumetric density of 30 to 40%.

8.5.2 Emission Sources: Displays and X-Ray Emitters

Samsung Display Innovation, a division of Samsung Corporation spent several years developing a field emission display (FED) using CNTS as field emission sources. While this project was discontinued before market introduction due to the emergence of LCD display technology, the design and manufacturing of the prototype CNT FED display serve as excellent examples of CNT integration for consumer electronics applications.

The Samsung CNT FED shown in Figure 8.10 was microfabricated on display glass and features a triode structure. A bias voltage is applied between the bottom (cathode, ITO) and middle (gate) electrodes of the triode structure, and the emission is focused by the top electrode (Choi et al., 2006). The CNT emitters are applied to the device by screen printing of a custom-made CNT paste or by direct growth of CNTs by chemical vapor deposition (CVD). In the former case, commercially available CNTs are mixed with

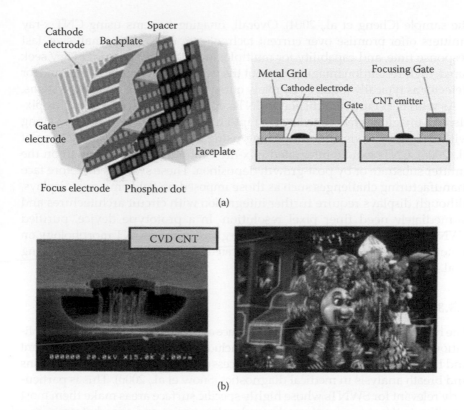

(a)

(b)

FIGURE 8.10
(See color insert.) Samsung CNT field emission display. (a) Display structure and triode pixel architecture. (b) SEM image of individual pixel with aligned CNTs grown directly on substrate, and image of prototype 15-inch diagonal display.

many ingredients and the paste is wiped through a screen mask placed over the device (Choi et al., 2004). In the latter case, a catalyst layer is deposited (e.g., sputtered invar) on the device and the CNTs are subsequently grown by thermal CVD of CO/H_2.

For screen printing of the paste, the maximum (and therefore limiting) process temperature is that of firing the paste after printing (<450°C); for CNT growth, the CVD temperature is limiting (lowest to date = 420°C; must be <500°C to prevent melting of glass). While isolated vertically aligned CNTs as grown by the CVD process are desired for field emission, the CNTs in the paste are generally tangled and parallel to the substrate. Therefore, CNTs as deposited by the paste method are aligned vertically by applying and peeling a tape from the substrate.

The use of CNTs as x-ray emission sources was pioneered by the group of Professor Otto Zhou at the University of North Carolina and is being commercialized by Xintex, Inc. In a CNT-based x-ray source, CNTs are placed in a triode configuration as a field electron emitter. The emitted electrons are incident upon a target (typically Mo) that in turn emits x-rays and is aimed toward the sample (Cheng et al., 2004). Overall, imaging systems using CNT x-ray emitters offer promise over current technology due to their small size, fast response time, and capability for multiplexing and array imaging. They seek rapid, high-resolution imaging without the need for moving the source and/or detector as typically required by many current x-ray and CT imaging systems.

As with field emission displays, CNT x-ray emitters can be arrayed as pixels, and imaging is facilitated by the rapid response time of the emitters along with frequency division multiplexing using a single flat detector (Zhang et al., 2006). CNTs can be integrated in x-ray emitters by direct growth on the emitter substrate or by post-growth deposition. These systems therefore face manufacturing challenges such as those imposed on field emission displays, although displays require further integration with circuit architectures and immediately need finer pixel resolution. In a prototype device, purified SWNTs were deposited from solution to give a tangled CNT morphology on a metal substrate that was then incorporated in a sealed emitter tube (Zhang et al., 2005).

8.5.3 Chemical Sensors

The high electrical conductivity and surface area of CNTs motivate their applications as chemical sensors. Examples include detection of trace contaminant and hazardous vapors for industrial process control and security applications and breath analysis in medical diagnostics (Snow et al., 2006). This is particularly relevant for SWNTs whose highly specific surface areas make them most sensitive to the presence of nearby molecular species and adsorbates.

Typically, a CNT-based chemical sensor consists of a tangled network of CNTs connected electrically to a substrate (Figure 8.11): in a resistive sensor, the CNTs are connected between electrode pairs such as interdigitated fingers; and in a capacitive sensor, the capacitance between the network and the substrate (isolated by a dielectric layer) is also monitored. Chemical adsorption on a CNT surface is driven by the partial pressure of the analyte relative to its vapor pressure (rather than relative to the ambient pressure); therefore, CNT sensors can detect small concentrations of low vapor pressure analytes such as explosives—a task many conventional sensors are unable to do.

Kong and colleagues (2000) demonstrated that adsorption of a molecule on the surface of a SWNT causes a transfer of charge (donation or withdrawal of an electron) to the SWNT and therefore changes its electrical conductivity. Many adsorbates bind strongly to CNTs, and the sensor was heated or exposed to ultraviolet light to remove the adsorbate from the CNT surfaces and "zero" the signal. Further, the sensitivity of a CNT to charge–transfer interactions (number of electrons transferred or withdrawn per adsorbed molecule) depends on the electronic structure (chirality) of the CNT and adsorbed molecule.

A general strategy to increase sensitivity (minimum detectable concentration) and/or specificity (response to a particular desired species) is to coat

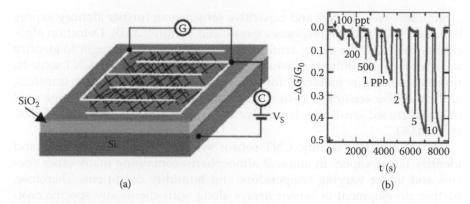

FIGURE 8.11
(a) Typical architecture of chemical sensor using CNT network. Tangled CNTs are in contact with interdigitated electrodes, and the conductance and capacitance of the network are measured via external circuitry. (From Snow, E.S., Perkins, F.K., and Robinson, J.A. 2006. *Chemical Society Reviews*, 35, 790–798. With permission.) (b) SWNT sensor conductance response to doses of NO_2. The sensor was reset by exposure to ultraviolet light between doses and exhibits sensitivity of 100 parts per trillion.

the CNTs with a polymer that selectively adsorbs a particular molecule or class of molecules. For example, coating a SWNT network sensor with polyethylenimine (PEI) allows detection of 100 ppt of NO_2 in a response time of 1000 seconds (Pengfei et al., 2003). As an aside, while the sensitivity of electrical conductivity to surface adsorbates can be helpful for CNTs as chemical sensors, it is important to consider these environmental effects in the context of stable operation of CNTs as other device elements such as transistors (Collins et al., 2000).

SWNT resistivity can also change due to collisions with gas molecules. This has been demonstrated by measuring a resistance increase across CNT network devices upon exposure to inert gases (Romero et al., 2005). This change is proportional to the molecular mass of the gas (as $M^{1/3}$), so a heavier molecule makes a larger "dent" in the sidewalls of the CNTs. The authors hypothesize that the collisions introduce a new channel that scatters conduction electrons travelling through the CNTs. Since this phenomenon depends on the ambient composition and temperature, it may require consideration for high-precision gas sensing using CNTs.

Adsorbed gases can also be sensed by measuring the capacitance of a CNT network; compared to resistive sensing, this method is faster and responsive to a wider variety of vapors. When an electric field is applied between a CNT network and the underlying "gate" electrode, molecules adsorbed on the CNT surface are polarized, and this increases the capacitance between the network and the gate (Snow et al., 2005). With a minimum (reversible) detectable capacitance change $\Delta C/C = 10^{-4}$, part-per-billion detection limits for nerve agent and explosive vapors are expected (Snow et al., 2006).

Simultaneous resistive and capacitive sensing can further identify vapors based on their relative responses (Snow and Perkins, 2005). Detection algorithms such as those using artificial neural networks are sought to identify analytes from the outputs of sensor arrays (Shi et al., 2006). For CNT sensors, resistive sensing is preferred for analytes that give large charge transfers, and capacitive sensing is better for analytes that transfer less charge as it offers increased sensitivity, larger dynamic range, and faster response (Snow et al., 2006).

For commercial viability, CNT sensor systems must reliably detect and identify target vapors in ambient atmospheres containing many other species and under varying temperature and humidity conditions. Therefore, further development of sensor arrays along with chemically specific coatings, preconcentration techniques, and signal analysis algorithms is necessary. More complex techniques such as mass spectrometry and ion mobility spectrometry will likely maintain superiority for chemical analysis and nonspecific agent detection, and CNTs offer opportunities in miniaturization of these systems as well (Figure 8.12).

8.5.4 Electromechanical Devices and Transducers

Nanoscale electromechanical transducers have attracted significant attention for advancing current device technology toward high-sensitive, low-power sensors. Further, as architectures continue to scale down, devices are approaching the quantum limit for detecting non-classical states of mechanical motion (Hierold et al., 2007), which holds significance for fundamental scientific research. In order to realize these scalability and performance factors, we require new materials with exceptional properties.

While CNTs exhibit many exciting properties, one of the most promising characteristics for small devices is the coupling between their electrical transport and mechanical deformation. This piezoresistive characteristic in conjunction with inherently high elasticity makes CNTs strong candidates for nanoscale electromechanical transducers because they can endure significant deformations yet return to their original states.

Based on these principles, several prototype devices with integrated CNTs for transduction have been developed (Hierold et al., 2007; Jungen et al., 2007a), including electromechanical sensors for measuring mass (Stampfer et al., 2007), force (Stampfer et al., 2006b), and pressure (Wood and Wagner, 2000; Sickert et al., 2006; Stampfer et al., 2006a). Figure 8.13 shows examples of these devices. Specifically, a single suspended SWNT was implemented as the transducer element in a microfabricated pressure sensor in which deflection of a suspended Al_2O_3 membrane strains the SWNT and thereby changes its I–V characteristics (Stampfer et al., 2006a). Gauge factors up to 1000 have been exhibited (Cao et al., 2003).

Other investigations were aimed at analyzing the transduction mechanism, especially how the electronic band structures of SWNTs are altered

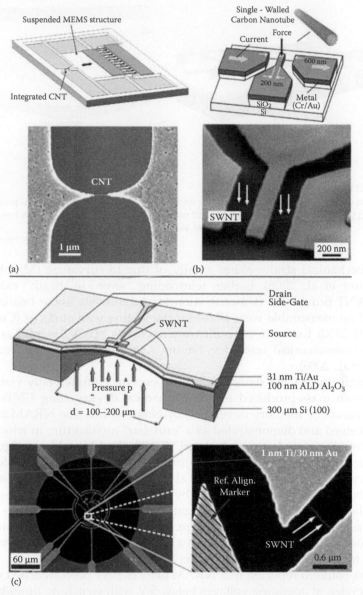

FIGURE 8.12
(See color insert.) (a) NEMS with integrated SWNT by direct growth and comb drive for sensing and actuation. SEM image shows a free-standing SWNT bridging two poly-Si tips. (b) SWNT-based nanoscale sensor system that may be used for measuring electrical responses to mechanical deformations applied at the center cantilever via an AFM tip. SEM image shows SWNT bridging two electrodes with the center cantilever exerting a force downward on the SWNT. (c) CNT-based pressure sensor consisting of an ultrathin Al_2O_3 membrane with SWNT adhering to it with electrodes. SEM images of the device show electrodes extending onto the membrane (black circle) and electrically contacting SWNT (left image), which is shown at higher magnification in the right image.

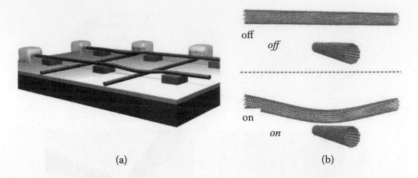

off

on

on

(a)   (b)

FIGURE 8.13

(a) Concept of non-volatile CNT memory device using bi-stable electromechanical junctions of suspended CNTs. (b) Renderings of on and off modes of CNT junction. (From Rueckes, T., Kim, K., Joselevich, E. et al. 2000. *Science*, 289, 94-97. With permission.)

under mechanical strain, either axially or due to curvature (Minot et al., 2003; Grow et al., 2005). Earlier, telescoping "sword-in-sheath" extension of a MWNT provided a demonstration for a nanoscale linear bearing that revealed no measurable wear between the sliding wall surfaces (Cumings and Zettl, 2000). Extending this idea, a telescoping MWNT is built as a tunable electromechanical resonator, having ≈200 MHz oscillation frequency (Jensen et al., 2006).

Although single CNT devices have not become commercially viable, the only known mass-produced electromechanical device using CNTs is the NRAM nanotube memory developed by Nantero, Inc. The NRAM concept was conceived and demonstrated as a "crossbar" architecture in which each bit is a cross junction between isolated SWNTs. This bi-stable configuration can be actuated electrostatically and interrogated by reading the electrical resistance of the junction (Rueckes et al., 2000). Because it has been impossible to fabricate a large crossbar array by either direct growth or post-growth manipulation of CNTs, commercialization of this idea realized the device using a suspended-ribbon architecture. A CNT membrane (1 to 2 nm thick) is electrostatically deflected to contact a bottom electrode, where it reversibly adheres to the electrode due to van der Waals interactions.

In 2006, Nantero reported that NRAM devices have been switched over 50 million cycles, at operating voltages below 5 V, with switching times below 3 ns. While flash memory has dominated the commercial market, the robustness of CNTs makes the NRAM architecture attractive for harsh environments.

A further important commercial aspect of Nantero's effort is the integration of CNTs into a CMOS process in which SWNTs are purified and dispersed in a solvent, and then the solvent is coated onto wafer substrates and can be subsequently patterned and etched using lithographic techniques. Within tangled CNT film geometries, such a CNT photoresist will be vital to CMOS-compatible fabrication of other devices including digital logic, reconfigurable antennas, and sensors without necessitating CNT growth; however, as emphasized earlier, direct deposition is unlikely to achieve highly ordered CNT assemblies.

8.5.5 Thermal Interface Materials

As the sizes of microelectronic circuits and devices continue to scale down and their power increases, efficient thermal management becomes more crucial. Since mating two rigid surfaces together is inherently limited by finite surface roughness (Figure 8.14a), an effective thermal interface material (TIM) is desirable for minimizing thermal contact resistance between the device and heat sink to maximize thermal dissipation. A good TIM therefore fills the gaps between contacting surfaces and provides high thermal conductivity (Gwinn and Webb, 2003).

These requirements are difficult to meet because materials having high thermal conductivity are typically hard solids, and it is difficult to maintain good contact between hard solids especially under the thermal cycling that occurs during microprocessor operation.

While the ideal TIM has yet to be discovered, CNTs are among the promising candidates (Huang et al., 2005; Tong et al., 2007) because they can potentially provide outstanding heat transport while CNT assemblies can be designed to maintain conformal contact against rough surfaces (Figure 14b).

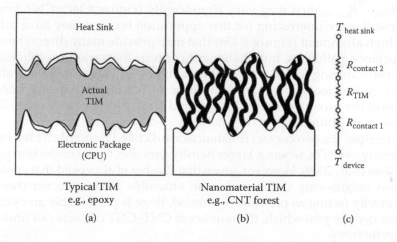

FIGURE 8.14
(a) Mating of two rough, rigid interfaces (device and heat sink) surrounding a thermal interface material. Note gaps between the TIM and rough surfaces. (From Gwinn, J.P. and Webb, R.L. 2003. *Microelectronics Journal*, 34, 215–222. With permission.) (b) Conceptual drawing of similar surfaces bridged by a CNT forest as a TIM. The compliance of the forest enables greater thermal contact between the TIM and the rough surfaces. (c) Thermal resistance model for these proposed systems as described by Equation (8.1).

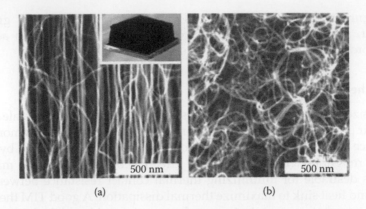

(a)   (b)

FIGURE 8.15
SEM images of typical CNT forest taken at (a) middle and (b) base of the sidewall. Note the distinct morphological change from top to bottom (beginning to end of growth). Inset is a photograph of a millimeter-tall forest adhered to its Si substrate. (From Meshot, E.R., Bedewy, M., Lyons, K.M. et al. 2010. *Nanoscale*, 2, 896–900. With permission.)

If the governing equation for the thermal resistance of an interface is expressed as

$$R = R_{contact1} + R_{TIM} + R_{contact2} \qquad (8.1)$$

and $R_{TIM} = R_{CNTs}$, then may vary significantly (Figure 8.14c). CNT forests in particular are interesting for this application because they have inherently high alignment (Figure 8.15a) that may provide many direct thermal conductance pathways, but challenges concern the varied morphologies that evolve during forest growth. For instance, in Figure 8.15b, we show the disordered morphology at the base of a CNT forest that correlates to the end of the growth process (Meshot and Hart, 2008; Bedewy et al., 2009; Meshot et al., 2010).

In principle, the two contact resistances are dictated primarily by the number density of CNTs, where a larger density provides more conduction pathways (Fan et al., 2009). However, given that Prasher et al. proved that contacts between neighboring CNTs within an ensemble hinder the net thermal conductivity (acting as parasitic resistance), there is presumably an optimal number density past which, the number of CNT–CNT contacts can limit the net conductivity.

Thus, for this application, vertically aligned forests of CNTs are highly advantageous since their orientations are predominantly parallel to the conduction pathway, with fewer CNT–CNT interactions than in a disordered CNT film. Still, considering that as-grown forests typically range from 1 to 5% density, even perfect scaling of individual CNT properties results in bulk conductivity estimates lower than those for Cu.

Experiments have shown that densifying a forest mechanically or by capillary forces increases the thermal conductivity from as low as 8 W/m-K to as high as 2369 W/m-K (Akoshima et al., 2009). Also, progress toward infusing a filler material (e.g., polymer) suggests that reducing the volume of air in the pores of the forest may enhance the thermal conductivity (Ivanov et al., 2006; Duong et al., 2009), but enhancements to date are only marginal. The limitations may arise from CNT–CNT contacts and interfaces between CNTs and filler materials, both of which will induce additional phonon scattering during thermal energy transport, leading to decreased conductance.

8.6 Conclusions and Outlook

Functional integration of CNTs in miniaturized systems requires control of the synthesis and organization of CNTs at hierarchical length scales spanning from the dimensions of individual CNTs to those of wafer substrates. Combined with constraints of microfabrication, including CMOS temperature limits and resistance to liquids and plasmas, it remains challenging to grow, place, and contact CNTs in a manner that meets manufacturing requirements.

While the exceptional properties of CNTs have been verified in all categories, prototype CNT devices provide many examples that reveal sacrifices in properties due to non-ideality of CNT quality, density, and alignment. Nevertheless, profitable demonstrations of CNTs as interconnects, non-volatile memories, inertial sensors, and thermal interfaces have put such devices on the path to commercialization. Further advances in low-temperature CNT growth, chirality separation and control, and directed self-assembly of 3D CNT structures will pave the way for new applications in the coming years, along with broad use of CNTs as new microfabrication materials. The future is certainly bright for continued innovations in the science and applications of CNTs.

Acknowledgments

Time spent preparing this chapter was supported by funds from the Nanomanufacturing Program of the National Science Foundation (CMMI-0800213, CMMI-0927364), the Office of Naval Research, the Belgian Fund for Scientific Research Flanders (FWO), and the University of Michigan.

References

Ajayan, P.M. and Iijima, S. 1993. Capillarity-induced filling of carbon nanotubes. *Nature*, 361, 333–334.

Akoshima, M., Hata, K., Futaba, D. et al. 2009. Thermal diffusivity of single-walled carbon nanotube forest measured by laser flash method. *Japan Journal of Applied Physics*, 48, 1–6.

Ashrafi, B., Hubert, P., and Vengallatore, S. 2006. Carbon nanotube-reinforced composites as structural materials for microactuators in microelectromechanical systems. *Nanotechnology*, 17, 4895–4903.

Atkinson, K.R., Hawkins, S.C., Huynh, C. et al. 2007. Multifunctional carbon nanotube yarns and transparent sheets: fabrication, properties, and applications. *Physica B-Condensed Matter*, 394, 339–343.

Awano, Y., Sato, S., Kondo, D. et al. 2006. Carbon nanotube via interconnect technologies: size-classified catalyst nanoparticles and low-resistance ohmic contact formation. *Physical Status Solidi A*, 203, 3611–3616.

Barber, A.H., Andrews, R., Schadler, L.S. et al. 2005. On the tensile strength distribution of multiwalled carbon banotubes. *Applied Physics Letters*, 87, 203106.

Barber, A.H., Cohen, S.R., and Wagner, H.D. 2003. Measurement of carbon nanotube polymer interfacial strength. *Applied Physics Letters*, 82, 4140–4142.

Baughman, R.H., Zakhidov, A.A., and de Heer, W.A. 2002. Carbon nanotubes: the route toward applications. *Science*, 297, 787–792.

Bedewy, M., Meshot, E.R., Guo, H.C. et al. 2009. Collective mechanism for the evolution and self-termination of vertically aligned carbon nanotube growth. *Journal of Physical Chemistry C*, 113, 20576–20582.

Biercuk, M.J., Llaguno, M.C., Radosavljevic, M. et al. 2002. Carbon nanotube composites for thermal management. *Applied Physics Letters*, 80, 2767–2769.

Cao, J., Wang, Q., and Dai, H.J. 2003. Electromechanical properties of metallic, quasimetallic, and semiconducting carbon nanotubes under stretching. *Physics Review Letters*, 90, 157601.

Chakrapani, N., Wei, B.Q., Carrillo, A. et al. 2004. Capillarity-driven assembly of two-dimensional cellular carbon nanotube foams. *Proceedings of the National Academy of Sciences of USA*, 101, 4009–4012.

Chen, Z., Appenzeller, J., Lin, Y.M. et al. 2006. An integrated logic circuit assembled on a single carbon nanotube. *Science*, 311, 1735.

Cheng, Y., Zhang, J., Lee, Y.Z. et al. 2004. Dynamic radiography using a carbon nanotube-based field-emission x-ray source. *Review of Scientific Instruments*, 75, 3264–3267.

Chhowalla, M., Teo, K.B.K., Ducati, C. et al. 2001. Growth process conditions of vertically aligned carbon nanotubes using plasma-enhanced chemical vapor deposition. *Journal of Applied Physics*, 90, 5308–5317.

Chiodarelli, N., Li, Y., Cott, D.J. et al. 2010. Integration and electrical characterization of carbon nanotube via interconnects. *Microelectronic Engineering*, 88, 837–843.

Chiu, H.Y., Deshpande, V.V., Postma, H.W.C. et al. 2005. Ballistic phonon thermal transport in multiwalled carbon nanotubes. *Physics Review Letters*, 95, 226101.

Choi, J.H., Zoulkarneev, A.R., Jin, Y.W. et al. 2004. Carbon nanotube field emitter arrays having an electron beam focusing structure. *Applied Physics Letters*, 84, 1022–1024.

Choi, S.U.S., Zhang, Z.G., Yu, W. et al. 2001. Anomalous thermal conductivity enhancement in nanotube suspensions. *Applied Physics Letters*, 79, 2252–2254.

Choi, Y.C., Jeong, K.S., Han, I.T. et al. 2006. Double-gated field emitter array with carbon nanotubes grown by chemical vapor deposition. *Applied Physics Letters*, 88, 263504.

Chung, J.Y., Lee, K.H., Lee, J.H. et al. 2004. Toward large-scale integration of carbon nanotubes. *Langmuir*, 20, 3011–3017.

Close, G.F. and Wong, H.S.P. 2007. Fabrication and characterization of carbon nanotube interconnects. *Proceedings of IEEE International Electronic Devices Meeting*, pp. 203–206.

Close, G.F., Yasuda, S., Paul, B., et al. 2008. A 1-Ghz integrated circuit with carbon nanotube interconnects and silicon transistors. *Nano Letters*, 8, 706–709.

Coiffic, J.C., Fayolle, M., Le Poche, H. et al. 2008. Realization of via interconnects based on carbon nanotubes. *Proceedings of IEEE International Interconnect Technology Conference*, pp. 153–155.

Collins, P.G., Bradley, K., Ishigami, M. et al. 2000. Extreme oxygen sensitivity of electronic properties of carbon nanotubes. *Science*, 287, 1801–1804.

Cooper, C.A., Cohen, S.R., Barber, A.H. et al. 2002. Detachment of nanotubes from a polymer matrix. *Applied Physics Letters*, 81, 3873–3875.

Correa-Duarte, M.A., Wagner, N., Rojas-Chapana, J. et al. 2004. Fabrication and biocompatibility of carbon nanotube-based 3D networks as scaffolds for cell seeding and growth. *Nano Letters*, 4, 2233–2236.

Cui, Y., Lauhon, L.J., Gudiksen, M.S. et al. 2001. Diameter-controlled synthesis of single-crystal silicon nanowires. *Applied Physics Letters*, 78, 2214–2216.

Cumings, J. and Zettl, A. 2000. Low-friction nanoscale linear bearing realized from multiwall carbon nanotubes. *Science*, 289, 602–604.

Dai, H. 2001. *Carbon Nanotubes: Synthesis, Structure, Properties, and Applications*. Berlin: Springer.

Dalton, A.B., Collins, S., Razal, J. et al. 2004. Continuous carbon nanotube composite fibers: properties, potential applications, and problems. *Journal of Material Chemistry*, 14, 1–3.

De Volder, M., Tawfick, S., Park, S.J. et al. 2010a. Directed assembly of complex and robust carbon nanotube microarchitectures by capillary forming. *Advanced Materials*, 22, 4384–4389.

Ding, F., Rosen, A., and Bolton, K. 2005. Dependence of SWNT growth mechanism on temperature and catalyst particle size: bulk versus surface diffusion. *Carbon*, 43, 2215–2217.

Dresselhaus, M.S. and Avouris, P. 2001. *Carbon Nanotubes: Synthesis, Structure, Properties, and Applications*. Berlin: Springer.

Dresselhaus, M.S., Dresselhaus, G., and Saito, R. 1995. Physics of carbon nanotubes. *Carbon*, 33, 883–891.

Dresselhaus, M.S., Lin, Y.M., Rabin, O. et al. 2003. Nanowires and nanotubes. *Materials Science and Engineering C*, 23, 129–140.

Duong, H., Yamamoto, N., Papavassiliou, D. et al., 2009. Inter-carbon nanotube contact in thermal transport of controlled-morphology polymer nanocomposites. *Nanotechnology*, 20, 155702.

Ebbesen, T.W., Lezec, H.J., Hiura, H. et al. 1996. Electrical conductivity of individual carbon nanotubes. *Nature*, 382, 54–56.

Endo, M., Hayashi, T., Kim, Y.A. et al. 2006. Development and application of carbon nanotubes. *Japan Journal of Applied Physics*, 45, 4883–4892.

Endo, M., Hayashi, T., Kim, Y.A. et al. 2004. Applications of carbon nanotubes in the twenty-first century. *Philosophical Transactions of the Royal Society*, 362, 2223–2238.

Erickson, L.M., Fan, H., Peng, H.Q. et al. 2004. Macroscopic, neat, single-walled carbon nanotube fibers. *Science*, 305, 1447–1450.

Fan, H., Zhang, K., and Yuen, M. 2009. The interfacial thermal conductance between a vertical single-wall carbon nanotube and a silicon substrate. *Journal of Applied Physics*, 106, 034307–034313.

Fan, S.S., Chapline, M.G., Franklin, N.R. et al. 1999. Self-oriented regular arrays of carbon nanotubes and their field emission properties. *Science*, 283, 512–514.

Fischer, J.E., Zhou, W., Vavro, J. et al. 2003. Magnetically aligned single-wall carbon nanotube films. *Journal of Applied Physics*, 93, 2157–2163.

Fornasiero, F., Park, H.G., Holt, J.K. et al. 2008. Ion exclusion by sub-2-nm carbon nanotube pores. *Proceedings of the National Academy of Sciences of USA*, 105, 17250–17255.

Futaba, D.N., Hata, K., Yamada, T. et al. 2006. Shape-engineerable and highly densely packed single-walled carbon nanotubes and their application as super-capacitor electrodes. *Nature Materials*, 5, 987–994.

Garcia, E.J., Hart, A.J., Wardle, B.L. et al. 2007. Fabrication and nanocompression testing of aligned carbon nanotube polymer nanocomposites. *Advanced Materials*, 19, 2151–2156.

Ghosh, S., Sood, A.K., and Kumar, N. 2003. Carbon nanotube flow sensors. *Science*, 299, 1042–1044.

Goel, A., Hebgen, P., Vander Sande, J.B. et al. 2002. Combustion synthesis of fullerenes and fullerenic nanostructures. *Carbon*, 40, 177–182.

Grow, R.J., Wang, Q., Cao, J. et al. 2005. Piezoresistance of carbon nanotubes on deformable thin-film membranes. *Applied Physics Letters*, 86, 93104–93107.

Guillorn, M.A., McKnight, T.E., Melechko, A. et al. 2002. Individually addressable vertically aligned carbon nanofiber-based electrochemical probes. *Journal of Applied Physics*, 91, 3824–3828.

Guillorn, M.A., Simpson, M.L., Bordonaro, G.J. et al. 2001. Fabrication of gated cathode structures using an in situ grown vertically aligned carbon nanofiber as a field emission element. *Journal of Vacuum Science and Technology B*, 19, 573–578.

Gwinn, J.P. and Webb, R.L. 2003. Performance and testing of thermal interface materials. *Microelectronics Journal*, 34, 215–222.

Hart, A.J., and Slocum, A.H. 2006. Force output, control of film structure, and microscale shape transfer by carbon nanotube growth under mechanical pressure. *Nano Lett*, 6, 1254–1260.

Hart, A.J., Slocum, A.H., and Royer, L. 2006. Growth of conformal single-walled carbon nanotube films from Mo/Fe/Al2o3 deposited by electron beam evaporation. *Carbon*, 44, 348–359.

Hayamizu, Y., Yamada, T., Mizuno, K. et al. 2008. Integrated three-dimensional microelectromechanical devices from processable carbon nanotube wafers. *Nature Nanotechnology*, 3, 289–294.

Hayashi, T., Kim, Y.A., Matoba, T. et al. 2003. Smallest free-standing single-walled carbon nanotube. *Nano Letters*, 3, 887–889.

Height, M.J., Howard, J.B., Tester, J.W. et al. 2004. Flame synthesis of single-walled carbon nanotubes. *Carbon*, 42, 2295–2307.

Hierold, C., Jungen, A., Stampfer, C. et al. 2007. Nano electromechanical sensors based on carbon nanotubes. *Sensors and Actuators A*, 136, 51–61.

Hinds, B.J., Chopra, N., Rantell, T. et al. 2004. Aligned multiwalled carbon nanotube membranes. *Science*, 303, 62–65.

Hofmann, S., Csanyi, G., Ferrari, A.C. et al. 2005. Surface diffusion: the low activation energy path for nanotube growth. *Physics Review Letters*, 95, 036101.

Holt, J.K., Noy, A., Huser, T. et al. 2004. Fabrication of a carbon nanotube-embedded silicon nitride membrane for studies of nanometer-scale mass transport. *Nano Letters*, 4, 2245–2250.

Holt, J.K., Park, H.G., Wang, Y.M. et al. 2006. Fast mass transport through sub-2-nanometer carbon nanotubes. *Science*, 312, 1034–1037.

Hone, J. 2001. *Carbon Nanotubes: Synthesis, Structure, Properties, and Applications.* Berlin: Springer.

Hone, J., Llaguno, M.C., Nemes, N.M. et al. 2000. Electrical and thermal transport properties of magnetically aligned single walled carbon nanotube films. *Applied Physics Letters*, 77, 666–668.

Horibe, M., Nihei, M., Kondo, D. et al. 2005. Carbon nanotube growth technologies using tantalum barrier layer for future vias with Cu/low-K interconnect processes. *Japan Journal of Applied Physics*, 44, 5309–5312.

Howard, J.B., Mckinnon, J.T., Makarovsky, Y. et al. 1991. Fullerenes C60 and C70 in flames. *Nature*, 352, 139–141.

Huang, H., Liu, C., Wu, Y. et al. 2005. Aligned carbon nanotube composite films for thermal management. *Advanced Materials*, 17, 1652–1656.

Huang, J.Y., Chen, S., Wang, Z.Q. et al. 2006. Superplastic carbon nanotubes. *Nature*, 439, 281.

Huang, S.M., Cai, X.Y., and Liu, J. 2003. Growth of millimeter-long and horizontally aligned single-walled carbon nanotubes on flat substrates. *Journal of the American Chemical Society*, 125, 5636–5637.

Hutchison, D.N., Aten, B., Turner, N. et al. 2009. High aspect ratio microelectromechanical systems: a versatile approach using carbon nanotubes. *Proceedings of IEEE Transducers*, pp. 1604–1607,

Iijima, S. 1991. Helical microtubules of graphitic carbon. *Nature*, 354, 56–58.

Iijima, S. and Ichihashi, T. 1993. Single-shell carbon nanotubes of 1-nm diameter. *Nature*, 363, 603–605.

Ivanov, I., Puretzky, A., Eres, G. et al. 2006. Fast and highly anisotropic thermal transport through vertically aligned carbon nanotube arrays. *Applied Physics Letters*, 89, 223110–223113.

Javey, A., Guo, J., Wang, Q. et al. 2003. Ballistic carbon nanotube field-effect transistors. *Nature*, 424, 654–657.

Jensen, K., Girit, C., Mickelson, W. et al. 2006. Tunable nanoresonators constructed from telescoping nanotubes. *Physical Review Letters*, 96, 215503.

Jeong, H.J., Jeong, S.Y., Shin, Y.M. et al. 2002. Dual-catalyst growth of vertically aligned carbon nanotubes at low temperature in thermal chemical vapor deposition. *Chemistry and Physics Letters*, 361, 189–195.

Jungen, A., Durrer, L., Stampfer, C. et al. 2007. Progress in carbon nanotube-based nanoelectromechanical systems synthesis. *Physics Status Solidi B*, 244, 4323–4326.

Kang, S.J., Kocabas, C., Ozel, T. et al. 2007. High-performance electronics using dense, perfectly aligned arrays of single-walled carbon nanotubes. *Nature Nanotechnology*, 2, 230–236.

Kim, P., Shi, L., Majumdar, A. et al. 2001. Thermal transport measurements of individual multiwalled nanotubes. *Physics Review Letters*, 8712.

Kim, S.G. 2006. Transplanting assembly of carbon nanotubes. *CIRP Annals: Manufacturing Technology*, 55, 15–18.

Kocabas, C., Hur, S.H., Gaur, A., et al. 2005. Guided growth of large-scale, horizontally aligned arrays of single-walled carbon nanotubes and their use in thin-film transistors. *Small*, 1, 1110–1116.

Kodambaka, S., Tersoff, J., Reuter, M.C. et al. 2006. Diameter-independent kinetics in the vapor–liquid–solid growth of Si nanowires. *Physics Review Letters*, 96, 096105.

Kong, J., Franklin, N.R., Zhou, C. et al. 2000. Nanotube molecular wires as chemical sensors. *Science*, 287, 622–625.

Koshino, M., Niimi, Y., Nakamura, E. et al. 2010. Analysis of the reactivity and selectivity of fullerene dimerization reactions at the atomic level. *Nature Chemistry*, 2, 117–124.

Kratschmer, W., Lamb, L.D., Fostiropoulos, K. et al. 1990. Solid C60: a new form of carbon. *Chemistry and Physics Letters*, 347, 354–358.

Kreupl, F., Graham, A.P., Duesberg, G.S. et al. 2002. Carbon nanotubes in interconnect applications. *Microelectronic Engineering*, 64, 399–408.

Krishnan, A., Dujardin, E., Ebbesen, T.W. et al. 1998. Young's modulus of single-walled nanotubes. *Physical Review B*, 58, 14013–14019.

Krupke, R., Hennrich, F., Kappes, M.M. et al. 2004. Surface conductance-induced dielectrophoresis of semiconducting single-walled carbon nanotubes. *Nano Letters*, 4, 1395–1399.

Lee, C.J., Son, K.H., Park, J. et al. 2001. Low temperature growth of vertically aligned carbon nanotubes by thermal chemical vapor deposition. *Chemistry and Physics Letters*, 338, 113–117.

Lee, S., Peng, J.W., and Liu, C.H. 2009. Probing plasma-induced defect formation and oxidation in carbon nanotubes by Raman dispersion spectroscopy. *Carbon*, 47, 3488–3497.

Li, H.J., Lu, W.G., Li, J.J. et al. 2005. Multichannel ballistic transport in multiwall carbon nanotubes. *Physics Review Letters*, 95, 086601.

Li, Y.L., Kinloch, I.A., and Windle, A.H. 2004. Direct spinning of carbon nanotube fibers from chemical vapor deposition synthesis. *Science*, 304, 276–278.

Liu, H., Li, S.H., Zhai, J. et al. 2004. Self-assembly of large-scale micropatterns on aligned carbon nanotube films. *Angewandte Chemie International*, 43, 1146–1149.

Madou, M.J. 2002. *Fundamentals of Microfabrication: The Science of Miniaturization*. Boca Raton, FL: CRC Press.

Magrez, A., Seo, J.W., Smajda, R. et al. 2010. Low-temperature, highly efficient growth of carbon nanotubes on functional materials by an oxidative dehydrogenation reaction. *ACS Nano*, 4, 3702–3708.

Majumder, M., Chopra, N., Andrews, R. et al. 2005. Nanoscale hydrodynamics: enhanced flow in carbon nanotubes. *Nature*, 438, 44–48.

Matweb Material Property Data. n.d. www.Matweb.com

Meitl, M.A., Zhu, Z.T., Kumar, V. et al. 2006. Transfer printing by kinetic control of adhesion to an elastomeric stamp. *Nature Materials*, 5, 33–38.

Melechko, A.V., Merkulov, V.I., Lowndes, D.H. et al. 2002. Transition between base and tip carbon nanofiber growth modes. *Chemistry and Physics Letters*, 356, 527–533.

Merkulov, V.I., Melechko, A.V., Guillorn, M.A. et al. 2001. Alignment mechanism of carbon nanofibers produced by plasma-enhanced chemical vapor deposition. *Applied Physics Letters*, 79, 2970–2972.

Meshot, E.R., Bedewy, M., Lyons, K.M. et al. 2010. Measuring lengthening kinetics of aligned nanostructures by spatiotemporal correlation of height and orientation. *Nanoscale*, 2, 896–900.

Meshot, E., Plata, D., Tawfick, S. et al. 2009. Engineering vertically aligned carbon nanotube growth by decoupled thermal treatment of precursor and catalyst. *ACS Nano*, 3, 2477–2486.

Meshot, E.R. and Hart, A.J. 2008. Abrupt self-termination of vertically aligned carbon nanotube growth. *Applied Physics Letters*, 92, 113107.

Minot, E.D., Yaish, Y., Sazonova, V. et al. 2003. Tuning carbon nanotube band gaps with strain. *Physics Review Letters*, 90, 156401.

Mora, E., Pigos, J.M., Ding, F. et al. 2008. Low-temperature single-wall carbon nanotube synthesis: feedstock decomposition limited growth. *Journal of the American Chemical Society*, 130, 11840–11841.

Motta, M., Li, Y.L., Kinloch, I. et al. 2005. Mechanical properties of continuously spun fibers of carbon nanotubes. *Nano Letters*, 5, 1529–1533.

Naeemi, A. and Meindl, J.D. 2007. Design and performance modeling for single-walled carbon nanotubes as local, semiglobal, and global interconnects in gigascale integrated systems. *IEEE Transactions on Electron Devices*, 54, 26–37.

Naeemi, A., Sarvari, R., and Meindl, J.D. 2005. Performance comparison between carbon nanotube and copper interconnects for gigascale integration. *IEEE Electron Device Letters*, 26, 84–86.

NAS 2005. High-Performance Structural Fibers for Advanced Polymer Matrix Composites, National Materials Advisory Board, National Academy of Sciences.

Nessim, G.D., Acquaviva, D., Seita, M. et al. 2010. The critical role of the underlayer material and thickness in growing vertically aligned carbon nanotubes and nanofibers on metallic substrates by chemical vapor deposition. *Advanced Functional Materials*, 20, 1306–1312.

Nessim, G.D., Seita, M., O'Brien, K.P. et al. 2009. Low temperature synthesis of vertically aligned carbon nanotubes with ohmic contact to metallic substrates enabled by thermal decomposition of the carbon feedstock. *Nano Letters*, 9, 3398–3405.

Nessim, G.D., Seita, M., Plata, D.L. et al. 2011. Precursor gas chemistry determines the crystallinity of carbon nanotubes synthesized at low temperature. *Carbon*, 49, 804–810.

Nihei, M., Horibe, M., Kawabata, A. et al. 2004. Simultaneous formation of multiwall carbon nanotubes and their end-bonded ohmic contacts to Ti electrodes for future UlSi interconnects. *Japan Journal of Applied Physics*, 43, 1856–1859.

Nihei, M., Kawabata, A., Kondo, D. et al. 2005. Electrical properties of carbon nanotube bundles for future via interconnects. *Japan Journal of Applied Physics*, 44, 1626–1628.

Peercy, P.S. 2000. The drive to miniaturization. *Nature*, 406, 1023–1026.

Pengfei, Q.F., Vermesh, O., Grecu, M. et al. 2003. Toward large arrays of multiplex functionalized carbon nanotube sensors for highly sensitive and selective molecular detection. *Nano Letters*, 3, 347–351.

Pint, C.L., Xu, Y.Q., Pasquali, M. et al. 2008. Formation of highly dense aligned ribbons and transparent films of single-walled carbon nanotubes directly from carpets. *ACS Nano*, 2, 1871–1878.

Plata, D.L., Hart, A.J., Reddy, C.M. et al. 2009. Early evaluation of potential environmental impacts of carbon nanotube synthesis by chemical vapor deposition. *Environmental Science and Technology*, 43, 8367–8373.

Pop, E., Mann, D., Wang, Q. et al. 2006. Thermal conductance of an individual single-wall carbon nanotube above room temperature. *Nano Letters*, 6, 96–100.

Prasher, R.S., Hu, X.J., Chalopin, Y. et al. 2009. Turning carbon nanotubes from exceptional heat conductors into insulators. *Physics Review Letters*, 102, 105901.

Reich, S., Li, L., and Robertson, J. 2005. Structure and formation energy of carbon nanotube caps. *Physical Review B*, 72, 165423.

Ren, Z.F., Huang, Z.P., Xu, J.W. et al. 1998. Synthesis of large arrays of well-aligned carbon nanotubes on glass. *Science*, 282, 1105-1107.

Romero, H.E., Bolton, K., Rosen, A. et al. 2005. Atom collision-induced resistivity of carbon nanotubes. *Science*, 307, 89–93.

Rotkin, S.V. and Subramoney, S. 2005. *Applied Physics of Carbon Nanotubes*. New York: Springer.

Rueckes, T., Kim, K., Joselevich, E. et al. 2000. Carbon nanotube-based nonvolatile random access memory for molecular computing. *Science*, 289, 94-97.

Sato, S., Nihei, M., Mimura, A. et al. 2006. Novel approach to fabricating carbon nanotube via interconnects using size-controlled catalyst nanoparticles. *Proceedings of IEEE International Interconnect Technology Conference*, pp. 230–232.

Shi, X.J., Wang, L.Y., Kariuki, N. et al. 2006. A multi-module artificial neural network approach to pattern recognition. *Sensors and Actuators B*, 117, 65–73.

Sickert, D., Taeger, S., Kuhne, I. et al. 2006. Strain sensing with carbon nanotube devices. *Physical Status Solidi B*, 243, 3542–3545.

Smith, B.W., Monthioux, M., and Luzzi, D.E. 1998. Encapsulated C60 in carbon nanotubes. *Nature*, 396, 323–324.

Snoeck, J.W., Froment, G.F., and Fowles, M. 1997. Filamentous carbon formation and gasification: Thermodynamics, driving force, nucleation, and steady-state growth. *J Catal*, 169, 240-249.

Snow, E.S., and Perkins, F.K. 2005. Capacitance and conductance of single-walled carbon nanotubes in the presence of chemical vapors. *Nano Lett*, 5, 2414–2417.

Snow, E.S., Perkins, F.K., Houser, E.J., Badescu, S.C., and Reinecke, T.L. 2005. Chemical detection with a single-walled carbon nanotube capacitor. *Science*, 307, 1942–1945.

Snow, E.S., Perkins, F.K., and Robinson, J.A. 2006. Chemical vapor detection using single-walled carbon nanotubes. *Chemical Society Reviews*, 35, 790–798.

Soh, H.T., Quate, C.F., Morpurgo, A.F. et al. 1999. Integrated nanotube circuits: controlled growth and ohmic contacting of single-walled carbon nanotubes. *Applied Physics Letters*, 75, 627-629.

Sood, A.K. and Ghosh, S. 2004. Direct generation of a voltage and current by gas flow over carbon nanotubes and semiconductors. *Physics Review Letters*, 93, 086601.

Stampfer, C., Guttinger, J., Roman, C. et al. 2007. Electron shuttle instability for nano electromechanical mass sensing. *Nano Letters*, 7, 2747–2752.

Stampfer, C., Helbling, T., Obergfell, D. et al. 2006a. Fabrication of single-walled carbon nanotube-based pressure sensors. *Nano Letters*, 6, 233–237.

Stampfer, C., Jungen, A., and Hierold, C. 2006b. Fabrication of discrete nanoscaled force sensors based on single-walled carbon nanotubes. *IEEE Sensors Journal*, 6, 613–617.

Takesue, I., Haruyama, J., Kobayashi, N. et al. 2006. Superconductivity in entirely end-bonded multiwalled carbon nanotubes. *Physics Review Letters*, 96, 057001.

Tans, S.J., Verschueren, A.R.M., and Dekker, C. 1998. Room-temperature transistor based on a single carbon nanotube. *Nature*, 393, 49–52.

Tawfick, S., O'Brien, K., and Hart, A.J. 2009. Flexible high-conductivity carbon nanotube interconnects made by rolling and printing. *Small*, 5, 2467–2473.

Teo, K.B.K., Singh, C., Chhowalla, M. et al. 2004. Catalytic synthesis of carbon nanotubes and nanofibres. In H. S. Nalwa (Ed.), *Encyclopedia of Nanoscience and Nanotechnology*, Stevenson Ranch, CA: American Scientific Publishers, Vol. 1, pp. 665–686.

Terranova, M.L., Sessa, V., and Rossi, M. 2006. The world of carbon nanotubes: overview of CVD growth methodologies. *Chemical Vapor Deposition*, 12, 315–325.

Terrones, M., Grobert, N., Olivares, J. et al. 1997. Controlled production of aligned nanotube bundles. *Nature*, 388, 52–55.

Thess, A., Lee, R., Nikolaev, P. et al. 1996. Crystalline ropes of metallic carbon nanotubes. *Science*, 273, 483–487.

Thostenson, E.T., Ren, Z.F., and Chou, T.W. 2001. Advances in the science and technology of carbon nanotubes and their composites. *Composite Science and Technology*, 61, 1899–1912.

Tong, T., Zhao, Y., Delzeit, L. et al. 2007. Dense, vertically aligned multiwalled carbon nanotube arrays as thermal interface materials. *IEEE Transactions on Components and Packaging Technologies*, 30, 92–100.

Tselev, A., Woodson, M., Qian, C. et al. 2008. Microwave impedance spectroscopy of dense carbon nanotube bundles. *Nano Letters*, 8, 152–156.

Vijayaraghavan, A., Blatt, S., Weissenberger, D. et al. 2007. Ultralarge scale directed assembly of single-walled carbon nanotube devices. *Nano Letters*, 7, 1556–1560.

Wagner, R.S. and Ellis, W.C. 1965. The vapor–liquid–solid mechanism of crystal growth and its application to silicon. *Transactions of Metallurgical Society of AIME*, 223, 1053–1064.

Walters, D.A., Ericson, L.M., Casavant, M.J. et al. 1999. Elastic strain of freely suspended single-wall carbon nanotube ropes. *Applied Physics Letters*, 74, 3803–3805.

Wang, D., Song, P.C., Liu, C.H. et al. 2008. Highly oriented carbon nanotube papers made of aligned carbon nanotubes. *Nanotechnology*, 19.

Williams, P.A., Papadakis, S.J., Falvo, M.R. et al. 2002. Controlled placement of individual carbon nanotube onto a microelectromechanical structure. *Applied Physics Letters*, 80, 2574–2576.

Wind, S.J., Appenzeller, J., and Avouris, P. 2003. Lateral scaling in carbon-nanotube field-effect transistors. *Physics Review Letters*, 91, 058301.

Wood, J.R. and Wagner, H.D. 2000. Single-wall carbon nanotubes as molecular pressure sensors. *Applied Physics Letters*, 76, 2883–2885.

Wu, Y.Y. and Yang, P.D. 2001. Direct observation of vapor–liquid–solid nanowire growth. *Journal of the American Chemical Society*, 123, 3165–3166.

Yamazaki, Y., Katagiri, M., Sakuma, N. et al. 2010. Synthesis of a closely packed carbon nanotube forest by a multi-step growth method. *Applied Physics Express*, 3, 055002.

Yang, D.J., Zhang, Q., Chen, G. et al. 2002. Thermal conductivity of multiwalled carbon nanotubes. *Physical Reviews B*, 66.

Yokoyama, D., Iwasaki, T., Ishimaru, K. et al. 2008. Electrical properties of carbon nanotubes grown at a low temperature for use as interconnects. *Japan Journal of Applied Physics*, 47, 1985–1990.

Yoon, D.M., Yoon, B.J., Lee, K.H. et al. 2006. Synthesis of carbon nanotubes from solid carbon sources by direct microwave irradiation. *Carbon*, 44, 1339–1343.

Zhang, J., Cheng, Y., Lee, Y.Z. et al. 2005. A nanotube-based field emission x-ray source for microcomputed tomography. *Review of Scientific Instruments*, 76, 094301.

Zhang, J., Yang, G., Lee, Y.Z. et al. 2006. Multiplexing radiography using a carbon nanotube-based x-ray source. *Applied Physics Letters*, 89, 064106.

Zhang, M., Atkinson, K.R., and Baughman, R.H. 2004. Multifunctional carbon nanotube yarns by downsizing an ancient technology. *Science*, 306, 1358–1361.

Zheng, L.X., O'Connell, M.J., Doorn, S.K. et al. 2004. Ultralong single-wall carbon nanotubes. *Nature Materials*, 3, 673–676.

Zhong, G., Hofmann, S., Yan, F. et al. 2009. Acetylene: a key growth precursor for single-walled carbon nanotube forests. *Journal of Physical Chemistry C*, 113, 17321–17325.

Zhong, G.F., Iwasaki, T., Honda, K. et al. 2005. Low temperature synthesis of extremely dense and vertically aligned single-walled nanotubes. *Japanese Journal of Applied Physics*, 44, 4A, 1558–1561.

Zhu, L.B., Sun, Y.Y., Hess, D.W. et al. 2006. Well-aligned open-ended carbon nanotube architectures: approach for device assembly. *Nano Letters*, 6, 243–247.

9

Heterogeneous Integration of Carbon Nanotubes on Complementary Metal Oxide Semiconductor Circuitry and Sensing Applications

Chia-Ling Chen, Sameer Sonkusale,
Michelle Chen, and Mehmet R. Dokmeci

CONTENTS

ABSTRACT Carbon nanotubes (CNTs) offer tremendous promise as emerging materials for sensing applications. However, we still lack a systematic approach for realization of functional nanodevices based on carbon nanotubes. An approach that produces carbon nanotubes with a conventional complementary metal oxide semiconductor (CMOS) technology will address such challenges. A simple methodology for integrating single-walled carbon nanotubes (SWNTs) onto CMOS integrated circuits is presented. The SWNTs are incorporated onto a CMOS chip between electrodes made with available metal layers from the CMOS process. For proof of concept, assembly of SWNTs serving as feedback resistors of a two-stage Miller compensated operational amplifier utilizing dielectrophoretic (DEP) assembly is demonstrated. The measured electrical properties from the integrated SWNTs yield ohmic behavior with a two-terminal resistance of ~37.5 KΩ. The measured small signal ac gain (about –2) from the inverting amplifier confirmed successful integration of CNTs onto the CMOS circuitry. Furthermore, the temperature response of the SWNTs integrated onto CMOS circuitry exhibited a temperature coefficient of resistance (TCR) of –0.4%/°C. Bare SWNTs were reported sensitive to various chemicals, and functionalization of SWNTs with biomolecular complexes further enhances their specificity and sensitivity. After decorating ss-DNA on SWNTs, the sensing response of the gas sensor is enhanced (up to ~300 and ~250% for methanol vapor and isopropanol alcohol vapor, respectively) compared with bare SWNTs. This methodology for integrating SWNTs onto CMOS technology is versatile, high yield, and paves the way to the realization of novel miniature CNT-based sensor systems.

9.1 Introduction

Single-walled carbon nanotubes (SWNTs), due to their miniature size, large surface area, and quasi one-dimensional electronic transport properties, are promising candidates for ultra-sensitive sensors. Chemical (Kong et al., 2000), biological (Cui and Lieber, 2001), and physical (Fung et al., 2004) sensors utilizing SWNTs as active materials have already been demonstrated. The main detection methods for SWNT sensors include resistive-, capacitive- and charge-based methods that often require additional read-out circuitry or measurement units for recognition of sensing responses. This limits the range of applications and also makes sensors prone to parasitics due to wire bonding pads and long interconnects.

Furthermore, for high density (>100) sensor arrays for detecting multiple targets, the area lost from wire pads and the signal lines would be intolerable for many applications. These problems are alleviated by monolithic integration of nanosensors with their measurement units. Another benefit of monolithic integration is the ability to perform signal detection, amplification, buffering, and storage on the same chip and possibly wireless transmission with on-chip coils.

Accordingly, CMOS technology provides an ideal platform for the integration of nanomaterials with built-in CMOS circuitry, offering numerous benefits such as direct detection, on-chip processing, storage, and even transmission of signals, enabling miniaturized, high performance sensing systems. Despite the immense potential, integration of nanotubes with CMOS circuitry is fairly challenging and hence is rarely implemented.

One of these challenges concerns the high growth temperatures for CNT synthesis. Well-known approaches for synthesizing carbon nanotubes include laser ablation (Guo et al., 1995), arc discharge (Iijima, 1991), and chemical vapor deposition (Li et al., 1996). All require elevated temperatures (>500°C) that are not compatible with CMOS technology. While lowering the nanotube synthesis temperatures is an area of active research, to date, the highest quality nanotubes are obtained only at growth temperatures exceeding 500°C.

A routine approach is to grow CNT devices on a substrate that can tolerate high temperatures and then to transfer print them onto other substrates (Hines et al., 2005). Another method is harvesting the nanotubes grown on a substrate followed by their purification, functionalization, and dispersion in an aqueous solution. Various systems for depositing nanotubes from aqueous solutions onto surfaces include drop casting (Chen et al., 2008), dip coating (Spotnitz et al., 2004; Xiong et al., 2009), inkjet printing (Song et al., 2008), and field assisted assembly (Khanduja et al., 2007).

One field-assisted manipulation method is dielectrophoretic (DEP) assembly. Classified as a bottom up technique, DEP is a versatile and low temperature technique for manipulating nanoscale materials (Pohl, 1978; Green et al., 1997; Hermanson et al., 2001). During the DEP process, nanomaterials suspended in a solution can be attracted to regions where the intensity of the field is maximum (positive DEP). This simple, low cost, high yield assembly process is also amenable with CMOS technology. Previous nanotube-based integration approaches utilized voltages >20 $V_{peak-peak}$ (V_{pp}) for DEP assembly, but such high voltages are not compatible with CMOS circuitry and also require additional photolithography steps (Close et al., 2008).

Another challenge in realizing nanotube based sensors is the requirement of good electrical contacts between the nanotubes and the metal electrodes. By adopting an electroless zincation process (Chen et al., 2009), good electrical contacts between the SWNTs and the metal electrodes can be achieved. Besides, by reducing the spacing (0.5 µm) between the assembly electrodes to a minimum (allowed by the commercial foundry process), the resulting low voltage (5 V_{pp}) DEP assembly process is compatible with most CMOS processes.

Furthermore, the top metal layer (M3) of the foundry CMOS process is ~1.7 µm higher than the rest of the substrate, and hence by assembling SWNT sensors on microelectrodes using an M3 layer enables the realization of suspended nanotube sensors that eliminate the influence of the substrate on the sensing performance critical for certain applications (Kim et al., 2001; Lu et al., 2006; Chen et al., 2010).

In this chapter, we first introduce the heterogeneous integration of CNTs onto CMOS circuitry. The CMOS chip, including an operational amplifier, is designed and fabricated using the AMI 0.5-µm CMOS technology. After successful integration of CNTs onto CMOS circuitry, two different sensor applications will be introduced. The first is a CNT-based thermal sensor which was reportedly capable of performing very low power thermal sensing and was about 1000 times lower than conventional micromachined polysilicon-based thermal sensors (Chan, Fung et al., 2004). The second is a single-stranded DNA (ss-DNA) decorated SWNT-based chemical sensor. Recent reports indicate that SWNTs decorated with a nanoscale layer of ss-DNA display remarkable chemical sensing capabilities, making them promising candidates for "electronic nose" applications (Staii et al., 2005). Both sensors based on SWNTs are integrated with CMOS circuitry by utilizing the DEP method which will be discussed along with the enhanced sensitivity obtained using DNA decoration of SWNT sensors.

9.2 Heterogeneous Integration

The approach for heterogeneous integration begins with a custom designed CMOS die (or wafer) that includes microelectrodes for nanotube assembly and operational amplifiers for sensor readout. Figure 9.1 shows a prototype of a CMOS chip composed of the microelectrodes for the assembly of SWNTs and the interface circuitry designed and fabricated using the AMI 0.5-µm CMOS process provided by MOSIS.

Next, a series of die-level mask-less post-CMOS fabrication steps are performed as explained in Figure 9.2 on two-dimensional (2D; between M3 and M3) and three-dimensional (3D; between M2 and M3) electrodes. SWNTs are assembled onto the microelectrodes realized by the top metal layer (M3) using dielectrophoresis. The integration of SWNTs onto CMOS circuitry is

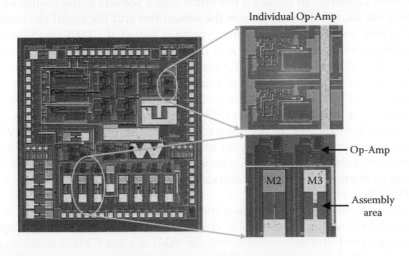

FIGURE 9.1
(See color insert.) CMOS chip designed and fabricated using the AMI 0.5-µm CMOS process provided by MOSIS.

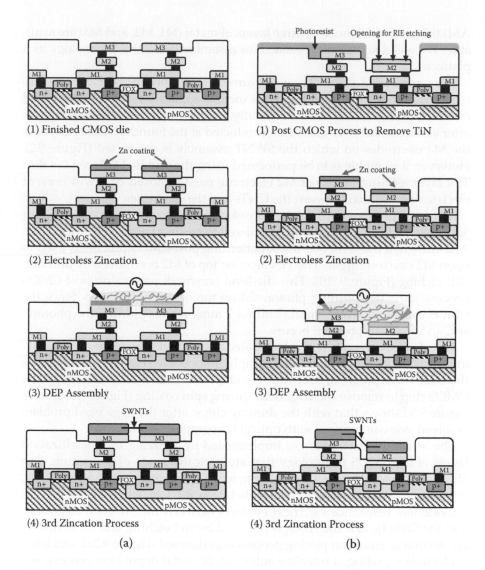

FIGURE 9.2
Fabrication processes for (a) two-dimensional electrodes; and, (b) three-dimensional electrodes.

thus realized utilizing a post-CMOS fabrication approach shown in the figure and described above.

9.2.1 Electroless Zincation

The SWNT-to-CMOS integration is achieved by a simple post-CMOS process. The CMOS chips include microelectrodes from the built-in metal layers to realize assembly of nanotubes. In our prototype implemented in the

AMI 0.5-μm CMOS process, three layers of metal (M1, M2, and M3) are available. We will discuss our approach for assembly using this technology as a platform.

The unpackaged chips, (2 mm × 2 mm size in our prototype; Figure 9.1) are received from the foundry. The chips include application-specific integrated circuitry (ASIC) that will be eventually used for readout of sensor signals after assembly. An over-glass etch conducted at the foundry readily exposes the M3 electrodes on which the SWNT assembly is conducted (Figure 9.2). However, if assembly is to be performed using the middle M2 layer, the thin TiN layer residing on top of M2 electrode must removed or it will prevent electrical connection between the CNTs and the electrodes.

In these experiments, M3 is the only electrode utilized for 2D assembly; for 3D assembly, both M2 and M3 are utilized to create a vertical electrical field. Accordingly, a die-level one-mask patterning post-CMOS process is used to open M2 electrode areas. The TiN layer on top of M2 is removed by utilizing RIE etching (Figure 9.2b1). This die-level one-mask patterning post-CMOS process starts by spinning photoresist on top of the CMOS chip. Since the CMOS chip is relatively small (2 mm × 2 mm), while spinning the photoresist, an edge bead problem occurs.

The photoresist is unevenly distributed and is thicker near the edges, making post-CMOS optical photolithography impossible (Figure 9.3a). To solve the edge bead problem, four dummy chips are placed on each side of the CMOS chip to remove the edge bead during spin coating (Figure 9.3b and c). Figure 9.3d shows that with the dummy chips, after the edge bead problem is solved, one can proceed with optical lithography.

The assembly electrodes are implemented by utilizing the metallization layers of the CMOS technology that are typically made of aluminum. The aluminum readily forms an insulating oxide upon exposure to atmosphere, resulting in poor electrical contacts between the CNTs and electrodes. To provide low resistance electrical contacts between the Al microelectrodes and the SWNTs, first the Al_2O_3 is removed by wet etching and immediately an electroless zincation plating process is performed (Figure 9.2a2 and b2).

Electroless plating, a selective autocatalytic metal deposition process, is a low-cost alternative deposition method compared to most vacuum deposition processes and does not require additional lithography steps. The process is fairly straightforward and only requires contact between the aqueous plating solution and the metal layers on the CMOS die (Lau, 1995).

The zincation process starts by immersing the CMOS chip in a 1M sodium hydroxide (NaOH) solution for 2 min to clean the surface of the chip, followed by a deionized water rinse. Next, the CMOS chip is placed into the zinc plating solution for 2 min to re-metalize (with zinc) the aluminum electrodes on the CMOS chip. The chip is then immersed in 30% nitric acid (HNO_3) for 15 sec to strip the granulated zinc deposits and form a more uniform zinc layer during the second zincation process.

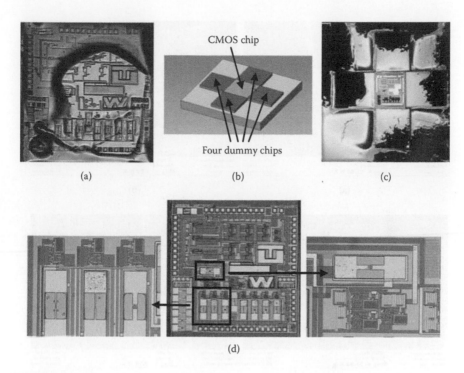

CMOS chip

Four dummy chips

(a)   (b)   (c)

(d)

FIGURE 9.3

(See color insert.) (a) Optical photograph of CMOS chip after spinning photoresist on top. Due to the edge bead problem, optical lithography on a CMOS chip is not possible. (b) Placing four dummy chips around the CMOS chip during photoresist application to solve the edge bead problem. (c) Optical photograph of CMOS chip surrounded by four dummy chips. (d) Optical photograph of CMOS chip after photolithography. Two inset images show close-ups of lithography patterns.

A deionized water rinse follows the second zincation process and completes the electroless zincation process. In this "displacement reaction" method, the top layer of the aluminum electrodes is replaced by the zinc ions, thus forming a thin zinc layer. This process prevents the underlying aluminum from being reoxidized (Datta and Merritt, 1999) and improves the contact resistance between the assembled SWNTs and aluminum microelectrodes.

9.2.2 Dielectrophoretic Assembly on CMOS Electrodes

After the electroless zincation process, SWNTs are assembled between the two zinc-coated metal electrodes utilizing the DEP assembly process. Commercially available aqueous suspensions of highly purified HIPCo-grown SWNTs (~1 to 2 nm diameter and 2 to 5 μm in length) are used with a concentration of 0.004 g/ml, which is diluted using deionized water. A 1-μL SWNT solution is first dispensed onto the microelectrodes of the CMOS chip and then an AC sinusoidal signal of 5 V_{pp} at a frequency of 10 MHz (Figures 9.2a.3

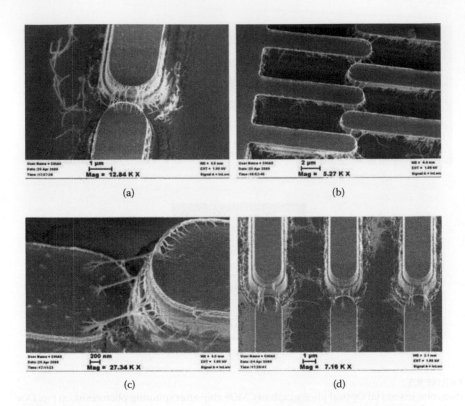

(a)   (b)

(c)   (d)

FIGURE 9.4
SEM micrographs of SWNTs assembled on two-dimensional single- (a) and multi-finger (b) electrodes and three-dimensional single- (c) and multi-finger (d) electrodes.

and b.3) is applied between the electrodes to perform the DEP assembly. After 30 sec of assembly, the solution is blow-dried with a nitrogen gun.

After assembly, the SWNTs are naturally attached onto the metal electrodes via weak van der Waals forces. The Figure 9.4 micrographs show SWNTs assembled on 2D and 3D electrodes. The two terminal resistance measurements from the SWNT bridges formed from this assembly are not stable.

To address the SWNT-to-metal-electrode contact problem, one approach is to utilize a focused ion beam (FIB) tool to selectively deposit metal patches onto areas where the SWNTs are attached to the metal electrodes. However, this is a serial and time consuming process. A third zincation process is used to stabilize the contact resistance between the SWNTs and the electrodes (Figures 9.2a.4 and b.4) and the zinc ions are directly deposited on top of the zinc-coated microelectrodes (Chen et al., 2009).

Due to the zincation process, stable readings are measured for 90 minutes as shown in Figure 9.5. The inset shows that the resistance of SWNTs assembled on to CMOS circuitry is about 27.86 ± 0.22 KΩ before and 12.56 ± 0.05

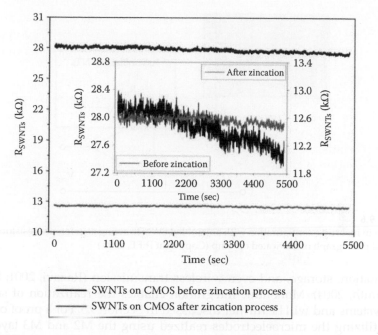

FIGURE 9.5

(See color insert.) Comparison of resistances of SWNTs assembled onto metal electrodes before and after the zincation process. The inset shows magnified resistance measurements from the SWNTs versus time. Before zincation, the resistance of the SWNTs assembled onto the CMOS circuitry was around 27.86 ± 0.22 KΩ measured for 90 min. After zincations, the measured resistance of SWNTs assembled onto CMOS circuitry was about 12.56 ± 0.05 KΩ during the same period. Therefore, the zincation process improves the SWNT-to-electrode contacts by decreasing contact resistance by about 54.9% and stabilizes the value with a smaller standard deviation.

KΩ after the zincation process. In short, the final zincation process reduces the contact resistance by about 54.9%, and this resistance value is found to be stable over time.

9.3 SWNT Thermal Sensors Implemented on CMOS Readout Circuitry

Measurement of temperature can be utilized to determine physical parameters such as heat energy or specific heat capacity. CNTs, due to their miniature size, low mass density, and fast response, are promising materials for temperature sensors.

The main advantage of monolithic integration in CMOS technology is the availability of circuitry that readily facilitates readout, signal processing,

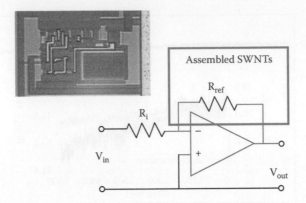

FIGURE 9.6
Op-amp in inverting configuration with assembled SWNTs acting as reference resistors. Inset is optical photograph of fabricated op-amp (Copyright IEEE, 2008.)

computation, storage, and even wireless transmission (Razavi, 2001; Sedra and Smith, 2004). Monolithic integration enables the realization of single-chip systems and will be further discussed in Section 9.5. For a proof of concept utilizing the microelectrodes realized using the M2 and M3 layers of the CMOS technology, the SWNTs are assembled onto these electrodes as feedback structures around the high gain Miller-compensated single-ended operational amplifier (Razavi, 2001). The SWNTs assembled onto these electrodes form the feedback resistance of the amplifier in the inverting configuration (R_{ref}) as shown in Figure 9.6.

9.3.1 Circuit Characterization

In the prototype, since the resistance of the assembled SWNTs is measured as ~44K Ω, an external resistor (R_i) of 22K Ω is used in the circuit to realize a gain of –2. Measured input and output signals from the inverting op-amp with SWNTs as the reference resistor are displayed in Figures 9.7a and b and the gains of the inverting amplifier are measured as about –1.95 and –1.98, respectively. This value agrees well with the calculated gain (gain = $-R_{ref}/R_i$ = –2).

Furthermore, to measure the response of the device, an output signal waveform corresponding to a unit step function is fed into the operational amplifier with the bundled SWNTs placed as the feedback resistor as shown in Figure 9.8. Using the approximate relationship between the time constant and cutoff frequency (Liu et al., 1999; Lai, Fung et al., 2006) is given as:

$$f_c = 1/(1.5 \times t_c) \tag{9.1}$$

where f_c is the cutoff frequency, t_c is the time constant of the response, and the cutoff frequency of the system is estimated to be ~9.5 kHz.

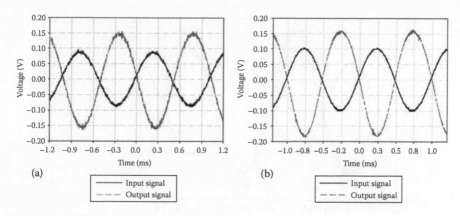

(a)

(b)

FIGURE 9.7

(See color insert.) Measured input and output signals from inverting op-amp with SWNTs as reference resistors. (a) Small-signal ac gain measurement from CNTs on two-dimensional electrodes. $R_i = 22$ KΩ; measured output gain is about –1.95. (b) Three-dimensional electrodes with small-signal ac gain measured about –1.98 (multi-finger electrodes). (Copyright IEEE, 2008.)

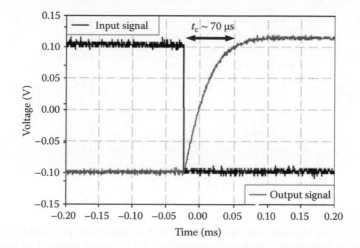

FIGURE 9.8

(See color insert.) Frequency response from inverting op-amp with SWNTs as feedback resistors. Cutoff frequency is estimated at 9.5 kHz. (Copyright IEEE, 4/2009.)

9.3.2 Thermal Sensor Applications

MWNT-based thermal sensors have been reported to be capable of performing very low power thermal sensing about 1000 times lower than conventional micromachined polysilicon-based thermal sensors (Chan et al., 2004). Utilizing a low voltage DEP assembly process, a SWNT based thermal sensor can be implemented on CMOS circuitry.

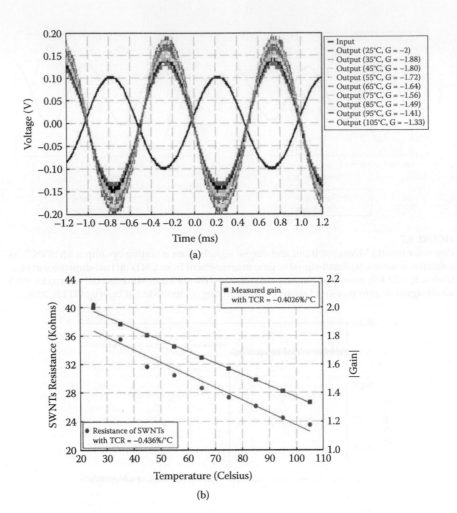

FIGURE 9.9

(See color insert.) (a) Measured output voltage of CMOS amplifier with integrated SWNT feedback resistor in response to variations in temperature. (b) Measured temperature response of SWNT thermal sensor. The measured temperature coefficient of resistance (TCR) calculated from the operational amplifier output is –0.4%/°C and agrees well with the measured TCR of –0.43%/°C obtained from assembled SWNTs on plain electrodes.

The SWNTs are assembled between the electrodes that serve as a feedback resistor of the inverting amplifier. Next, the CMOS chip is placed on a probe station with a heatable chuck connected to external biasing resistors that are maintained outside the chuck at ambient temperature. The temperature of the chuck is then swept from 25 to 105°C in increments of 10°C. The input signal and the measured output voltage of the amplifier with respect to temperature are shown in Figure 9.9a.

As reported by Nguyen and colleagues (2006), SWNTs are sensitive to temperature and their resistance decreases corresponding to an increase in temperature. As expected, increasing the temperature of the chip caused the resistance of the SWNTs to decrease (Figure 9.9a) and led to a reduction in amplifier gain compared to the value measured at room temperature.

In these experiments, a bundle of SWNTs are also assembled onto a set of electrodes (not connected to the amplifier) as a control group. At each temperature, the resistance of the SWNTs attached to the electrodes is measured first. Then the voltage gain of the inverting amplifier with an integrated SWNT resistor is measured and both data are plotted in Figure 9.9b. Again, as the temperature of the chip increased, the measured output voltage of the amplifier (Figure 9.9b) decreased—agreeing with Nguyen's results. The temperature coefficient of resistance (TCR) of the SWNT thermal sensors is calculated from the gain of the amplifier that had a value of –0.4%/°C and agreed well with the measured TCR of –0.43%/°C from the assembled SWNTs with no amplifier connections.

9.4 ss-DNA Decorated SWNTs for Chemical Sensor Applications

Real-time monitoring of environmental gases is becoming increasingly important for analyzing local and global trends and the complicated side effects of air pollution. Accordingly, miniature, low power and ultra-sensitive gas sensing systems are urgently needed.

Carbon nanotubes (CNTs) are excellent materials for ultra-sensitive gas and chemical sensors in environmental monitoring and low-power lab-on-a-chip systems due to their hollow geometry at the nanoscale with large surface area-to-volume ratios that give rise to very high gas absorptive capacity and electrical mobility. Moreover, functionalization of CNTs with polymers and biomolecular complexes such as single-stranded DNA (ss-DNA) is shown to enhance the specificity and sensitivity of the CNT-based sensors.

Among nucleic acid biomolecules, ss-DNA is an intriguing candidate as a molecular recognition layer because it can bind to SWNTs through non-covalent π-π stacking interactions (Zheng et al., 2003a; Johnson et al., 2008), and can be engineered to achieve affinity to a variety of molecular targets (Patel et al., 1997; Breaker, 2004).

According to some reports, SWNTs decorated with nanoscale layers of ss-DNA display remarkable chemical sensing capabilities, making them promising for "electronic nose" applications (Staii et al., 2005). Label-free detection of DNA hybridization can also be achieved optically (Jeng et al., 2006) and electrically (Star et al., 2006; Tang et al., 2006; Martinez et al., 2009) on ss-DNA functionalized SWNTs.

Furthermore, ss-DNA can be used to sort chirality of SWNTs (Zheng et al., 2003b), and the DNA SWNT may provide a means for ultrafast DNA sequencing (Meng et al., 2007). Additionally, most current CNT sensors are attached to substrates (Wong et al., 1997; Peng and Cho, 2000; Snow et al., 2005). Sensors based on suspended CNTs maximize the active surface area for molecular functionalization and also eliminate nanotube–substrate interactions that are desirable for various nanotube-based sensing devices (Kim et al., 2001; Lu et al., 2006).

The integration of SWNTs onto functional electronic circuitry via a post-CMOS fabrication process is described in the previous section. Double zincation is performed to remove the Al_2O_3 layer prior to DEP assembly of SWNTs onto CMOS circuitry. Then a third zincation process is performed to ensure stable electrical contacts between the SWNTs and metal electrodes (Chen et al., 2009). In this section, successful decoration of ss-DNA onto SWNTs assembled on CMOS circuitry is described. Enhanced gas sensing response due to the decoration of ss-DNA on SWNTs with both passive (simple electrodes with no amplifier connections) and active (connected to amplifier) integration onto CMOS circuitry is then detailed. Finally, the sensitivity enhancement versus different ss-DNA sequences (Chen et al., 2010) is explored. The ss-DNA sequences chosen for the experiments are:

Sequence 1   5′ to 3′ GAG TCT GTG GAG GAG GTA GTC

Sequence 2   5′ to 3′ GTG TGT GTG TGT GTG TGT GTG TGT

Sequence 3   5′ to 3′ CTT CTG TCT TGA TGT TTG TCA AAC

Sequences 1 and 3 were chosen based on previous fluorescence (White and Kauer, 2004) and electrical (Staii et al., 2005) measurements showing sensitivity to various volatile compounds. Sequence 2 was chosen based on a previous report that it can readily wrap around SWNTs and result in effective sorting of the SWNTs (Staii et al., 2005).

The oligonucleotides were obtained from Invitrogen (Carlsbad, CA) and diluted in deionized water to make a stock solution of 100 μmoles. After odor responses of the bare SWNT sensors are measured, a 2-μL drop of ss-DNA solution is applied to SWNTs for 45 min in a humid environment and then dried with a nitrogen stream (Staii et al., 2005).

Figure 9.10a is a schematic of assembled ss-DNA decorated SWNTs on CMOS circuitry. The I–V measurements (HP 4155A Semiconductor Parameter Analyzer) from the assembled SWNTs on the CMOS chip before (34.67 KΩ) and after (54.44 KΩ) ss-DNA decoration are shown in Figure 9.10b.

The ss-DNA decoration on SWNTs is found to increase the resistance of SWNTs by about 57.02%. This may be due to weak carrier scattering by the

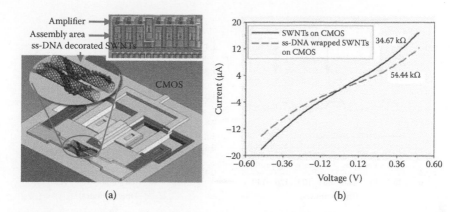

(a)　　　　　　　　　　　　　　　　　(b)

FIGURE 9.10
(See color insert.) (a) Optical photograph of CMOS chip and schematic of DNA decorated SWNT sensors integrated onto CMOS circuitry. (b) I–V characterization of SWNTs on CMOS circuitry before and after ss-DNA decoration. The decoration was found to increase the resistance while maintaining sufficient conduction.

molecular coating and displacement of a small number of SWNTs during the decoration of ss-DNA and/or the drying process.

Increasing interest surrounds assemblies of suspended nanostructures for sensing applications since they expose more sensor surface. Selecting the top metal layer (M3) as the integration layer allows the realization of suspended structures since M3 layer is 1.7 μm above the substrate. Figures 9.11a and b are typical side view micrographs of suspended SWNTs integrated on CMOS circuitry before and after ss-DNA decoration, respectively.

Chemical sensing tests are performed on both bare SWNTs and ss-DNA decorated SWNTs assembled onto microelectrodes on CMOS chips. These microelectrodes are stand-alone test structures and not connected to the on-chip amplifiers; hence they are utilized for the initial characterization of the sensors.

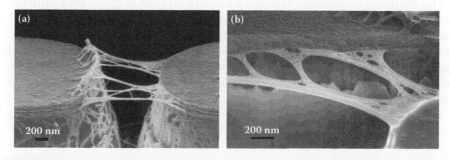

FIGURE 9.11
(a) Typical side and top view SEMs of suspended SWNTs assembled between microelectrodes realized by the M3 layer of the CMOS chip. (b) Close-up of ss-DNA decorated SWNTs on CMOS circuitry.

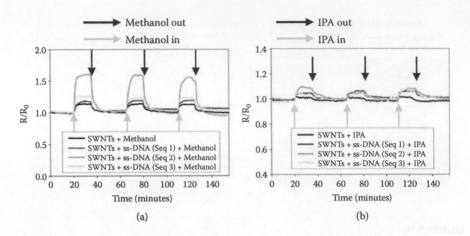

FIGURE 9.12

(See color insert.) Change in sensor resistance upon chemical vapor exposure. Resistances are normalized to the value when exposed to air. (a) Bare SWNTs respond to methanol vapor (black line) with about 13.41 ± 1.03% increase in resistance. The SWNTs decorated with ss-DNA show enhanced response to methanol (red, green and yellow lines for sequences 1, 2, and 3, respectively) with resistance increases of about 18.43 ± 0.81, 58.02 ± 3.36, and 24.7 ± 1.34%, respectively. (b) Bare SWNTs respond to isopropanol alcohol vapor (black line) with about 3.23 ± 0.50% increase in resistance. The same SWNTs decorated with ss-DNA show enhanced response to isopropanol alcohol (red = sequence 1; green = sequence 2; yellow = sequence 3) with resistance increases of about 5.65 ± 0.30%, 11.25 ± 0.33%, and 7.38 ± 0.49%.

In the experiments, two chemical vapors (methanol and isopropanol alcohol) are used to characterize the responses of the SWNT gas sensors. The saturation vapor pressure at 20°C is 97.48 torr for methanol and 33 torr for isopropanol alcohol. The experiments are performed by measuring their resistances under a probe station (SUSS, MicroTec PM5) using a multimeter (HP 34401A) with LabView control.

The chemical responses of the bare SWNTs on CMOS chips are evaluated first. The resistance of the sensors is measured for 20 min under ambient conditions to reveal the stability of the devices. Then the sensors are exposed to chemical vapors for 15 min, during which their resistance increases and reaches a stable value. Finally, the vapor source is removed and the sensors are allowed to rest in ambient conditions for 30 min for their resistance to reach the initial value prior to exposure to chemical vapors.

The three cycles shown in Figure 9.12 demonstrate the reproducibility and stability of the sensing response. The resistances are normalized to the value when exposed to air (R_0 ~20 kΩ in this case). Under methanol vapor (Figure 9.12a [black data]), the increase in the measured resistance of SWNTs is about 13.41 ± 1.03%, suggesting that the methanol vapor near the SWNTs changed the electrostatics in a way that decreased the hole carrier density in the SWNTs (Philip et al., 2003).

Next, three different sequences of ss-DNA are decorated onto the SWNTs to enhance their chemical sensing properties. Exposing ss-DNA decorated SWNT sensors to methanol vapor increased their resistance by about 18.43 ± 0.81%, 58.02 ± 3.36%, and 24.7 ± 1.34% for sequences 1, 2, and 3, respectively, as shown in Figure 9.12a. The enhanced chemical sensing property of ss-DNA-functionalized SWNT sensors compared with that of bare SWNTs was measured at about 37.43, 332.66, and 84.19% for sequences 1, 2, and 3, respectively, under methanol vapor.

Next, bare and ss-DNA decorated SWNTs are exposed to isopropanol vapor. When bare SWNTs are exposed to isopropanol alcohol vapor, an increase in resistance of about 3.23 ± 0.5% is measured (Figure 9.12b [black data]). The change in resistance of bare SWNTs under isopropanol alcohol vapor is less than that when they are exposed to methanol vapor. This observation is due to the differences in the chemical nature of the solvents and the different saturation vapor concentrations of the two chemicals.

Three different sequences of ss-DNA decorated SWNTs are exposed to isopropanol alcohol vapor and their responses are measured. Measurements shown in Figure 9.12b indicate that under isopropanol alcohol vapor, the resistance of the sample increased by about 5.65 ± 0.3%, 11.25 ± 0.33%, and 7.38 ± 0.49% and the measured chemical sensing response of ss-DNA functionalized SWNTs sensors are enhanced by about 73.84, 248.30, and 128.48% for sequences 1, 2, and 3, respectively.

Both data show that the ss-DNA decorated SWNT sensors reach a stable resistance value within a few minutes after exposure to the analyte vapors. Similarly, after removal of the vapor source, the resistance of the sensors drops down to half (50%) of the highest resistance value within a minute before taking a longer time to return to its initial value. Comparing our sensor's response with other investigators' results, we found that our response and recovery times were in the order of a few minutes—longer than the few seconds reported by Staii et al. (2005).

One possible reason is that the sensing response and recovery time include the time required for the chemical analytes to be delivered to the sensors. In our experimental setup, the chemical vapors are delivered to the SWNTs through diffusion in the chamber; where Staii et al. delivered the chemical analytes to the devices through a carrier gas with a 0.1 mL/sec flow rate. The dimensions of our chamber are about 10 cm × 10 cm × 2.5 cm and the chemical analyte inside a container is placed in the center of the cell. The diffusion coefficient is about 0.2 cm^2/sec for methanol (Mrazek et al., 1968) and 0.0959 cm^2/sec for IPA (http://www.gsi-net.com/en/publications/gsi-chemical-database/single/324.html), respectively.

The diffusion time in one dimension can be estimated by

$$L = 2\sqrt{Dt}$$

where L is the diffusion length, D is the diffusion coefficient, and t is the diffusion time (calculated around 31.2 sec and 65.17 sec for methanol and IPA, respectively). From this calculation, it is clear that the diffusion time for the chemical analytes to reach the walls of the cell is in the range of 30 to 60 sec in one dimension. For a three-dimensional cell, the diffusion time should be longer.

A second possible reason is that these devices consist of small bundles of SWNTs and it is reasonable to assume that it would take a longer time for the chemical analytes to reach the saturation sensing responses for all tubes.

Based on the time-constant calculations, the time constants for different DNA sequences are sequence dependent. For instance, sequence 2 (highest response) has a higher time constant compared to sequences 1 and 3, possibly because the binding between DNA sequence 2 and gas molecules is stronger than the bindings between sequences 1 and 3.

Under methanol vapor, the changes in sensing response are 18.43, 58.02, and 24.7% for sequences 1, 2, and 3, respectively and the calculated time constants are 1.03, 2.36, and 1.74, respectively. These data suggest that the binding between sequence 2 and gas molecules is the strongest among the three ss-DNA sequences used.

A similar trend is observed when the decorated SWNT sensors are exposed to IPA vapour; the SWNTs decorated with sequence 2 exhibited the maximum response (Table 9.1). Because of its maximum response, several experiments comparing sensor response versus vapor pressure were conducted and the measurements are shown in Figure 9.13.

To demonstrate SWNT gas sensors with active circuitry, we next assembled ss-DNA decorated SWNTs onto microelectrodes connected to operational-amplifiers (op-amps) on CMOS chips. The SWNTs are assembled onto the feedback path (R_{ref}) of a Miller-compensated single-ended op-amp (Razavi, 2001) configured as an inverting amplifier with an external resistor (R_i) connected to the circuit. Measured input and output signals from

TABLE 9.1

Vapor Testing of ss-DNA Decorated SWNTs

Odor		Bare SWNTs	SWNTs + Sequence 1	SWNTs + Sequence 2	SWNTs + Sequence 3
Methanol (97.48 Torr)	$\Delta R/R_0$ (%)	13.41 ± 1.03	18.43 ± 0.81	58.02 ± 3.36	24.7 ± 1.34
	Improvement (%)	–	37.43	332.66	84.19
	Time constant (min)	2.85	1.03	2.36	1.74
IPA (33 Torr)	$\Delta R/R_0$ (%)	3.23 ± 0.50	5.65 ± 0.30	11.25 ± 0.33	7.38 ± 0.49
	Improvement (%)	–	73.84	248.30	128.48
	Time constant (min)	3.02	2.56	10.03	2.42

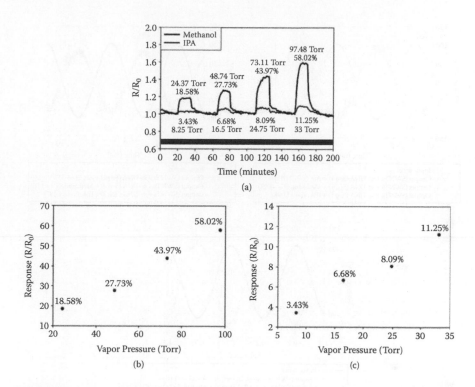

FIGURE 9.13
Measurements of sensor responses versus vapor pressures of ss-DNA (sequence 2) decorated SWNTs upon exposure to IPA and methanol. (a) Time versus response measurement. (b) Vapor pressure versus response measurement for methanol. (c) Vapor pressure versus response measurement for IPA.

the inverting op-amp with three different sequences of ss-DNA decorated SWNTs in the feedback path is displayed in Figure 9.14. The gains of the inverting amplifiers were –1.02, –1.00, and –1.09 for sequences 1, 2, and 3, respectively.

The SWNT sensors on the CMOS chip were next exposed to various gaseous vapor conditions and their gas sensing properties characterized. We expected that as the sensors were exposed to vapors of different gases, their resistance would increase and lead to an increase in amplifier gain compared to the value measured under ambient conditions. Figure 9.14 shows the responses of the ss-DNA decorated SWNT sensors to gas vapors (methanol and isopropanol alcohol) measured from the output of the operational amplifier.

Table 9.2 shows that the measured gains of the inverting amplifiers when exposed to methanol vapor are –1.23, –1.55, and –1.43 for sequences 1, 2, and 3 (20.71, 55.00, and 31.19% higher than values measured under ambient conditions), respectively. A similar trend was observed during chip exposure

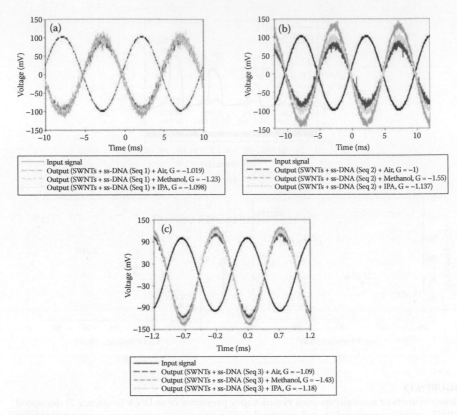

FIGURE 9.14

(See color insert.) Measured ac amplifier gain in response to variations in gas vapors. Corresponding to different gas vapors, the gain decreased according to $-R_{SWNT}/R_i$. (a) For sequence 1 ss-DNA decorated SWNT sensors, during exposure to methanol vapor, the gain of the inverting amplifier increased by about 20.71%. During exposure to isopropanol alcohol vapor, the gain increased by 7.75%. (b) For sequence 2 ss-DNA decorated sensors, during exposure to methanol vapor, the gain of the inverting amplifier increased by about 55.00%. During exposure to isopropanol alcohol vapor, the gain is increased by 13.70%. (c) For sequence 3 ss-DNA decorated sensors, during exposure to methanol vapor, the gain of the inverting amplifier increased by about 31.19%. During exposure to isopropanol alcohol vapor, the gain increased by 8.25%.

to isopropanol alcohol vapor. Under IPA vapor, the measured gains of the inverting amplifiers were found to be −1.10, −1.14, and −1.18 for sequences 1, 2, and 3 (7.75, 13.70, and 8.25% higher compared to measurements from the same sample under ambient conditions), respectively. The amplifier measurements agree well with the measurements obtained from the SWNT sensors attached to plain electrodes not connected to active amplifiers. These data validate successful integration of the ss-DNA decorated SWNT sensors onto CMOS circuitry.

TABLE 9.2

Gas Sensing Results from CNTs Connected to On-Chip Amplifiers

		SWNTs + Sequence 1			SWNTs + Sequence 2			SWNTs + Sequence 3		
Odor	Vapor Pressure (Torr)	Gain	% G/G_0	% R/R_0	Gain	% G/G_0	% R/R_0	Gain	% G/G_0	% R/R_0
Air		−1.02	–	–	−1.00	–	–	−1.09	–	–
Methanol	97.48	−1.23	20.71	18.43	−1.55	55.00	58.02	−1.43	31.19	24.7
IPA	33	−1.10	7.75	5.65	−1.14	13.70	11.25	−1.18	8.25	7.38

9.5 CMOS Integrated Circuits: Scaling and Possibilities

The operational amplifier implemented in the prototype is only one example of the possibilities afforded by monolithic integration for readout. It demonstrated access to a single electrode site where the SWNTs are assembled. However one can envision a single-chip solution for dense sensor arrays of functionalized SWNTs for a true "single-chip electronic nose."

Figure 9.15 is a conceptual block diagram of a CMOS sensor array of $2^N \times 2^M$ interface electrodes. The architecture is similar to that of a DRAM (dynamic random access memory) architecture used in a digital memory. It consists of a row and column address decoder, timing control logic, read and write logic, sense amplifiers, function generators, and microcontrollers (or DSPs) with built-in wireless transmitters and receivers.

Unlike the DRAM cell, each cell in the sensor array operates on analog signals. The cell is nothing but electrodes on which SWNTs are assembled, with two transistors for read and write access controlled by the row and column decoders. The row decoder takes N-bit address words as inputs and provides access to any one of the 2^N rows at any given time. The column decoder enables simultaneous read and write operation to L of 2^M columns. Timing control logic updates the address input to the row decoder for a time-multiplexed and sequential access to all rows in an array. It also provides pulse-width control to set the exposure time of each electrode in an array.

The frame rate of the sensor array is the maximum rate at which an electrode array is read. An L:1 analog multiplexer at the output of the column array provides a serial output fed to a correlated double sampling (CDS) switched capacitor amplifier. The CDS amplifier removes the background fixed-pattern noise and provides programmable gain control for amplification of the electrode signal (Johns and Martin, 1997; Razavi, 2001). A high resolution delta-sigma A/D converter then digitizes this output and sends it to an on-board microcontroller or DSP for further analysis and processing. The resolution of the A/D converter (ADC) depends on the desired

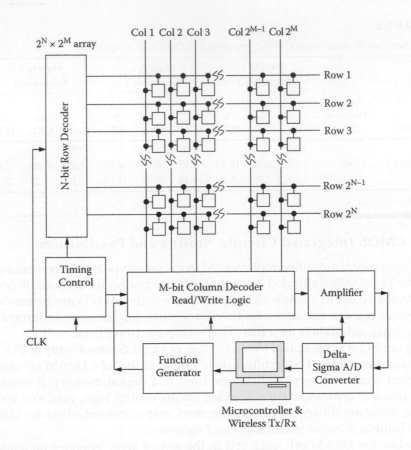

FIGURE 9.15
Conceptual single-chip realization of electronic nose, showing dense arrays of electrode sites for SWNT assembly with built-in CMOS circuitry for row and column decoder, amplification, A/D conversion, microcontroller with wireless transmitter and receiver, and function generator. Monolithic integration with CMOS technology benefits from rapid scaling.

signal-to-noise ratio (SNR) from the system. To realize high SNR, a high order sigma-delta ADC can be used. For example, a typical 16-bit sigma-delta ADC operating at a 100-KHz sampling rate provides a SNR of more than 90 dB.

The function generator controlled by the microcontroller/DSP provides controlled voltage stimulus for different amplitudes and frequencies to a given sensor site. These function generators can provide simultaneous stimulus to a group of sites in a selected row or column to facilitate parallel assembly. A given row is selected by setting an appropriate row address, and a group of L electrodes within the row can be selected by setting an appropriate column address. Timing logic controls the pulse width of the stimulus signal to control exposure time. More electrodes can be accessed simultaneously by increasing L at the cost of increased area and power consumption.

Thus a chip that facilitates both assembly and readout of sensor output from a dense array is possible. Continuous scaling of the CMOS technology makes it possible to achieve even higher density of sensor sites on silicon. Moreover, creative circuit ideas for low power dissipation such as using transistors in subthreshold regions (Trakimas and Sonkusale, 2008a) for amplification and novel architectures for A/D conversion (Agarwal et al., 2005; Trakimas and Sonkusale, 2008b) promise smart, adaptive, integrated sensor devices for a variety of application needs.

9.6 Conclusions and Future Directions

In summary, a key enabling technology for the heterogeneous integration of SWNTs onto CMOS integrated circuits for the next generation of nanosensors has been described in this chapter. The technique is simple, low cost, and utilizes a low-voltage dielectrophoretic assembly process directly on CMOS chips.

Novel three-step electroless zincation plating after assembling SWNTs is described and stable electrical contacts are achieved between the SWNTs and the metal electrodes on CMOS chips. We demonstrate a robust CNT temperature sensor utilizing such a platform on which temperature sensing is accurately measured by an amplifier whose gain depends on the temperature coefficient of the SWNTs.

Successful integration of ss-DNA decorated SWNTs onto functional CMOS circuitry for chemical sensing applications is also presented. The chemical sensing properties are tested with both bare SWNTs and ss-DNA attached SWNTs on CMOS chips with and without amplifier connections. Bare SWNTs were sensitive to the vapors of methanol and isopropanol; increases in resistance of 13.41 ± 1.03% and 3.23 ± 0.5%, respectively, were measured. The ss-DNA decoration enhanced the sensing responses of nanotube sensors to methanol by 37.43, 332.66, and 84.19% for three different DNA sequences. With DNA decoration, the enhancements in the sensing response of the nanotube sensors to isopropanol were 73.84, 248.30, and 128.48% for specific DNA sequences.

The initial responses and the recovery rates of the bare and ss-DNA decorated devices were fairly fast—within a few minutes of vapor exposure and removal. Moreover, ss-DNA decorated SWNTs successfully integrated onto CMOS circuitry and their sensing responses with and without connections to on-chip amplifiers were in agreement with each other.

The gain of the inverting amplifier integrated with ss-DNA decorated SWNTs increased by about 20.71, 55.00, and 31.19% for sequences 1, 2, and 3, respectively, when the devices were exposed to methanol. The results were 7.75, 13.70, and 8.25%, respectively after exposure to isopropanol vapors.

We described ss-DNA decoration and integration of SWNT sensors onto CMOS circuitry and included specific details for sensing applications. The methodology presented is simple and versatile with potential applications to achieve high sensitivity and low power CNT-based biological and chemical sensor arrays for miniature environmental and health monitoring applications.

Acknowledgments

This work was supported by the National Science Foundation Nanoscale Science and Engineering Center (NSEC) for High-Rate Nanomanufacturing (NSF Grant 0425826 and NSF EAGER Grants (0947874 and 0947781). The authors would like to thank Nantero Inc. for supplying us with CMOS grade SWNT solution and Professor Jeffrey A. Hopwood of Tufts University for valuable suggestions.

References

Agarwal, A., Kim, Y.B., and Sonkusale, S. 2005. Low power current mode ADC for CMOS sensor IC. Paper presented at IEEE International Symposium on Circuits and Systems.

Breaker, R.R. 2004. Natural and engineered nucleic acids as tools to explore biology. _Nature_, 432, 838–845.

Chan, R.H.M., Fung, C.K.M., and Li, W.J. 2004. Rapid assembly of carbon nanotubes for nanosensing by dielectrophoretic force. _Nanotechnology_, 15, S672–S677.

Chen, C.L., Agarwal, V., Sonkusale, S. et al. 2009. The heterogeneous integration of single-walled carbon nanotubes onto complementary metal oxide semiconductor circuitry for sensing applications. _Nanotechnology_, 20, 225–302.

Chen, C.L., Lopez, E., Jung, Y-J. et al. 2008. Mechanical and electrical evaluation of parylene-C encapsulated carbon nanotube networks on a flexible substrate. _Appl. Phys. Lett._, 93, 93–109.

Chen, C.L., Yang, C.F., Agarwal, V. et al. 2010. DNA-decorated carbon-nanotube-based chemical sensors on complementary metal oxide semiconductor circuitry. _Nanotechnology_, 21, 095504.

Close, G.F., Yasuda, S., Paul, B. et al. 2008. A 1-GHz integrated circuit with carbon nanotube interconnects and silicon transistors. _Nano Lett._, 8, 706–709.

Cui, Y. and Lieber, C.M. 2001. Functional nanoscale electronics devices assembled using silicon nanowire building blocks. _Science_, 291, 851–853.

Datta, M. and Merritt, S.A. 1999. Electroless remetallization of aluminum bond pads on CMOS driver chip for flip chip attachment to vertical cavity surface emitting lasers and VSCELs. *IEEE Trans. Compon. Packag. Technol.*, 22, 229–236.

Fung, C.K.M., Wong, V.T.S., Chan, R.H.M. et al. 2004. Dielectrophoretic batch fabrication of bundled carbon nanotube thermal sensors. *IEEE Trans. Nanotechnol.*, 3, 395–403.

Green, N.G., Morgan, H., and Milner, J.J. 1997. Manipulation and trapping of submicron bioparticles using dielectrophoresis. *J. Biochem. Biophys. Methods*, 35, 89–102.

Guo, T., Nikolaev, P., Rinzler, A.G. et al. 1995. Self assembly of tubular fullerenes. *J. Phys. Chem.*, 99, 10694–10697.

Hermanson, K.D., Lumsdon, S.O., Williams, J.P. et al. 2001. Dielectrophoretic assembly of electrically functional microwires from nanoparticle suspensions. *Science*, 294, 1082–1086.

Hines, D.R., Mezhenny, S., Breban, M. et al. 2005. Nanotransfer printing of organic and carbon nanotube thin-film transistors on plastic substrates. *Appl. Phys. Lett.*, 86, 163101.

Iijima, S. 1991. Helical microtubules of graphitic carbon *Nature*, 354, 56–58.

Jeng, E.S., Moll, A.E., Roy, A.C. et al. 2006. Detection of DNA hybridization using the near-infrared band-gap fluorescence of single-walled carbon nanotubes. *Nano Lett.*, 6, 371–375.

Johns, D.A. and Martin, K. 1997. *Analog Integrated Circuit Design*. New York: John Wiley & Sons.

Johnson, R.R., Johnson, A.T., and Klein, M.L. 2008. Probing the structure of DNA-carbon nanotube hybrids with molecular dynamics. *Nano Lett.*, 8, 69–75.

Khanduja, N., Selvarasah, S., Chen, C.L. et al. 2007. Three-dimensional controlled assembly of gold nanoparticles using a micromachined platform. *Appl. Phys. Lett.*, 90, 83–105.

Kim, P., Shi, L., Majumdar, A. et al. 2001. Thermal transport measurements of individual multiwalled nanotubes. *Phys. Rev. Lett.*, 87, 215502.

Kong, J., Franklin, N.R., Zhou, C. et al. 2000. Nanotube molecular wires as chemical sensors. *Science*, 287, 622–625.

Lai, K.W.C., Fung, C.K.M., Wong, V.T.S. et al. 2006. Development of an automated microspotting system for rapid dielectrophoretic fabrication of bundled carbon nanotube sensors. *IEEE Trans. Automat. Sci Eng.*, 3, 218–227.

Lau, J. H. 1995. *Flip Chip Technologies*. New York: McGraw-Hill.

Li, W.Z., Xie, S.S., Qian, L.X. et al. 1996. Large-scale synthesis of aligned carbon nanotubes. *Science*, 274, 1701–1703.

Liu, C., Huang, J.B., Zhu, Z. et al. 1999. A micromachined flow shear-stress sensor based on thermal transfer principle. *J. Microelectromech. Syst.*, 8, 90–99.

Lu, J., Kopley, T., Dutton, D. et al. 2006. Generating suspended single-walled carbon nanotubes across a large surface area via patterning self-assembled catalyst-containing block copolymer thin films. *J. Phys. Chem B*, 110, 10585–10589.

Martínez, M.T., Tseng, Y.C., Ormategui, N. et al. 2009. Label-free DNA biosensors based on functionalized carbon nanotube field effect transistors. *Nano Lett.*, 9, 530–536.

Meng, S., Maragakis, P., Papaloukas, C. et al. 2007. DNA nucleoside interaction and identification with carbon nanotubes. *Nano Lett.*, 7, 45–50.

Mrazek, R.V., Wicks, C.E., and Prabhu, K.N.S. 1968. Dependence of diffusion coefficient on composition in binary gaseous systems. *J. Chem. Eng. Data*, 13, 508–510.

Nguyen, H.Q. and Huh, J.S. 2006. Behavior of single-walled carbon nanotube-based gas sensors at various temperatures of treatment and operation. *Sensors Actuators B*, 117, 426–430.

Patel, D.J., Suri, A.K., Jiang, F. et al. 1997. Structure, recognition and adaptive binding in RNA aptamer complexes. *J. Mol. Biol.*, 272, 645–664.

Peng, S. and Cho, K. 2000. Chemical control of nanotube electronics. *Nanotechnology*, 11, 57–60.

Philip, B., Abraham, J.K., Chandrasekhar, A. et al. 2003. Carbon nanotube/PMMA composite thin films for gas-sensing applications. *Smart Mater. Struct.*, 12, 935–939.

Pohl, H.A. 1978. *Dielectrophoresis: The Behavior of Neutral Matter in Nonuniform Electric Fields*. New York: Cambridge University Press.

Razavi, B. 2001. *Design of Analog CMOS Integrated Circuits*. New York: McGraw-Hill.

Sedra, A. S. and Smith, K.C. 2004. *Microelectronic Circuits*, 5th ed. New York: Oxford University Press.

Snow, E.S., Perkins, F.K., Houser, E.J. et al. 2005. Chemical detection with a single-walled carbon nanotube capacitor. *Science*, 307, 1942–1945.

Song, J.W., Kim, J., Yoon, Y.H. et al. 2008. Inkjet printing of single-walled carbon nanotubes and electrical characterization of the line pattern. *Nanotechnology*, 19, 095702.

Spotnitz, M.E., Ryan, D., and Stone, H.A. 2004. Dip coating for the alignment of carbon nanotubes on curved surfaces. *J. Mater. Chem.*, 14, 1299–1302.

Staii, C., Johnson, A.T., Jr., Chen, M. et al. 2005. DNA-decorated carbon nanotubes for chemical sensing. *Nano Lett.*, 5, 1774–1778.

Star, A., Tu, E., Niemann, J. et al. 2006. Label-free detection of DNA hybridization using carbon nanotube network field effect transistors. *PNAS*, 103, 921–926.

Tang, X., Bansaruntip, S., Nakayama, N. et al. 2006. Carbon nanotube DNA sensor and sensing mechanism. *Nano Lett.*, 6, 1632–1636.

Trakimas, M. and Sonkusale, S. 2008a. A 0.8-V asynchronous ADC for energy constrained sensing applications. Paper presented at Custom Integrated Circuits Conference, San Jose, CA.

Trakimas, M. and Sonkusale, S. 2008b. A 0.5-V bulk input OTA with improved common-mode feedback for low-frequency filtering applications. *Analog Integr. Circuits Signal Process.*, 59, 83–89.

White, J.E. and Kauer, J.S. 2004. Intelligent electro-optical nucleic acid-based sensor array and method for detecting volatile compounds in ambient air. U.S. Patent 7,062,385.

Wong, E.W., Sheehan, P.E., and Lieber, C.M. 1997. Nanobeam mechanics: elasticity, strength and toughness of nanorods and nanotubes. *Science*, 277, 1971–1975.

Xiong, X., Chen, C.L., Ryan, P. et al. 2009. Directed assembly of high density single-walled carbon nanotube patterns on flexible polymer substrates. *Nanotechnology*, 20, 295–302.

Zheng, M., Jagota, A., Semke, E.D. et al. 2003a. DNA-assisted dispersion and separation of carbon nanotubes. *Nat. Mater.*, 2, 338–342.

Zheng, M., Jagota, A., Strano, M.S. et al. 2003b. Structure-based carbon nanotube sorting by sequence-dependent DNA assembly. *Science*, 302, 1545–1548.

10

NEMS-Based Ultra Energy-Efficient Digital ICs: Materials, Device Architectures, Logic Implementation, and Manufacturability

Hamed F. Dadgour and Kaustav Banerjee

CONTENTS

ABSTRACT   The search for steep subthreshold slope devices has been intensified during the past few years due to the significant increases in the subthreshold leakage currents of CMOS transistors. Nanoelectromechanical switches (NEMS) are very promising steep subthreshold devices that offer the prospect of improved energy efficiency in digital circuits. Their near-zero subthreshold leakage and unique device architectures provide exciting circuit design opportunities. NEMS devices also allow high-temperature (>500°C) operation (Lee et al., 2010) and are relatively immune to radiation damage that can cause logical errors in charge-based devices (George, 2003). However, certain concerns must be addressed, before NEMS-based circuits can be implemented in main-stream digital chips; reliability issues, performance limitations, and scalability challenges must be investigated and resolved. This chapter presents a brief overview of NEMS operation and methods for modelling, analysis, and design of NEMS-based digital circuits. It also provides a critical analysis of the promises and challenges of NEMS technology. Preliminary circuit implementations and simulations indicate that clever design of NEMS-based circuits can offer better energy efficiency than their CMOS counterparts. Therefore, NEMS-based circuits are expected to find wide usage in low-power applications where longer battery lives are necessities.

10.1 NEMS: Opportunities and Challenges

10.1.1 Prospects for Near-Zero Leakage Circuits

While aggressive transistor scaling offers performance improvements as per Moore's law (Moore, 1965), it has also resulted in higher power dissipation especially due to a substantial increase in subthreshold leakage. This is because of weaker control of the gate terminal over the channel in scaled CMOS devices, which makes the transition between the on and off states less abrupt.

Thus, the subthreshold swing (S) of devices that essentially indicates the amount of gate voltage (V_{gs}) reduction necessary to reduce the subthreshold current (I_{ds}) by one decade ($S = dV_{gs}/dlogI_{ds}$) is higher for scaled transistors (Taur and Ning, 1998). Although improving the subthreshold performance (lowering the swing) of CMOS devices has been extensively researched, a significant improvement seems unlikely. The reason is that the subthreshold swing for bulk CMOS transistors has a fundamental lower limit of 60mV/decade (Taur and Ning, 1998). Hence, achieving extreme steep subthreshold operations requires a different approach from drift-diffusion based conduction mechanisms employed by CMOS devices (Dadgour and Banerjee, 2009).

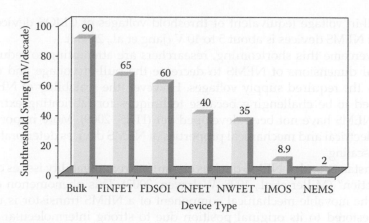

FIGURE 10.1

Minimum subthreshold swing values reported in the literature for various emerging device structures. The value reported here for the carbon nanotube FET corresponds to the tunneling-CNFETs that have the lowest subthreshold swings.

Various research groups have attempted to develop non-classical steep subthreshold semiconductor devices. For comparison purposes, Figure 10.1 summarizes minimum subthreshold swings reported in the literature. For bulk CMOS devices, in reality, subthreshold swing values are typically around 90 mV/decade (it is difficult to achieve the theoretical lower limit). Non-classical CMOS-based transistors such as fully depleted SOI (FDSOI) (Wouters et al., 1990) and FinFET (Liu et al., 2003) devices offer lower subthreshold swing values that are closer to the theoretical limit.

Furthermore, tunnelling type carbon nanotube transistors (CNFETs; Appenzeller et al., 2004), nanowire-based transistors (NWFETs; Lin et al., 2006), and impact ionization-based MOS (IMOS; Gopalakrishnan et al., 2002) show subthreshold slopes of 40, 35, and 8.9 mV/decade, respectively. While these proposed devices offer low subthreshold swing values, they suffer from certain limitations (Dadgour and Banerjee, 2009).

As shown in Figure 10.1, experiments have shown that electromechanical devices can exhibit incredibly low subthreshold swings of ~2 mV/decade (Abelé et al., 2005). As a result of their extraordinary subthreshold characteristics, NEMS have generated a great deal of interest especially for integration in future ultra low-power digital IC design applications (Dadgour and Banerjee, 2007; Zhou et al., 2007; Chong et al., 2009).

10.1.2 Scalability and Reliability Challenges

While near-zero leakage properties of NEMS are impressive, widespread usage of NEMS transistors in logic applications is hindered by their need for high (10 to 20 V) supply voltages (Ekinci and Roukes, 2005). This is because the pull-in voltage (equivalent of threshold voltages of CMOS devices) for current NEMS devices is about 5 to 10 V (Jang et al., 2008a).

To overcome this shortcoming, researchers are attempting to reduce the physical dimensions of NEMS to decrease the pull-in voltage and consequently the required supply voltages. However, the scalability of NEMS is predicted to be challenging because techniques for fabricating extremely small NEMS have not been developed yet (ITRS, 2009). More importantly, some electrical and mechanical properties of NEMS devices deteriorate with device scaling.

For instance, scaled NEMS devices can suffer from reliability issues caused by "stiction" (Miller et al., 1998; Walraven, 2003). This phenomenon occurs when the movable mechanical component of a NEMS transistor is unable to be restored to its original position due to strong intermolecular forces between mating surfaces (Tabor, 1976; Allameh, 2003).

Recently, various important characteristics of NEMS technology have been evaluated by the International Technology Roadmap for Semiconductors (ITRS, 2009) for NEMS memories and devices as shown in Figure 10.2. The higher values indicate a more desirable condition. It can be observed that, for NEMS memories shown in Panel (a), scalability, performance, and operational reliability scores are below average and hence require more in-depth investigation and technological improvement. Similarly, for NEMS devices, the same aspects are identified as challenging issues. However, NEMS exhibit high potential in terms of energy efficiency, CMOS technological and architectural compatibilities, and operational temperature and thus demonstrate sufficient incentives to offset their challenges. This chapter will attempt to evaluate the prospects of employing NEMS technology for the implementation of energy efficient digital circuits.

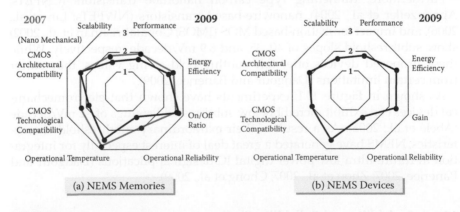

FIGURE 10.2
(See color insert.) Technology performance evaluation predicted by the ITRS (2009). (a) NEMS memories; the blue and red lines correspond to the data available for years 2007 and 2009, respectively. (b) NEMS devices. Various properties are indicated with numbers; 1, 2, and 3 correspond to poor, average, and excellent, respectively.

10.1.3 Scope of Chapter

This chapter is aimed at providing an overview on the feasibility of designing energy-efficient digital circuits using NEMS devices and the challenges surrounding this technology. In Section 10.2, the existing NEMS device structures are discussed. Section 10.3 covers the impact of material choice on the performance and reliability of scaled NEMS devices. Section 10.4 provides a summary of the existing compact NEMS-based logic gate designs. The implications of these compact gates for logic minimization and optimization are investigated in Section 10.5.

10.2 Device Structures and Operational Principles

Microelectromechanical systems (MEMS) have been fabricated for decades and used for aerospace, electronics, automotive, communications, and biotechnology applications. In recent years, more advanced fabrication techniques have miniaturized MEMS into the nanoscale regime and the miniature devices are called NEMS. These systems can act as switching devices in integrated circuits where the flow of electrical current between source and drain terminals can be controlled using simultaneous interactions of electrical and mechanical means. NEMS are usually composed of a mechanically moveable beam that can deform in response to an applied electrical bias; thereby aiding or blocking the flow of electrical current.

Various device structures can implement NEMS transistors. The most common NEMS structures are cantilever (beam)-based relays or fixed-base–fixed-base devices (Jang et al., 2005; Kam et al., 2005; Abelé et al., 2006b; Gong et al., 2007; Dadgour et al., 2010a).

10.2.1 Fixed–Fixed Structures

One example of fixed–fixed structures is the suspended gate MOSFET (SG-MOSFET) proposed by Ionescu et al. (2002) and then fabricated by Abelé et al. (2005). Figure 10.3a is a scanning electron micrograph (SEM) of the fabricated SG-MOSFET device. The gate is suspended over the channel area much like a bridge. This device is considered a fixed–fixed structure because the suspended gate is mechanically clamped at both ends. To better explain the device operation, Figure 10.3b and Figure 10.3c illustrate the position of the suspended gate in both on and off states, respectively, as viewed from cross-sections AA′ and BB′ in Figure 3a.

As shown in 10.3b and 10.3c, in the absence of a gate bias (off state), an air gap exists between the gate material and the gate dielectric layer. Hence, the source and drain are separated by a non-conducting semiconductor material

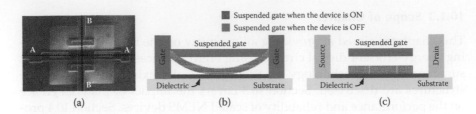

(a)   (b)   (c)

FIGURE 10.3
(See color insert.) Suspended gate (SG) transistor structure. (a) SEM of fabricated device. (b) Cross-sectional view along AA′ line when the device is in the on and off states. (c) Cross-sectional view along BB′ line when the device is in the on and off states. (From Abelé, N., Fritschi, R., Boucart, K. et al. 2005. *Proceedings of International Electron Devices Meeting*, pp. 479–481. With permission.)

(substrate) and there is no conduction channel. However, when a sufficiently high gate voltage is applied to the gate terminal (on state), opposite charges appear on the suspended gate and the substrate, generating an electrostatic force between these two components.

Due to this electrostatic attraction, the suspended gate deflects from its original straight position and touches the underlying dielectric layer. Once the suspended gate is pulled all the way down, the structure acts similar to a standard MOS device in which a conducting channel is formed between the source and the drain in the underlying silicon substrate.

10.2.2 Vertically Actuated Cantilever Structures

An alternative approach for fabricating NEMS switches is a cantilever (beam)-based relay (Kinaret et al., 2003). The electromechanical principle of operation is similar to that of a SG-MOSFET. In these structures (Figure 10.4a), instead of a movable gate, the conducting channel between the source and the drain is movable (Lee et al., 2004). In Figure 10.4b, the source terminal is connected to a conductive bendable cantilever (beam) suspended over the

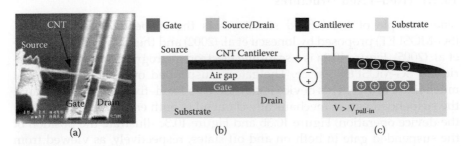

(a)   (b)   (c)

FIGURE 10.4
(See color insert.) Vertically actuated cantilever-based NEMS relay. (a) SEM of fabricated device (From Lee S., Lee, D., Morjan, R. et al. 2004. *Nano Lett.*, 4, 2027–2030. With permission.). (b) Schematic of device in off state. (c) Schematic of device in on state.

gate terminal. This is different from the fixed–fixed architecture, because the beam is fixed at one end (source) and free at the other end (drain).

In the off state, as shown in Figure 10.4b, the drain and source terminals have no connection. However, if sufficient voltage is applied between the gate and source (V > $V_{pull-in}$), the cantilever can deform and make an electrical connection between the source and drain terminals (Figure 10.4c). Various materials are use to fabricate the movable beams, for example, metals and carbon nanotubes (CNTs). The device shown in Figure 10.4a (Lee et al., 2004) is a CNT-based relay. The choice of material affects the mechanical and electrical properties of relays and will be discussed later in this chapter.

10.2.3 Laterally Actuated Cantilever Structures

Another alternative is to design a cantilever-based NEMS to facilitate the movement of the beam in the lateral direction instead of the vertical fashion (Dadgour et al., 2010a). Figure 10.5a is a SEM of a laterally actuated NEMS device. The top view schematic (Figure 10.5b) corresponds to the SEM. The sketch (10.5c) provides a cross-sectional view of the transistor.

This device has two gate terminals (Gates A and B in Figure 10.5a) that are controlled independently. The basic operation of the device is illustrated in Figure 10.6. When a bias voltage is applied between one of the gates (for example, Gate A) and the source (Gate B is biased at the same voltage as the source), opposite charges appear on the beam and the corresponding gate terminal, generating an electrostatic force.

If the gate voltage is smaller than a threshold value ($V_{pull-in}$), as shown in Figure 10.6a, the beam bends slightly, but does not touch the drain terminal. However, if the bias voltage is larger than $V_{pull-in}$, the beam deflects sufficiently to touch the drain and hence, creates a conduction path from the source to the drain as shown in Figure 10.6b.

Figure 10.6c is a sketch of a typical I_{DS}-V_{GS} characteristic of such a device. I_{DS} denotes the source drain current and V_{GS} refers to the gate source voltage

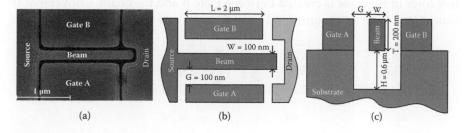

(a)   (b)   (c)

FIGURE 10. 5
Laterally actuated double-gate NEMS device. (a) SEM of fabricated device (Dadgour, H.F., Hussain, M.M., Casey, S. et al. 2010a. *Proceedings of ACM Design Automation Conference*, pp. 893–896. With permission.) (b) Top view schematic of device in off state. (c) Side schematic of device in off state.

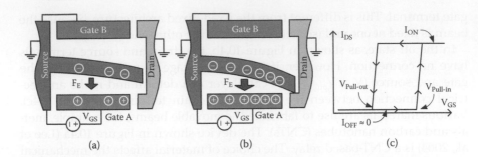

FIGURE 10.6
Basic operation of laterally actuated NEMS devices. (a) VGS < $V_{pull-in}$. (b) VGS > $V_{pull-in}$. (c) IDS VGS characteristics. (From Dadgour, H.F., Hussain, M.M., Casey, S. et al. 2010a. *Proceedings of ACM Design Automation Conference*, pp. 893–896 With permission.).

difference. It can be observed that the device exhibits a hysteresis behavior because of the surface forces that prevent the beam from being restored to its original position after it is pulled down (Elwenspoek and Wiergerink, 2001). As a result of this hysteresis, if V_{GS} is increased gradually from zero, there is virtually zero off current for V_{GS} < $V_{pull-in}$ and the device turns on when V_{GS} > $V_{pull-in}$. In contrast, while decreasing the V_{GS} value from $V_{pull-in}$, for V_{GS} > $V_{pull-out}$, the device remains on and turns off only for V_{GS} < $V_{pull-out}$.

Figure 10.7 illustrates the operation of a laterally actuated NEMS device under all possible input combinations. It is assumed that the source terminal is always connected to the ground (GND). It is also assumed that VDD is larger than $V_{pull-in}$. When Gate A is connected to the power supply (Gate A = VDD) and Gate B is tied to the ground (Gate B = GND), an attractive electrostatic force (F_E) will occur between the beam and Gate A (Figure 10.7a) and the source and drain will be connected.

Since the source and Gate B are at the same voltage (GND), there is no attractive force between them. A similar scenario occurs when Gate A = GND and Gate B = VDD (Figure 10.7b). The only difference is that the attractive force in this case is created between Gate B and the beam. Conversely, if Gate A and Gate B are connected to VDD, both gate terminals create equally

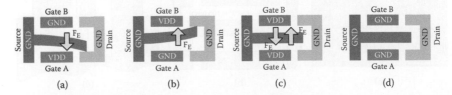

FIGURE 10.7
Operation of NEMS under various bias conditions. (a) Gate A = VDD and Gate B = GND. (b) Gate A = GND and Gate B = VDD. (c) Gate A = VDD and Gate B = VDD. (d) Gate A = GND and Gate B = GND. (From Dadgour, H.F., Hussain, M.M., Casey, S. et al. 2010a. *Proceedings of ACM Design Automation Conference*, pp. 893–896. With permission.)

strong electrostatic forces. Since these two forces are in the opposite directions, the beam will not be deflected (Figure 10.7c).

Finally, when Gate A = GND and Gate B = GND, the beam will not be deflected due to the absence of electrostatic force between the beam and the gate terminals.

10.2.4 Comparison of Structures

The choice of switch architecture affects important device properties such as $V_{pull-in}$ and oscillation frequency (f_0). The pull-in voltage is the voltage that must be applied to the gate of a NEMS transistor to switch it from the off to on state. The resonance frequency is the fastest rate at which the NEMS can operate, as determined by the delay of the mechanical movement of the beam. Generally, lower $V_{pull-in}$ and higher f_0 values are preferable.

Since the fixed–fixed structures are stiffer (more force is required to deflect them), they offer higher f_0 values. However, for the same reason, the $V_{pull-in}$ for a fixed–fixed beam is higher than that of a same-size cantilever structure. As a result, for these structures, achieving reasonable $V_{pull-in}$ values comparable to the threshold voltages of CMOS devices is possible only by employing extremely small air gaps. Since reliably fabricating such small air gaps is difficult with today's technology, this chapter will consider only cantilever-based structures for further investigation and analysis.

As discussed earlier, the two possibilities for implementing cantilever-based structures are (1) vertically and (2) laterally actuated beams. While the vertically actuated structures are widely used for the implementation of NEMS (Tilmans et al., 1994; Osterberg, 1995; O'Mahony et al., 2003; Chowdhury et al., 2005), they suffer from impact bouncing and a long settling time due to release vibrations (Gorthi et al., 2006) as illustrated in Table 10.1 (Dadgour et al., 2010b).

The reason can be explained by considering an equivalent mechanical model of the device that consists of the beam mass (m), a damping factor (c), and a spring constant (k). Note that the damping factor represents the viscosity of the air molecules that reside between the beam and the substrate or gate. The spring constant represents the elasticity of the beam. The settling time of the vertical NEMS is longer due to the smaller damping factor (c) compared to the lateral NEMS (2c). This is because the vibrations of the lateral beam are dampened twice as fast by the air molecules that exist between the beam and the two gates.

Another advantage of the laterally actuated NEMS is the availability of two gate terminals. Such a structure provides one more degree of freedom that may be exploited for the design of compact logic circuits as discussed later in this chapter.

TABLE 10.1

Device Switching Behaviors of Vertically and Laterally Actuated NEMS

NEMS	Device Schematic	Mechanical System	Release Vibrations
Vertical NEMS	Beam C_C Substrate	c $\lessgtr$ k / m	y(t)
Lateral NEMS	Gate A C_C Beam C_C Gate B Substrate	k/2 — m — k/2 / c — m — c / OR / k' = k — m — c' = 2c	x(t)

10.2.5 High-Frequency Operation of NEMS Devices

Interestingly, the scaling down of mechanical systems into the nanometer regime has very important implications for ICs: the mechanical switching speeds of NEMS can match those of CMOS transistors.

The reason is that the intrinsic delay of a mechanical switch is limited by the mass of the moving segment of the NEMS device (that scales as the cube of the dimensions), whereas the switching speed of a CMOS device is restricted by the time required for the charge to be relocated inside the device to form the conduction channel (that scales as square of the dimensions). In other words, the intrinsic delays of NEMS devices decrease at a higher rate than those of CMOS devices with scaling. Therefore, GHz operations are achievable using NEMS devices when the dimensions are scaled down to the nanometer range (Dadgour and Banerjee, 2009).

Mathematically, the mechanical delay of the beam is equal to the inverse of resonant frequency of the beam, which in turn equals

$$\sqrt{k/m}$$

where k is the spring constant of the suspended beam (inverse of its stiffness) and m is its mass. The spring constant (k) can be expressed as $k = (16\ Ewh^3/l^3)$, where E is the Young's modulus; l, h, and w are the length, thickness, and width of the beam, respectively. Also, the mass of the beam can be expressed as $m = (lwh) \times \rho$, where ρ is the mass density of the beam. Substituting these values in the resonant frequency formula

$$f = \sqrt{k/m}$$

one arrives at

$$f = \frac{h}{l^2}\sqrt{16E/}$$

According to this formula, the resonant frequency is proportional to $1/(l^2)$; it increases dramatically as the device is scaled down. Since both k and m are proportional to the width of the beam (w), the resonant frequency of the beam (inverse of its intrinsic delay) is independent of its width. Figure 10.8 plots the resonant frequency of suspended-gate type NEMS devices with different beam lengths (l), where thickness (h) is assumed to be 5 and 10 nm.

10.3 Choices of Materials and Fabrication Steps

10.3.1 Impact of Beam Material on Performance

The characteristics of NEMS depend on several parameters including the choice of material, architecture, and geometry (Akarvardar et al., 2008). For instance, the resonance frequency (f_0) is equal to $(1/2\pi) \times (k/m)^{1/2}$ where k is the spring constant of the gate (inverse of its stiffness) and m is the mass of

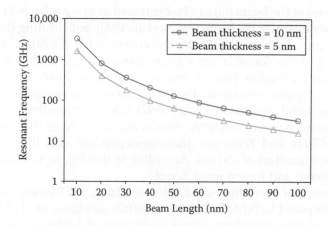

FIGURE 10.8

Resonant frequency of NEMS increases dramatically by scaling and hence GHz operation is feasible for nanoscale mechanical switches. Young's modulus (E) and mass density (ρ) of silicon are assumed to be 150 GPa and 2330 Kg.m^{-3}, respectively. The beam thickness (h) is 5 nm (triangle) for one set curves and 10 nm (square) for the other. (From Dadgour, H.F. and Banerjee, K. 2009. *IET Computers and Digital Techniques*, 3, 593–608. With permission.)

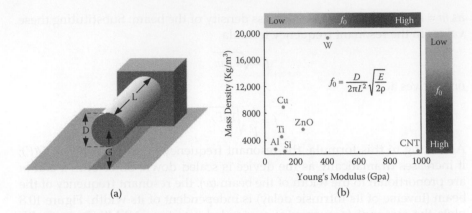

(a)   (b)

FIGURE 10.9
Impact of beam material on resonance frequency (f_0) of NEMS devices. (a) Schematic of cylindrical cantilever-based NEMS and its key geometrical parameters. (b) Resonance frequency assuming different beam materials. CNT-based devices offer high resonance frequency due to their very low mass density (ρ) and high Young's modulus (E). (From, Dadgour, H.F., Cassell, A.M., and Banerjee, K. 2008. *Proceedings of International Electron Devices Meeting*, pp. 529–532. With permission.)

the beam. The spring constant k depends on the cross-sectional shape and the material properties of the beam.

For example, considering a cylindrical beam model shown in Figure 10.9a, the spring constant is $k = 8EI/L^3$ (Dequesnes et al., 2002). L is the length of the beam, E is the Young's modulus, and I is the moment of inertia. The moment of inertia for a cylinder is $I = \pi D^4/12$ where D is the diameter of the beam. Also, the mass of the beam (m) can be expressed as $m = \rho \times L \times (\pi D^2/4)$ where ρ is the mass density of the beam (Frank et al., 1998). Substituting these values in the resonance frequency formula, one arrives at $f_0 = (D/2\pi L^2) \times (E/2\rho)^{1/2}$.

According to this formula, for a given geometry, f_0 is higher if the beam material is stiffer (higher Young's modulus E) and has lower mass density ρ. For instance, Figure 10.9b shows the Young's modulus versus the mass density of some metal elements (IUPAC, 1997), CNTs (Dresselhaus et al., 2004) and ZnO-based nanowires (NWs; Stan et al., 2007). Note that the Young's moduli of CNTs and NWs are diameter-dependent and their values are reported for diameters of ≈50 nm. According to this figure, CNTs exhibit the highest stiffness and lowest mass density.

This shows that assuming identical geometries, CNT-based NEMS offer higher f_0 compared to NEMS of other materials (Dadgour et al., 2008). Note, however, that this advantage comes at the cost of higher $V_{pull-in}$ for CNT-based devices because CNTs are extremely stiff and hence higher electrostatic forces (and higher voltages) are needed to deflect them. Another disadvantage for CNT-based NEMS is that their integration in the standard CMOS fabrication may be more challenging than incorporation of the metal- or polysilicon-based NEMS devices.

This is an important issue. Although NEMS devices have extremely low off currents, they do not offer as high on currents as CMOS transistors do. As a result, circuits that only employ NEMS devices cannot achieve very high performance. Therefore, a hybrid NEMS-CMOS technology was proposed recently to combine near-zero leakage characteristics of NEMS with high on currents of CMOS transistors (Dadgour and Banerjee, 2007 and 2009).

10.3.2 Impact of Contact Material on Performance

Another important criterion in the selection of a beam material is the performance and reliability of the transitory contacts between the beam and drain terminal when the device turns on (Dadgour et al., 2011) (Figure 10.6b). Figure 10.10a is a SEM of a fabricated device that includes enlarged pictures of its contact area. Figures 10.10b and c depict the contact area when the device is off and on, respectively.

Although it seems that the beam and the drain terminal form a perfect contact, the effective contact area is less than the surface area determined by surface irregularities. That is because real metal surfaces, if considered in the nanoscale regime, are rather rough and not atomically smooth (Figure 10.11a). This means that when two such rough films come into physical contact, the actual area of conduction is much smaller than the total surface area since electrical contacts occur exclusively at asperities (a spots) or local maxima of the film surfaces (Greenwood and Williamson, 1966; Tabor, 1976).

The contact area for two sphere-shaped asperities pressed against each other can be calculated using well known methods (Hertz, 1881; Chang et al., 1987; Johnson et al., 1971) as shown in Figure 10.11b. These models indicate that the contact radius r and its electrical resistance are functions of the mechanical properties of the mating surfaces. This is because, under an identical applied force, the extent of deformation differs for various materials. Softer materials deform more and hence create larger contact areas.

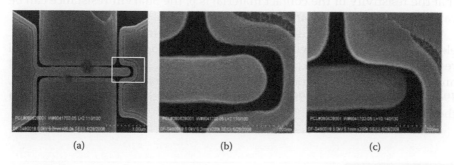

(a)   (b)   (c)

FIGURE 10.10
SEMs of contact area. (a) NEMS device [area inside box is enlarged and shown in (b) and (c) panels]. (b) Contact area when device is off. (c) Contact area when device is on. (From Dadgour, H.F., Hussain, M.M., Cassell, A. et al. 2011. *IEEE International Reliability Physics Symposium*, pp. 280–289.)

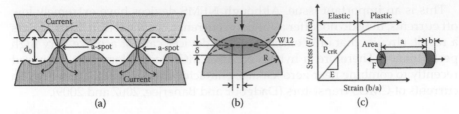

FIGURE 10.11
Contacting surfaces and their deformations. (a) Surface roughness and current conduction at a-spots. (b) Deformation of two half sphere a-spots under external force (F). (c) Stress–strain curve for a typical metal. At stress levels below or above a critical stress (P_{crti}), the deformation is elastic or plastic.

A typical relation between stress (force per area) and strain (longitudinal deformation), called the stress–strain curve, is shown in Figure 10.11c. F, A, and E denote the applied force, cross-sectional area, and Young's modulus of the metal bar, respectively. At low levels of stress (elastic region), the strain increases linearly with stress. The slope of the stress–strain curve in this region is equal to the Young's modulus of the material as shown in Figure 10.11c. When the stress is higher than a threshold value (yield strength) of a metal, the result is called plastic deformation.

The choice of contact material also is important because it determines the resistivity at the asperities. The flow of current through each asperity is limited by two mechanisms: (1) the resistance due to the lattice scattering mechanism, which is called Maxwell resistance (R_M; Holm, 1965; Jansen et al., 1980) and (2) the resistance caused by the boundary scattering of electrons, which is referred to as the Sharvin resistance (R_S; Sharvin, 1965).

The Sharvin resistance is dominant when the contact radius (r) is small compared to the electron mean free path (λ). In this case, electrons can be projected ballistically through the contact area without scattering. Assuming that the resistivity of the contact material is ρ, the Sharvin resistance can be calculated as $R_S = 4\,\rho\lambda/(3\,\pi r^2)$.

When the contact radius is larger than the electron mean free path, the electrical current is limited by the lattice scattering mechanism (Maxwell resistance) which can be calculated as $R_M = \rho/(2\,r)$. Note that both Maxwell and Sharvin resistances contribute to the overall contact resistance (Wexler, 1966; Nikolic and Allen, 1999) and both are proportional to the resistivity of the contact material (ρ).

10.4 Logic Gate Design Using NEMS

Several attempts have been made to design and implement NEMS-based digital circuits. These implementations differ mainly by employing various device structures. The two major digital design areas where NEMS have been employed are (1) logic gate and (2) memory design. Many interesting device architectures and system design approaches are proposed for NEMS-based memory designs (Rueckes et al., 2000; Abelé et al., 2006a; Jang et al., 2008b; Choi et al., 2008; Han et al., 2009). Moreover, a chip-level implementation of NEMS-base memory units was by Belov et al. (2009).

The literature on NEMS-based memories is particularly rich but beyond the scope of this chapter. Therefore, this section will provide an overview on innovative NEMS-based logic gate designs proposed in the literature.

10.4.1 Suspended Channel Architecture

In Chen et al. (2010), the device structure shown in Figure 10.12 is used to demonstrate several monolithically integrated NEMS-based basic building blocks and circuits. The schematic of the NEMS device used in that work is presented in Figure 10.12a. For better understating of the device operation, the cross-sections along AA' are also shown in Figure 10.11b for the off state (top) and the on state (bottom).

In the off state, the moveable gate electrode is suspended over the body electrode beneath it and no electrical conduction path exists between the drain and source. When a bias voltage larger than the $V_{\text{pull-in}}$ is applied to the gate with respect to its body, the suspended gate moves all the way down. Then, the channel (attached to the underside of the gate, with an intermediary gate oxide layer) will form contacts with the drain and source electrodes. To turn the switch off, the bias voltage must be reduced below the $V_{\text{pull-out}}$ of the device.

Figure 10.13a presents a schematic of such a NEMS-based carry-generation circuit for a one-bit full-adder and its truth table. A and B are the inputs of the carry-generation circuit; C_{in} and C_{out} are the input and output carry signals, respectively. According to the truth table of a full adder, C_{out} must equal

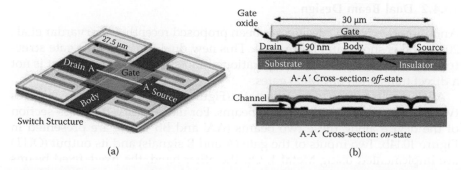

(a)

(b)

FIGURE 10.12

See color insert. Schematic of NEMS device. (a) Top view. (b) Cross sections along AA' in the off state (top) and on state (bottom). (From Chen, F., Spencer, M., Nathanael, R. et al. 2010. *Proceedings of International Solid-State Circuits Conference*, pp. 26–28. With permission.)

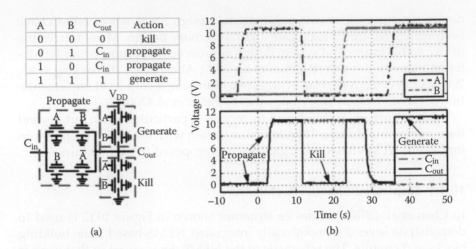

FIGURE 10.13
NEMS-based carry-generation circuit. (a) Schematic along with simplified truth table. (b) Measured signal waveforms (Chen *et al.*, 2010). (From Chen, F., Spencer, M., Nathanael, R. et al. 2010. *Proceedings of International Solid-State Circuits Conference*, pp. 26–28. With permission.)

C_{in} when A and B have different logic values (or A_x or B = 1). This functionality can be achieved using the proposed circuit because C_{in} is passed on to C_{out} through the propagate module while the generate and kill modules act as open circuits. On the other hand, when A = B= 0, the propagate and generate subcircuits are open and the kill module ensures that C_{out} = 0. Similarly, when A = B = 1, the propagate and kill subcircuits are open and the generate module forces C_{out} = 1. The signal waveforms obtained from measurements shown in Figure 10.13b validate the accurate operation of the proposed circuit (Chen et al., 2010).

10.4.2 Dual Beam Design

An innovative NAND design has been proposed recently (Akarvardar et al., 2007) and is shown in Figure 10.14. This new dual-beam NAND gate structure is based on a principle of operation unique to NEMS devices that is not a direct translation of CMOS gates.

As shown in the top view schematic (Figure 10.14a), this structure features two doubly clamped (fixed–fixed) beams. For more clarity, the cross-section of the device along these two beams (AA' and BB' lines) are presented in Figure 10.14b. Two inputs of the gate (A and B signals) and its output (OUT) are implemented using Metal 1. On the other hand, the fixed–fixed beams are realized using Metal 2.

Note that one of the beams is biased to VDD (to connect OUT to VDD when A or B =0) and the other beam is biased to GND (to connect OUT to VDD when A = B = 1). The NAND functionality is achieved by intentionally

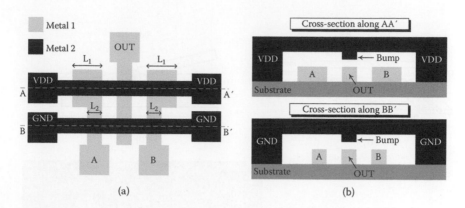

(a)   (b)

FIGURE 10.14

(See color insert.) NEMS-based NAND. (a) Top view diagram of logic gate. (b) Cross sections of structure along its beams [AA′ (top) and BB′ (bottom) lines]. (From Akarvardar et al. 2007. With permission.)

making the GND beam overlaps with the inputs (L_2) smaller than those of the VDD beam (L_1). The overlapping length L_1 is such that the GND beam does not collapse unless A= B = VDD. On the other hand, L_2 is long enough to pull in the VDD beam when A or B = 0, resulting in a NAND functionality.

10.4.3 Seesaw NAND and NOR Gates

Tsai et al. (2008) proposed and fabricated a similar design for NAND and NOR gates. The concept of the proposed NEMS logic gates is shown clearly in Figures 10.15a and b. The logic gate consists of a moveable plate on the top that can incline around its axis located in the middle, much like a seesaw.

The right and left segments of this plate are electrically isolated and biased at Vcc– (0) and Vcc+ (1), respectively. Each of the logic gate inputs (A and B) are connected to two electrodes: a large one on the right and a small one on the left. The output terminal of the logic gate (OUT) is composed of two connected metal lines that reside on both sides of the structure. The seesaw motion of the moveable plate connects the output terminal on each end to export the corresponding output voltage to either Vcc– or Vcc+.

The logic function of the structure is facilitated by different voltages and dimensions of the input pads (A_l and A_r in Figure 10.15a). This is because the motion of the movable plate is determined by the net attractive force from the input pads, which depends on their bias voltages and their sizes. The four cases that correspond to four possible input combinations are shown in Figure 10.15c and explained here.

A = Vcc+ and B = Vcc– — The seesaw electrode tilts to the right and exports the output voltage Vcc. This is because the attraction force between the right portion of the seesaw (biased at Vcc–) and the large pad connected to A (= Vcc+) is larger than the attraction force between the left section of the seesaw

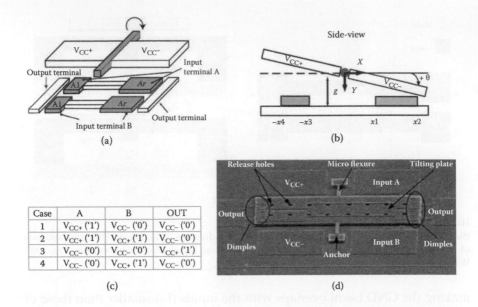

Case	A	B	OUT
1	V_{CC+} ('1')	V_{CC-} ('0')	V_{CC-} ('0')
2	V_{CC+} ('1')	V_{CC+} ('1')	V_{CC-} ('0')
3	V_{CC-} ('0')	V_{CC-} ('0')	V_{CC+} ('1')
4	V_{CC-} ('0')	V_{CC+} ('1')	V_{CC-} ('0')

(c)   (d)

FIGURE 10.15
NEMS-based NOR gate architecture. (a) Top view diagram of logic gate. (b) Side view of structure. (c) Truth table. (d) SEM of fabricated device. (From Tsai, C.Y., Kuo, W.T., Lin, C.B. et al. 2008. *Journal of Micromechanics and Microengineering*, 18, 045001. With permission.)

(biased at Vcc+) and the small pad connected to B (= Vcc–). Note that there is no electrostatic attraction between parts of the structure that are biased at the same voltage.

A = Vcc+ and B = Vcc+ — Two attraction forces created by the two input pads pull down the moveable electrode toward the right and the output terminal will be connected to Vcc–.

A = Vcc– and B = Vcc– — Two small pads create enough pull-in force to tilt the seesaw to the left and connect the output terminal to Vcc+. Note that both the large input pads are biased at the same voltage as the right segment of the moveable electrode and hence there is no electrostatic attraction between them.

A = Vcc– and B = Vcc+ — This is the same situation as in the first case. This structure has been fabricated as shown in Figure 10.14d using a CMOS-compatible process flow based on a surface micromachining method (Tsai et al., 2008).

In addition to the advantage of no leakage current, the proposed design offers two more advantages. First, the output terminal connects the fixed electrodes on the top or bottom in all combinations of input signals. Therefore, this logic device does not have undefined states (Taur and Ning, 1998)—an issue for some logic devices composed of solid state transistors. Note that this is also valid for the suspended-channel and dual-beam designs.

Second, when reversing the bias voltages on the top and bottom electrodes, the logic function of this device switches from a NAND gate to a NOR gate. Therefore, the mechanical NOR gate and NAND gate can have the same mechanical structure with different electrical interconnects (Tsai et al., 2008).

10.4.4 Laterally Actuated Double-Gate Devices

An alternative approach for implementing NEMS is to employ laterally actuated double-gate devices (Dadgour et al., 2010b). These structures offer several advantages over vertically actuated devices as discussed earlier in this chapter.

The existence of two gate terminals that may be controlled independently offers higher flexibility and provides unique and exciting circuit design opportunities. For instance, as shown in Figure 10.16, an XOR/XNOR logic gate can be implemented using only two such NEMS transistors. Note that the standard CMOS implementations of XOR/XNOR gates require at least eight transistors. The circuits shown in Figures 10.16a and b represent two alternative implementations of an XOR gate. Figures 10.16c and d depict two possible XNOR architectures. In these figures, A and B denote the input signals and the output nodes are identified by their corresponding Boolean symbols (A ⊕ B). The output nodes are formed by connecting the drain terminals of pull-up and pull-down NEMS.

The source terminal of the pull-up and pull-down NEMS is connected to VDD/GND. Note that the only difference between the two XOR designs is that in Figure 10.16a, the gate terminals of the pull-down NEMS are connected to A and $\overline{B}$ and in Figure 10.16b, they are tied to B and $\overline{A}$. Also, in the XNOR

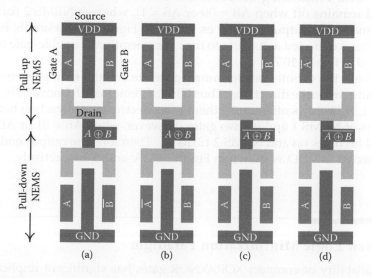

FIGURE 10.16
Two equivalent designs for NEMS-based XOR gates (a) and (b) and XNOR gates (c) and (d).

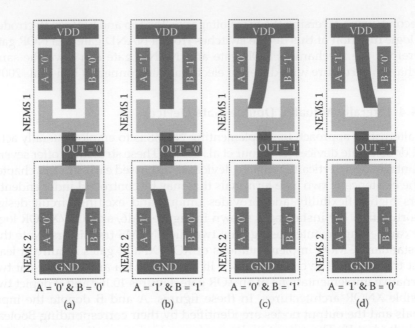

FIGURE 10.17
Operation of XOR gate shown in Figure 10.16a under various input combinations.

implementation in Figure 10.16c, the gate terminals of the pull-up NEMS are connected to A and $\overline{B}$ while in Figure 10.16d, they are tied to B and $\overline{A}$.

Figure 10.17 illustrates the basic operation of the XOR gate proposed in Figure 10.16a considering all possible input combinations. In this circuit, NEMS 1 remains off when AB = 00 or AB = 11, whereas NEMS 2 turns on and connects the output to GND as shown in Figures 10.17a and b, respectively. This is expected according to the basic operation of double-gate NEMS presented in Figure 10.7.

When A = B = 0, both gate terminals generate equally strong electrostatic forces but in opposite directions. Thus NEMS 1 remains off. Conversely, when A = B= 1, NEMS 1 is off because there is no electrostatic attraction between the beam of NEMS 1 and its two gates. However, when AB = 01 or AB = 10, NEMS 1 becomes on and NEMS 2 turns off. Therefore, the output node will be connected to VDD as shown in Figures 10.17c and d, respectively.

10.5 New Logic Minimization Paradigm

The availability of compact XOR/XNOR gates has significant implications for minimizing Boolean functions (Dadgour et al., 2010b). In order to appreciate these implications, the Karnaugh maps (K-maps) logic minimization

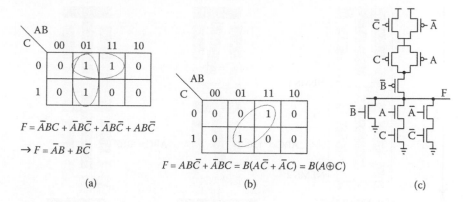

FIGURE 10.18

Diagonal grouping of 1s. (a) K-map. (b) Boolean function. (c) Transistor-level implementation. (From Dadgour, H.F., Hussain, M.M., and Banerjee, K. 2010b. *Proceedings of International Symposium on Low Power Electronics and Design*, pp. 7–12. With permission.)

technique is used as an example. In the conventional K-map technique, 1s are grouped only horizontally and/or vertically. Figure 10.18a shows a simple example in which three minterms (1s) are organized as a horizontal and vertical pair. This simplification results in a Boolean function of the form $F = \bar{A}B + B\bar{C}$.

On the other hand, the K-map presented in Figure 10.18b cannot be simplified using the standard approach because such a simplification results in a Boolean function of the form $F = B(A \oplus C)$ that requires an XOR operation. Since CMOS-based XOR implementations are not power- and area-efficient, the optimal approach is to implement such a Boolean function without simplification as shown Figure 10.18c.

10.5.1 New Paradigm: Diagonal Grouping of 1s in K-Maps

The diagonal grouping of 1s (as shown in Figure 10.18b) is possible if circuits are designed using laterally actuated double-gate NEMS devices due to the availability of power- and area-efficient XOR/XNOR gates (Dadgour et al., 2010b). This unique approach can significantly reduce the number of transistors required to implement Boolean functions.

For example, the Boolean function $F = B(A \oplus C)$ can be successfully designed using only four NEMS devices as illustrated in Figure 10.19a. In this circuit, the pull-up network is composed of NEMS 1 and NEMS 2 in series; therefore, the pull-up network connects the output (F) to VDD only when B and $(A \oplus C)$ are simultaneously 1. NEMS 1 connects the internal node (INT) to VDD when AB is either equal to 10 or 01 (XOR function). Then, assuming that INT is already connected to VDD, NEMS 2 connects INT to F only when B = 1. This is because when B = 0, the beam of NEMS 2 is attracted by two equal but opposing electrostatic forces generated by the gate terminals.

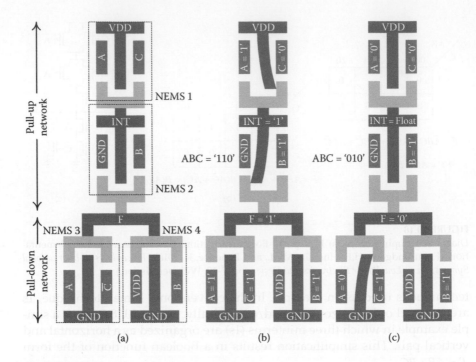

FIGURE 10.19
Design of Boolean function shown in Figure 10.18a. (a) NEMS-based implementation and its operation assuming (b) ABC = 110 and (c) ABC = 010.

However, when B = 1, the beam is attracted only by the gate terminal connected to the ground (GND). Note that in Figure 10.19a, the pull-down network is the dual of the pull-up network to connect the output node (F) to GND when it is not connected to VDD.

Figures 10.19b and c illustrate the operation of the circuit shown in Figure 10.19a when inputs (ABC) are 110 and 010, respectively. When ABC = 110, only NEMS 1 and NEMS 2 turn on and thus F becomes connected to VDD. In contrast, when ABC = 010, only NEMS 3 turns on and F will be connected to GND. A comparison of the NEMS-based design (Figure 10.19a) and the CMOS-based circuit (Figure 10.18c) reveals that the former requires fewer devices to implement an identical Boolean function.

The four-variable K-map (Figure 10.20a) is the most extreme case in which NEMS-based designs can offer maximum advantage over their standard CMOS-based counterparts (Dadgour et al., 2010b). The implementation of such a K-map requires three XOR operations [$F = ((A \oplus B) \oplus C) \oplus D$]. This Boolean function can be implemented using three cascaded NEMS-based XOR gates as shown in Figure 10.20b. In this design, NEMS 1 and NEMS 2 generate $A \oplus B$ which is used by NEMS 3 and NEMS 4 to create $((A \oplus B) \oplus C)$. Finally, NEMS 5 and NEMS 6 are employed at the last stage to generate the desired Boolean function ($F = ((A \oplus B) \oplus C) \oplus D$). Note that the NEMS-based

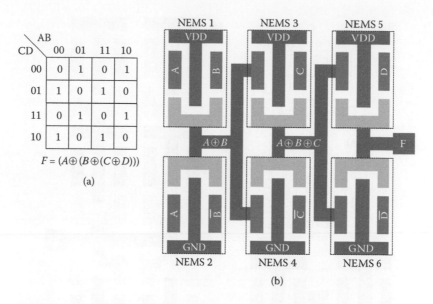

FIGURE 10.20
(a) Four-variable K-map that can be created using three XOR operators. (a) Truth table. (b) Its NEMS-based implementation.

circuit requires only 6 devices whereas its CMOS-based counterpart needs at least 44 transistors. Therefore, an improvement of more than 600% is achieved in terms of the number of devices required to implement this particular Boolean function.

It should be noted that the advantage of employing the diagonal grouping varies for different Boolean functions. Therefore, the actual benefits of the proposed approach may be identified only if practical design problems such as the implementation of a full adder are investigated.

10.5.2 NEMS-Based Full-Adder Design

Full-adder circuits are the building blocks of arithmetic modules in advanced microprocessors. The novel design paradigm proposed in this section allows the design of NEMS-based full adders using fewer devices due to possibility of the diagonal grouping of 1s (Dadgour et al., 2010b). The K-maps associated with carry-out (C_{out}) and sum (Sum) of a full adder are shown in Figures 10.21a and b, respectively.

A and B are the input bits and C_{in} denotes the carry-in signal. The K-map corresponding to C_{out} is simplified as shown in Figure 10.21a which results in $C_{out} = BC_{in} + A(B \oplus C_{in})$. The first term is a result of the horizontal grouping of minterms ABC = 011 and ABC = 111 while the second term represents the diagonal grouping of minterms ABC = 110 and ABC = 101. Similarly, Figure 10.21b represents the K-map for Sum, simplified as $Sum = C_{in} \oplus (A \oplus B)$.

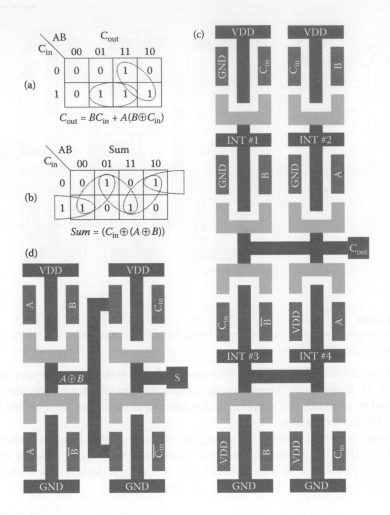

FIGURE 10.21
(a) NEMS-based full adder. (b) K-maps for C_{out} and Sum. (c) and d) NEMS-based implementations for C_{out} and Sum, respectively.

The transistor-level implementations of C_{out} and Sum are demonstrated in Figures 10.21c and d, respectively. The pull-up network associated with C_{out} is composed of two parallel branches. The one on the left implements and the other produces $A(B \oplus C_{in})$ similar to Figure 10.19a. Naturally, the pull-down network is designed to be the dual of the pull-up network.

Figure 10.21d represents the circuit implementation of Sum which is composed of two cascaded NEMS-based XOR gates. The output of the first XOR gate ($A \oplus B$ in the figure) is connected to the inputs of the second XOR to generate $Sum = A \oplus (B \oplus C_{in})$ at its output. Note that the NEMS-based design only requires 12 devices compared to 28 transistors needed by its CMOS counterpart.

10.5.3 Power and Performance Comparison Using Circuit Simulations

10.5.3.1 Energy-Efficiency Enhancement

Using the proposed design paradigm, extremely energy-efficient logic circuits can be implemented. Two main reasons support such an improvement. First, NEMS devices have inherently low power due to their steep subthreshold slopes. This implies that the power consumption of a NEMS-based circuit due to subthreshold leakage is negligible compared to that of its CMOS counterpart. Therefore, assuming an identical level of dynamic power consumption, a NEMS-based circuit will offer superior energy efficiency.

Second, the proposed design methodology enables circuit designers to implement various Boolean functions in a more compact fashion. By reducing the number of devices, the dynamic power consumption of the circuit also decreases due to fewer internal nodes that must be charged or discharged.

10.5.3.2 Compact Circuit Models

To investigate the performance and power of NEMS-based full adders, the compact circuit models presented in Dadgour et al. (2010b) are used. These compact models are illustrated in Figure 10.22. The coupling capacitances between the beam and the Gate A and Gate B are denoted as C_A and C_B, respectively. Furthermore, the voltage differences between Gate A and Gate B are labeled V_A and V_B, correspondingly. If the device is in the off state, the values of C_A and C_B depend on V_A and V_B as indicated in Figures 10.22a and b because the curvature of the beam (and hence, the coupling capacitance) is a function of gate bias.

Using the models proposed by Dadgour et al. (2010b), one can take into account the dependency of C_A and C_B on V_A and V_B. While both C_A and C_B are functions of the gate bias in the off state, these variables are independent of the gate voltage when the device is on because, in the on state, the

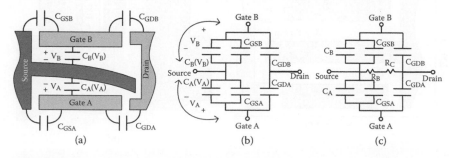

FIGURE 10.22
Compact circuit models for laterally actuated double-gate NEMS. (a) Schematic of device in off state along with all capacitances. (b) Compact model in off state. (c) Compact model in on state.

shape of the bended beam is invariant; therefore, the values of C_A and C_B are independent of the gate biases (as indicated in Figure 10.22c).

The parasitic capacitances between the gate and the source (C_{GSA} and C_{GSB}) and the gate and the drain (C_{GDA} and C_{GDB}) are also included in these compact models. Since these parasitic capacitances exist between stationary parts of the device, their values can be easily calculated using the parallel plate capacitance model.

The source and the terminals are electrically isolated when the transistor is off (Figure 10.22b). A conduction path is formed when the device turns on (Figure 10.22c). Note that contact resistance will occur at the interface of the beam and the drain terminal when they come into contact. Therefore, this electrical connection can be modeled with a series combination of beam resistance (R_B) and contact resistance (R_C). The values of R_B and R_C can be calculated using existing models (Bromley and Nelson, 2001; Chen et al., 2008) in the literature or extracted from the I_{DS} and V_{GS} characteristics of the devices.

10.5.3.3 Circuit Simulation Results

The power and performance of the proposed full adder is compared to its CMOS counterpart at 65-nm technology node as described here. Assuming a power supply of 1 V, the physical dimensions of the NEMS devices are designed to result in $V_{\text{pull-in}} = 0.5$ V. These dimensions are shown in Figure 10.23. It is assumed that reliable methods are available for the fabrication of small gap sizes.

Assuming that the cantilever beam is made of titanium (resistivity $\rho = 0.42\ \mu\Omega/m$), the resistance of the beam is calculated as $R_B \approx 210\ \Omega$. Using a methodology proposed by Bromley and Nelson (2001) and Chen et al. (2008) based on the contact theory, the contact resistance (R_C) is estimated at ≈ 2.2 kΩ. Moreover, the gate capacitances C_A and C_B vary in the range of 0.2 to 1

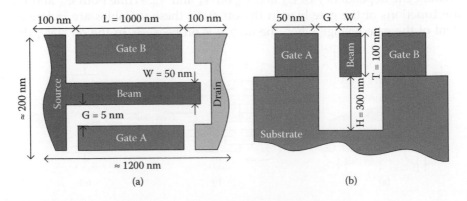

(a)   (b)

FIGURE 10.23
Laterally actuated double-gate NEMS device used in circuit simulations. (a) Top view schematic. (b) Side view schematic. The different dimensions of the structure are indicated.

fF, depending on the corresponding gate bias. Other parasitic capacitances (C_{GSA}, C_{GSB}, C_{GDA}, and $_{CGDB}$) are estimated to be ≈ 0.05 fF.

Note that the delay of a NEMS device consist of two components: (1) mechanical delay or the time needed to deflect the beam and form a contact, and (2) electrical delay or the time required to charge or discharge the output load. The mechanical delays of NEMS devices can also be integrated in the circuit model (not shown in Figure 10.22) using available analytical models (Dadgour et al., 2010b). The electrical delays of the NEMS transistors and circuits can be determined using commercial circuit simulators and the model proposed in Figure 10.22.

A NEMS-based 32-bit carry–save adder (Taur and Ning, 1998) was simulated using the NEMS model (Dadgour et al., 2010b) and its average delay was determined to be ≈ 1.92 ns. This means that the circuit can reach a maximum performance of 200 MOPS (million operations per second) at a clock rate of 500 MHz. This adder consumes 0.92 mW under full load condition (activity factor = 100%) mostly due to its dynamic power consumption. The power consumption caused by the mechanical movement of the beam is estimated to be as low as ~6% of the total power (Dadgour et al., 2010b). When the adder is idle (activity factor = 0%), it consumes negligible leakage power as a result of the near-zero subthreshold leakage of its NEMS devices.

A 32-bit CMOS adder is also designed to serve as a reference for power and performance comparisons. Assuming an identical silicon area for both NEMS- and CMOS-based adders, the sizes of CMOS devices are estimated to be 750 nm × 65 nm (all transistors are sized equally for simplicity). Considering these dimensions, the 32-bit CMOS adder is simulated using BPTM models (BPTM online) and its average delay is measured at ≈ 975 ps. This means that the circuit can perform 1000 MOPS at a clock rate of 1 GHz. This adder consumes 1.98 mW under full load condition (activity factor = 100%) mainly because of its dynamic power dissipation. When the adder is idle (activity factor = 0%), it consumes 11.6 mW as a result of the subthreshold leakage of its CMOS devices.

The power–performance comparison of these two adders is summarized in Figure 10.24. The horizontal axis represents the total power consumption of the 32-bit adders and the vertical axis denotes the logarithms of their performances (throughputs) in MOPS. For medium levels of performance (<134 MOPS), the NEMS-based adder can offer matching performances with superior energy efficiency (region A) because the CMOS-based adder always consumes power even when it is idle. As a result, in this region, the NEMS-based adder can offer the same level of performance with optimal energy efficiency.

For higher performances (134 to 205 MOPS), the CMOS adder becomes more power efficient (regions B and C). Note that NEMS adder is not able to offer very high performance values (>205 MOPS) due to the combined effect of high mechanical delays and charge–discharge delays (region C). Therefore, Figure 10.24 indicates that the NEMS-based adder is a significantly more energy-efficient choice for medium levels of performance.

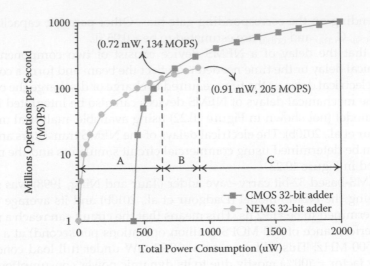

FIGURE 10.24
Power–performance analysis for CMOS and NEMS based 32-bit adders. (From Dadgour, H.F., Hussain, M.M., and Banerjee, K. 2010b. *Proceedings of International Symposium on Low Power Electronics and Design*, pp. 7–12. With permission.)

10.6 Summary

Amid increasing concerns about the leakage power consumption of CMOS transistors, the semiconductor industry is motivated to explore emerging energy efficient nanotechnologies. NEMS transistors represent one such emerging technology. These transistors are attractive because they offer unbeatable subthreshold characteristics (energy efficiency) compared to all other emerging solid state transistors. NEMS devices can also be CMOS-compatible, which means that NEMS and CMOS devices can be integrated on the same wafer.

Although NEMS devices can in principle solve the leakage problem, they still present challenges. Therefore, this chapter provided a comprehensive discussion of the reliability issues, performance limitations, and scalability challenges that must be investigated and resolved. Furthermore, the implications of employing various NEMS devices on digital circuit design are explored.

A new class of devices called laterally actuated double-gate NEMS transistors is highlighted and analyzed. Such devices can be employed to implement highly energy efficient and ultra-compact XOR gates that are the key building blocks of more complex computational units. The lateral NEMS devices also create new opportunities in Boolean logic minimization and are promising for high-performance arithmetic modules (such as adders). A comprehensive scaling analysis of the NEMS devices was conducted to identify the key challenges that must be overcome before such transistors can be incorporated in mainstream IC technologies.

The logical future direction for this research is fabrication of the proposed complex arithmetic components. This new research can lead to implementation of a modern nanoscale "Babbage machine" using NEMS. While simple circuits have been fabricated using NEMS, the use of NEMS-based circuits for practical applications has not yet been demonstrated. Furthermore, the implications of NEMS devices for the design of sequential circuits and their timing issues are not addressed in the literature. Since most realistic VLSI circuits require some types of sequential components such as flip-flops, exploring NEMS-based design opportunities for sequential components is critical.

It is also important to demonstrate the feasibility of reliable scaling of NEMS devices. Improving the existing fabrication methods to enable creation of air gaps in the nanometer range is essential. This research requires development of in-depth physical insights into the scalability issues of NEMS devices. Although this chapter covers some of those challenges, several other areas need the prompt attention of researchers. For example, contact reliability degrades with scaling and must be thoroughly studied. Such research attempts can assist in optimal selection of fabrication methods and materials that can enable further scaling of NEMS devices.

References

Abelé, N., Fritschi, R., Boucart, K. et al. 2005. Suspended-gate MOSFET: bringing new MEMS functionality into solid-state MOS transistor. *Proceedings of International Electron Devices Meeting*, pp. 479–481.

Abelé, N., Villaret, A., Gangadharaiah, A. et al. 2006a. 1T MEMS memory based on suspended gate MOSFET. *Proceedings of International Electron Devices Meeting*, pp. 509–512.

Abelé, N., Séguéni, K., Boucart, K. et al. 2006b. Ultra-low voltage MEMS resonator based on RSG MOSFET. *Proceedings of 19th IEEE International Conference on MEMS*, pp. 882– 885.

Akarvardar, K., Eggimann, C., Tsamados, D. et al. 2008. Analytical modeling of the suspended-gate FET and design insights for low power logic. *IEEE Transactions on Electronic Devices*, 55, 48–59.

Akarvardar, K., Elata, D., Parsa, R. et al. 2007. Design considerations for complementary nanoelectromechanical logic gates. *Proceedings of International Electron Devices Meeting*, pp. 299–302.

Allameh, S.M. 2003. An introduction to mechanical properties-related issues in MEMS structures. *Journal of Material Science*, 38, 4115–4123.

Appenzeller, J., Lin, Y.M., Knoch, J. et al. 2004. Band-to-band tunneling in carbon nanotube field-effect transistors. *Physical Review Letters*, 93, 196805.

Belov, N., Adams, D., Ascanio, P. et al. 2009. Nanochip: a MEMS-based ultra-high data density memory device. *Sensors and Transducers Journal*, 7, 34–46.

BPTM online, Berkeley Predictive Technology Model. http://www.eas.asu.edu/ptm/.

Bromley, S.C. and Nelson, B.J. 2001. Performance of microcontacts tested with a novel MEMS device. *IEEE Holm Conference on Electrical Contacts*, pp. 122–127.

Chang, W.R., Etsion, I., and Bogy, D.B. 1987. An elastic-plastic model for the contact of rough surfaces. *Journal of Tribology*, 109, 257–263.

Chen, F., Kam, H., Markovic, D. et al. 2008. Integrated circuit design with NEM relays. *Proceedings of IEEE/ACM International Conference on Computer-Aided Design*, pp. 750–757.

Chen, F., Spencer, M., Nathanael, R. et al. 2010. Demonstration of integrated micro-electro-mechanical (MEM) switch circuits for VLSI applications. *Proceedings of International Solid-State Circuits Conference*, pp. 26–28.

Choi, W.Y., Osabe, T., and King Liu, T.J. 2008. Nano-electro-mechanical nonvolatile memory (NEMory) cell design and scaling, *IEEE Transactions on Electron Devices* 55, 3482–3488.

Chong, S., Akarvardar, K., Parsa, R. et al. 2009. Nanoelectromechanical (NEM) relay integrated with CMOS SRAM for improved stability and low leakage. *Proceedings of IEEE/ACM International Conference on Computer-Aided Design*, pp. 478–484.

Chowdhury, S., Ahmadi, M., and Miller, W.C. 2005. Pull-in voltage calculations for MEMS sensors with cantilevered beams. *NEWCAS*, pp. 143–146.

Dadgour, H.F. and Banerjee, K. 2007. Design and analysis of hybrid NEMS-CMOS circuits for ultra low-power applications. *Proceedings of ACM Design Automation Conference*, pp. 306–311.

Dadgour, H.F., Cassell, A.M., and Banerjee, K. 2008. Scaling and variability analysis of CNT-based NEMS devices and circuits with implications for process design. *Proceedings of International Electron Devices Meeting*, pp. 529–532.

Dadgour, H.F. and Banerjee, K. 2009. Hybrid NEMS–CMOS integrated circuits: a novel strategy for energy-efficient designs. *IET Computers and Digital Techniques*, 3, 593–608.

Dadgour, H.F., Hussain, M.M., Casey, S. et al. 2010a. Design and analysis of compact ultra low-power logic gates using laterally actuated double-electrode NEMS. *Proceedings of ACM Design Automation Conference*, pp. 893–896.

Dadgour, H.F., Hussain, M.M., and Banerjee, K. 2010b. A new paradigm in the design of energy-efficient digital circuits using laterally actuated double-gate NEMS. *Proceedings of International Symposium on Low Power Electronics and Design*, pp. 7–12.

Dadgour, H.F., Hussain, M.M., Cassell, A. et al. 2011. Impact of scaling on the performance and reliability degradation of metal-contacts in NEMS devices. *IEEE International Reliability Physics Symposium*, pp. 280–289.

Dequesnes, M., Rotkin, S.V., and Aluru, N. R. (2002). Calculation of pull-in voltages for carbon-nanotube-based nanoelectromechanical switches. *Nanotechnology*, 13, 120–131.

Dresselhaus, M.S., Dresselhaus, G., Charlier, J.C. et al. 2004. Electronic, thermal and mechanical properties of carbon nanotubes. *Philosophical Transactions of the Royal Society*, 362, 2065–2098.

Ekinci, K.L. and Roukes, M.L. 2005. Nanoelectromechanical systems. *Review of Scientific Instruments*, 76, 61101–61112.

Elwenspoek, M. and Wiergerink, R.J. 2001, *Mechanical Microsensors,* Springer, Berlin.

Frank, S., Poncharal, P., Wang, Z.L. et al. 1998. Carbon nanotube quantum resistors. *Science,* 280, 1744–1746.

Gong, J.F., Xiao, Z.Y., and Chan, P. 2007. Integration of an RF MEMS resonator with a bulk CMOS process using a low-temperature and dry-release fabrication method. *Journal of Micromechanics and Microengineering,* 17, 20–25.

Gopalakrishnan, K., Griffin, P.B., and Plummer, J.D. 2002. I-MOS: a novel semiconductor device with a subthreshold slope lower than kT/q. *Proceedings of International Electron Devices Meeting,* pp. 289–292.

Gorthi, S., Mohanty, A., and Chatterjee, A. 2006. Cantilever beam electrostatic MEMS actuators beyond pull-in. *Journal of Micromechanics and Microengineering,* 16, 1800–1810.

George, T. 2003. Overview of MEMS/NEMS technology development for space applications at NASA/JPL. *Proceedings of SPIE,* 5116, 136–148.

Greenwood, A. and Williamson, J.P. 1966. Contact of nominally flat surfaces. *Proceedings of Royal Society (London),* A295, 300–319.

Han, J.W., Ahn, J.H., Kim, M.W. et al. 2009. Monolithic integration of NEMS-CMOS with a fin flip-flop actuated channel transistor (FinFACT). *Proceedings of International Electron Devices Meeting,* pp. 621–624.

Hertz, H. 1881. The contact of elastic solids. *Die Journal für die Reine und Angewandte Mathematik,* 92, 156–171.

Holm, R. 1967. *Electric Contacts: Theory and Applications,* 4th ed. Springer, Berlin.

International Technology Roadmap for Semiconductors (ITRS), 2009 Edition, http://www.itrs.net/Links/2009ITRS/Home2009.htm

Ionescu, A.M., Pott, V., Fritschi, R. et al. 2002. Modeling and design of a low-voltage SOI suspended-gate MOSFET (SG-MOSFET) with a metal-over-gate architecture. *Proceedings of IEEE International Symposium on Quality Electronic Design,* pp. 496–501.

IUPAC. 1997. *Compendium of Chemical Terminology,* 2nd ed. Blackwell Scientific, Oxford.

Jang, J.E., Cha, S.N., Choi, Y. et al. 2005. Nanoelectromechanical switches with vertically aligned carbon nanotubes. *Applied Physics Letters,* 87, 163114.

Jang, W.W., Lee, J.O., Yoon, J.B. et al. 2008a. Fabrication and characterization of a nanoelectromechanical switch with 15-nm thick suspension air gap. *Applied Physics Letters,* 92, 103–110.

Jang, W.W., Lee, J.O., Yang, H.H. et al. 2008b. Mechanically operated random access memory (MORAM) based on an electrostatic microswitch for nonvolatile memory applications. *IEEE Transactions on Electron Devices,* 55, 2785–2789.

Jansen, A.G.M., van Gelder, A.P., and Wyder, P. 1980. Point-contact spectroscopy in metals, *Journal of Physics C,* 13, 6073–6118.

Johnson, K.L., Kendall, K., and Roberts, A.D. 1971. *Proceedings of Royal Society A,* 324, 301–313.

Kam, H., Lee, D., Howe, R.T. et al. 2005. A new nano-electro-mechanical field effect transistor (NEMFET) design for low-power electronics. *Proceedings of International Electron Devices Meeting,* pp. 463–466.

Kinaret, J.M., Nord, T., and Viefers, S. 2003. A carbon nanotube based nanorelay, *Applied Physics Letters,* 82, 1287–1289.

Lee S., Lee, D., Morjan, R. et al. 2004. A three-terminal carbon nano-relay. *Nano Letters,* 4, 2027–2030.

Lee, T.H., Bhunia, S., and Mehregany, M. 2010. Electromechanical computing at 500°C with silicon carbide. *Science*, 329, 1316–1318.

Lin, H.C., Lee, M.H., Su, C.J. et al. 2006. Fabrication and characterization of nanowire transistors with solid-phase crystallized poly Si channels. *IEEE Transactions on Electron Devices*, 53, 2471–2477.

Liu, Y., Ishii, K., Tsutsumi, T. et al. 2003. Fin-type double-gate metal-oxide-semiconductor field-effect transistors fabricated by orientation-dependent etching and electron beam lithography. *Japanese Journal of Applied Physics*, 42, 4142-4146.

Miller, S.L., Rodgers, M.S., LaVigne, G. et al. 1998. Failure modes in surface micromachined microelectromechanical actuators. *Proceedings of IEEE International Reliability Physics Symposium*, pp. 17–25.

Moore, G.E. 1965. Cramming more components onto integrated circuits, *Electronics*, 38, 114–117.

Nikolic, B. and Allen, P.B. 1999. Electron transport through a circular constriction, *Physical Review B*, 60, 3963–3969.

O'Mahony, C., Hill, M., Duane, R. et al. 2003. Analysis of electro-mechanical boundary effects on the pull-in of micromachined fixed beams. *Journal of Micromechanics and Microengineering*, 13, 75–80.

Osterberg, P.M. 1995. Electrostatically actuated microelectromechanical test structures for material property measurements. PhD Dissertation, Massachusetts Institute of Technology, Cambridge, pp. 52–87.

Rueckes, T., Kim, K., Joselevich, E. et al. 2000. Carbon nanotube-based nonvolatile random access memory for molecular computing. *Science*, 289, 94–97.

Sharvin, Y.V. 1965. A possible method for studying Fermi surfaces. *Soviet Physics JETP*, 21, 655–656.

Stan, G., Ciobanu, C.V., Parthangal, P.M. et al. 2007. Diameter-dependent radial and tangential elastic moduli of ZnO nanowires. *Nano Letters*, 7, 3691–3697.

Tabor, D. 1976. Surface forces and surface interactions, *Journal of Colloid and Interface Science*, 58, 2–13.

Taur, Y. and Ning, T. 1998. *Fundamentals of Modern VLSI Devices*, Cambridge University Press, New York.

Tilmans, H.A.C. and Legtenberg, R. 1994. Electrostatically driven vacuum-encapsulated polysilicon resonators. II: Theory and performance, *Sensors and Actuators A*, 45, 67–84.

Tsai, C.Y., Kuo, W.T., Lin, C.B. et al. 2008. Design and fabrication of MEMS logic gates, *Journal of Micromechanics and Microengineering*, 18, 045001.

Walraven, J.A. 2003. Future challenges for MEMS failure analysis. *Proceedings of International Test Conference*, pp. 805–855.

Wexler, G. 1966. The size effect and the non-local Boltzmann transport equation in orifice and disk geometry. *Proceedings of Physical Society*, 89, 927–941.

Wouters, D.J., Colinge, J.P., and Maes, H.E. 1990. Subthreshold slope in thin-film SOI MOSFETs. *IEEE Transactions on Electron Devices*, 37, 2022–2033.

Zhou, Y., Thekkel, S., and Bhunia, S. 2007. Low power FPGA design using hybrid CMOS-NEMS approach. *International Symposium on Low Power Electronics and Design*, pp. 14–19.

11

Carbon Nanotube Y-Junctions

Prabhakar R. Bandaru

CONTENTS

ABSTRACT Nonlinear carbon nanostructures such as Y-shaped carbon nanotube morphologies are interesting in that they represent new paradigms of nanostructures and potential applications. These forms have been synthesized through the use of certain catalysts that promote branching on linear nanotube structures. In a shining example in which "function follows form," the Y-shaped structures have been used for three terminal switching devices. Additionally, the varying diameters of the individual branches can be designed to play a role in current blocking that may be used for space charge-based devices integrated with linear electronics. While not all the parameters responsible for their formation have been elucidated, it is presumed that such details are pertinent for observing interesting variants of electrical behavior. This chapter considers these implications in terms of their nanostructure property relationships. The overall goal is to investigate possibilities for obtaining novel functionalities at nanoscale level that may lead to new device paradigms.

11.1 Introduction

In recent years, carbon nanotubes (CNTs) have emerged as the foremost manifestations of nanotechnology and extensive research has been expended in probing their various properties. While many desirable attributes in terms of electrical, mechanical, and biological properties have been attributed to CNTs, many obstacles remain before their widespread, practical application (Baughman et al., 2002) becomes feasible.

Some of the foremost hindrances are (1) variation of properties from one nanotube to another, partly due to unpredictability in synthesis and the random occurrence of defects, and (2) lack of a tangible method for widescale synthesis (Kaul and Bandaru, in press). Generally, (1) is a natural consequence of nanoscale structures and could be difficult to solve, at least in the short term. It would then seem that fundamentally new ideas might be needed. Some interesting viewpoints are also being considered, for example, defect manipulation could be used on purpose (Nichols et al., 2007).

Wide scale synthesis methods, for example, by aligning the nanotubes with the underlying crystal orientation (Kang et al., 2007) have recently proved successful, but it is still not clear whether such methods would allow for practical implementation, say at the scale of silicon microelectronics.

In this context, it would be pertinent to take a pause and consider the rationale for the use of nanotubes, especially in the context of electronic characteristics and devices. Carbon-based nanoelectronics technologies (McEuen, 1998) promise greater flexibility compared to conventional silicon electronics, for example, the extraordinarily large variety of carbon- based organic structures. It would then be interesting to look into the possible implications of this large variety, particularly with respect to morphology and associated properties. Such an outlook gives rise to the possibility of examining nonlinear forms, some examples of which are depicted in Figure 11.1.

While Y-junctions can be used as three-terminal switching devices or diodes, helical nanostructures can give rise to nanoscale inductors or, more interestingly, a sequence of metallic and semiconducting junctions (Castrucci et al., 2004). It is to be noted at the outset that we now seek to

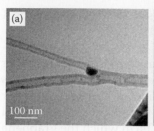 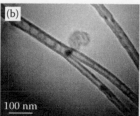 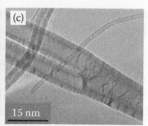

FIGURE 11.1

Nanoengineering of carbon nanotubes (CNTs) to produce nonlinear structures is manifested most clearly through the Y-junction morphology. Such structures can be prepared by adding carbide forming elements such as Ti, Zr, and Hf to ferrocene-based thermal CVD at different branching angles (i.e., a versus b) and spatial locations of catalyst particles (i.e., a versus c) by varying the growth conditions.

explore completely novel forms of electronics, as cited, for example, in the ITRS (International Technology Roadmap for Semiconductors) recommendations on emerging research devices:

> The dimensional scaling of CMOS devices and process technology, as it is known today, will become much more difficult as the industry approaches 16 nm (6 nm physical channel length) around the year 2019 and will eventually approach an asymptotic end. Beyond this period of traditional CMOS scaling, it may be possible to continue functional scaling by integrating alternative electronic devices onto a silicon platform. These alternative electronic devices include 1D structures such as CNTs ....

It is the purpose of investigating novel nanotube morphologies, then, to demonstrate many of the compelling attributes noted in the ITRS roadmap that include (1) room temperature operation, (2) functional scalability by orders of magnitude, and (3) energy dissipation per functional operation substantially less than CMOS. We will show specifically, how the exploration of Y-junction topologies would help lay the foundation for an entirely new class of electronic and optical devices.

11.2 Carbon Nanotubes

We commence with an overview of the underlying constituents of Y-shaped structures—linear nanotubes. Several comprehensive expositions (Dai, 2002) of the fundamental aspects are extant in literature (Ajayan, 1999; Dai, 2002; McEuen et al., 2002; Dresselhaus et al., 2004; Bandaru, 2007).

Nanotubes are essentially graphene sheets rolled up into varying diameters (Saito et al., 1998) and are attractive from both scientific and technological perspectives as they are extremely robust (Young's modulus approaching 1 TPa) and at least in pristine forms, chemically inert.

By varying the nature of wrapping of a planar graphene sheet and consequently its diameter, a nanotube can be constructed to function as either semiconductors or metals (Yao et al., 1999) that can be used in electronics (Collins and Avouris, 2000). The literature describes a variety of tubular structures composed of carbon and referred to as nanotubes (single-walled and multi-walled nanotubes or SWNTs and MWNTs) in which the graphene walls are parallel to the axis of the tube. Other configurations are called nanofibers, e.g., the graphene sheets are at an angle to the tube axis.

The electrical and thermal conductivity (Hone et al., 2000) properties of both SWNTs (Tans et al., 1997) and MWNTs have been well explored. While SWNTs (diameter ~1 nm) can be described as quantum wires due to the ballistic nature of electron transport (White and Todorov, 1998), the transport in MWNTs (diameter in the range 10 to 100 nm) is found to be diffusive or quasi-ballistic (Delaney et al., 1999; Buitelaar and Bachtold, 2002). Quantum dots can be formed in both SWNTs (Bockrath et al., 1997) and MWNTs (Buitelaar and Bachtold, 2002) and the Coulomb blockade and quantization of the electron states can be used to fabricate single-electron transistors (Tans et al., 1998).

Several electronic components based on CNTs such as single-electron transistors or SETs (Tans et al., 1998; Freitag et al., 2001; Postma et al., 2001), nonvolatile random access memory or RAM (Rueckes et al., 2000; Radosavljevic et al., 2002), field effect transistors or FETs (Radosavljevic et al., 2002), and logic circuits (Bachtold et al., 2001; Martel et al., 2002; Javey et al., 2003) have also been fabricated. However, most of these devices involve conventional lithography schemes and electronics principles, using nanotubes as conducting wires or modifying them along their length via atomic force microscopy (AFM)-based techniques (Postma et al., 2001).

While extremely important in elucidating fundamental properties, the above experiments used external electrodes made through conventional lithographic processes to contact the nanotubes and do not represent true nanoelectronic circuits. Additionally, in the well known MOSFET (metal oxide semiconductor field effect transistor) architecture, the nanotube serves as the channel between the electrodes (source and drain), and a SiO_2/Si-based gate modulates the channel conductance. In other demonstrations, cumbersome AFM manipulations (Postma et al., 2001) may be needed.

It would, therefore, be more attractive to propose new nanoelectronic elements to harness new functionalities peculiar to novel CNT forms such as nanotubes with bends or Y-junctions (Bandaru et al., 2005). One can also envision a more ambitious scheme and circuit topology in which both interconnect and circuit elements are all based on nanotubes, realizing true nanoelectronics (Figure 11.2). For example, nanotube-based interconnects do not suffer from the problems of electromigration that plague copper-based lines due to the strong carbon–carbon bonds. They can thus support higher

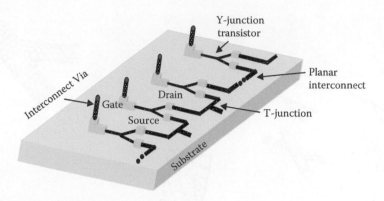

FIGURE 11.2
(See color insert.) Conceptual view of a possible CNT technology platform including Y-junction devices, interconnect vias, and directed nanotube growth. The overall objective is to create nanodevices with novel functionalities that go beyond existing technologies.

current densities (Collins et al., 2001b), for example, ~10 μA/nm^2 or 10^9 A/cm^2 versus 10 nA/nm^2 or 10^6 A/cm^2 for noble metals such as Ag. Additionally, the predicted large thermal conductivity (Kim et al., 2001) from ~3000 W/mK at 300 K up to an order of magnitude higher than copper may help alleviate the problem of heat dissipation in ever-shrinking devices.

In addition to miniaturization and lower power consumption, nanotube-based devices could also allow us to exploit the advantages of inherently quantum mechanical systems for practical devices such as ballistic transport and low switching voltages (Wesstrom, 1999), ~26 mV at room temperature ≡ $k_B T/e$).

11.2.1 Branched Carbon Nanostructures: Initial Work

At the very outset, any deviation from linearity, say in a branched Y-junction, must be accompanied by the disruption of the regular hexagonal motif. This can be accomplished in the simplest case by the introduction of pentagons and heptagons to account for the curvature (Iijima et al., 1992; Figure 11.3a). Since the charge distribution is likely to be nonuniform in these regions, the interesting possibility of localized scattering centers can be introduced. For example, rectification behavior was posited due to different work functions of contacts with respect to metal (M) and semiconductor (S) nanotubes (electrostatic doping) on either side of the bend (Yao et al., 1999). We will see later that how this can be exploited for more interesting device electronics.

Nanotube junctions were formed through high energy (1.25 MeV) electron beam exposure (in a transmission electron microscope) welding of linear SWNTs at high temperature (800°C) to form X-, Y-, or T- junctions (Terrones et al., 2002; Figures 11.3b and c). The underlying mechanism invoked was primarily the "knock-off" of carbon atoms and in situ annealing. Molecular

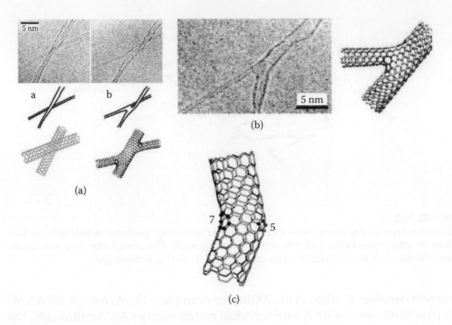

FIGURE 11.3
(See color insert.) (a) Bend in a nanotube introduces regions of positive and negative curvature. The associated heptagons and pentagons can have local excess or deficit of charge and can be used as scattering centers for nanoelectronics. (From From Yao, Z. et al. 1999. *Nature*, 402, 273. With permission.) (b) X-shaped and (c) Y-shaped nanotube molecular junctions can be fabricated by irradiating crossed single-walled nanotube junctions with high energy (~1.25 MeV) and beam intensity 10 A/cm²) electron beams. (From Terrones, M. et al. 2002. *Physics Review Letters*, 89, 075505. With permission.)

dynamics simulations intimate that vacancies and interstitials play a role. However, the purposeful synthesis of branched morphologies can be accomplished through chemical vapor deposition (CVD) methods.

Preliminary work on individual Y-junctions, grown through CVD in branched nanochannel alumina templates (Li et al., 2001) resulted in the observation of nonlinear I–V characteristics at room temperature through ohmic contact (Papadapoulos et al., 2000) and tunneling conductance (Satishkumar et al., 2000) measurements. From an innate synthesis view, nanotubes with T-, Y-, L-, and more complex junctions (resembling those in Figure 11.1) were initially observed in arc discharge-produced nanotubes (Zhou and Seraphin, 1995). Beginning in 2000, a spate of publications reported on the synthesis of Y-junctions through the use of organometallic precursors such as nickelocene and thiophene, in CVD (Papadapoulos et al., 2000; Satishkumar et al., 2000). Y-junction (Li et al., 2001) and multi-junction carbon nanotube networks (Ting and Chang, 2002) were also synthesized through the pyrolysis of methane over Co catalysts and through growth on roughened Si substrates. However, the mechanism of growth was not probed adequately.

11.3 Controlled Carbon Nanotube Y-Junction Synthesis

It is to be noted at the very outset that the Y-junctions synthesized are quite different in form and structure compared to crossed nanotube junctions (Fuhrer et al., 2000; Terrones et al., 2002) in which the nanotubes are individually placed and junctions produced through electron irradiation (Terrones et al., 2002).

Significant control in the growth of Y-junction nanotubes (Gothard et al., 2004) on bare quartz or SiO_2/Si substrates through thermal CVD was accomplished through the addition of Ti containing precursor gases to the usual nanotube growth mixture. In one instance, a mixture of ferrocene ($C_{10}H_{10}Fe$), xylene ($C_{10}H_{10}$), and a Ti containing precursor gas ($C_{10}H_{10}N_4Ti$) was decomposed at 750°C in the presence of flowing argon (~600 sccm) and hydrogen (75 sccm) carrier gases. The two-stage CVD reactor consisted of (1) a low temperature (~200°C) preheating chamber for the liquid mixture vaporization followed by (2) a high temperature (~750°C) main reactor. A yield of 90% MWNT Y-junction nanotubes that grow spontaneously on quartz substrates in the main reactor was obtained.

The mechanism for the Y-junction growth was hypothesized to depend on the carbide forming ability of Ti as measured by its large heat of formation (ΔH_f of –22 Kcal/g-atom). The Ti-containing Fe catalyst particles seeded nanotube nucleation by a root growth method in which carbon was absorbed at the root and then ejected to form vertically aligned MWNTs (Figure 11.4a). As the supply of Ti-containing Fe catalyst particles continues (Figure 11.4b), some of the particles (Fe and Ti) attach onto the sidewalls of the growing nanotubes (Figure 11.4c). The catalysts on the side then promote the growth of a side branch (Figure 11.4d) that, when further enhanced, forms a full-fledged Y-junction.

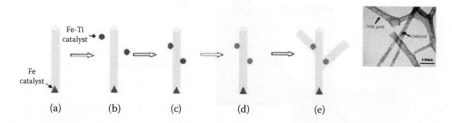

FIGURE 11.4
(See color insert.) Postulated growth sequence of Y-junction nanotube. (From Gothard, N. et al. 2004. *Nanoletters*, 4, 213. With permission.) (a) Initial seeding of straight nanotube through conventional catalytic synthesis. (From Teo, K.B.K. et al. 2004. In Nalwa, H.S., Ed., *Encyclopedia of Nanoscience and Nanotechnology*. Stevenson Ranch, CA: American Scientific Publishers. With permission.) (b) Ti-doped Fe catalyst particles (from ferrocene and $C_{10}H_{10}N_4Ti$) attach (c) to sidewalls and nucleate (d) the side branches (e).

The correlation of the carbide forming ability to branch formation was also supported by Y-junction synthesis in Hf-, Zr-, and Mo-doped Fe catalyst particles (Choi and Choi, 2005; Gothard et al., 2004) that also have large ΔH_f (HfC –26 Kcal/g-atom and ZrC –23 Kcal/g-atom) It was generally found that the use of Zr and Hf catalysts yielded larger diameter Y-junctions.

The ratio of the Ti precursor gas and the feedstock gases may be adjusted to determine the growth of the side branches at specific positions (Figure 11.5). For example, a decreased flow of the xylene gas at a point of time would halt the growth of the nanotube, while preponderance of the Fe-Ti precursor gas catalyst particles would nucleate the branch. The Y-junction formation has also been found to be sensitive to temperature, time, and catalyst concentration. The optimal temperature range is between 750 and 850°C; below 750°C, the yield is very low and temperatures greater than 850°C produce V-shaped nanotube junctions. Y-junction CNTs with minimal defects at the junction

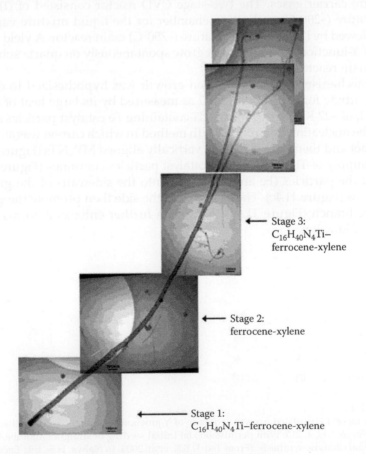

Stage 3:
$C_{16}H_{40}N_4Ti-$
ferrocene-xylene

Stage 2:
ferrocene-xylene

Stage 1:
$C_{16}H_{40}N_4Ti-$ferrocene-xylene

FIGURE 11.5
Controlled addition of Ti (see stage 3) can induce branching in linear nanotubes. (From Gothard, N. et al. 2004. *Nanoletters*, 45, 213. With permission.)

region were obtained when the atomic composition of Fe:Ti:C was in the ratio of 1:3:96.

The growth of the Y-junctions essentially seems to be a nonequilibrium phenomenon and various other methods have been found to be successful in proliferating branches, such as sudden reduction of temperature during a normal tip growth process (Teo et al., 2004) in which over-saturation by the carbon feedstock gas causes a surface energy-driven splitting of the catalyst particle and branch nucleation. Other catalyst particles such as Ca and Si have also been found to nucleate side branches (Li et al., 2001). The location of the junction is in any case a point of structural variation (Ting and Chang, 2002), the control of which seems to determine the formation of Y-junctions and their subsequent properties. It should also be possible to undertake a rigorous thermodynamic analysis (Bandaru et al., 2007) to rationalize the growth of nonlinear forms.

11.4 Electrical Characterization of Y-Junction Morphologies

Initial work has focused mainly on the electrical characterization of the nano-structures. In a large part, it was motivated by the possible influence of topology on electrical transport and supported through theoretical predictions on SWNT junctions. Current theoretical explanations of electrical behavior in Y-junctions are mainly based on SWNT Y-junctions, and the experimental demonstrations detailed in this paper were made on both SWNT and MWNT Y-junctions (Papadapoulos et al., 2000; Satishkumar et al., 2000). It is speculated that the relatively low-temperature CVD methods used to date may not be adequate to reliably produce SWNT based Y-junctions with larger energies (Terrones et al., 2002).

While SWNTs have been extensively studied theoretically (Ajayan, 1999; Dai, 2001), MWNTs have been relatively less scrutinized. An extensive characterization of their properties is found in the literature (Forro and Schonenberger, 2001; Bandaru, 2007). MWNTs are generally found to be metal-like (Forro and Schonenberger, 2001) with possibly different chiralities for the constituent nanotubes. Currently, there is some understanding of transport in straight MWNTs; it has been shown that electronic conduction mostly occurs through the outermost wall (Bachtold et al., 1999) and interlayer charge transport in the MWNT is dominated by thermally excited carriers (Tsukagoshi et al., 2004).

While the outer wall dominates in the low-bias regime (<50 mV), at higher bias many shells can contribute to the conductance with an average current-carrying capacity of 12 μA/shell at room temperature (Collins et al., 2001b). In contrast to SWNTs with micrometer coherence lengths, the transport in MWNTs is quasi-ballistic (Buitelaar and Bachtold, 2002) with mean free paths <100 nm. Based on the above survey of properties in straight MWNTs, we could extend the hypothesis in that noncoherent electronic transport dominates the Y-junctions and other branched morphologies.

11.5 Carrier Transport in Y-Junctions: Electron Momentum Engineering

The progenitor of a Y-junction topology for electronic applications was basically derived from an electron-wave Y-branch switch (YBS; Palm and Thylen, 1992) in which a refractive index change of either branch through electric field modulation can affect switching.

This device, demonstrated in the GaAs/AlGaAs (Worschech et al., 2001) and InP/InGaAs-based (Hieke and Ulfward, 2000; Lewen et al., 2002) two-dimensional electron gas (2-DEG) system, relies on ballistic transport and was proposed for low power, ultra-fast (THz) signal processing. It was derived theoretically (Wesstrom, 1999) and proven experimentally (Shorubalko et al., 2003) that based on the ballistic electron transport, nonlinear and diode-like I–V characteristics were possible.

These devices based on III–V materials, while providing proof of concept, were fabricated through conventional lithography. It was also shown in 2-DEG geometry (Song et al., 1998) with artificially constructed defects and barriers that the defect topology can affect the electron momentum and guide the current to a pre-determined spatial location independent of input current direction. This type of rectification involves a new principle of electron momentum engineering in contrast to the well known band engineering.

Nanotubes provide a more natural avenue to explore such rectification behavior. It was theoretically postulated (Andriotis et al., 2001) that switching and rectification (Figure 11.6) could be observed in symmetric (e.g., no change in chirality from stem to branch) Y-junction SWNTs, assuming quantum conductivity of electrons in which rectification could be determined (Andriotis et al., 2003) by the:

1. Formation of a quantum dot asymmetric scattering center (Song et al,. 1998) at the location of the Y-junction
2. Finite length of the stem and branches connected to metallic leads
3. Asymmetry of the bias applied or the potential profile (Tian et al., 1998) across the nanotube
4. Strength of the nanotube–metal lead interactions and influence of the interface (Meunier et al., 2002)

Some of the possibilities in which the nature of the individual Y-junction branches determines the electrical transport are illustrated in Figure 11.7.

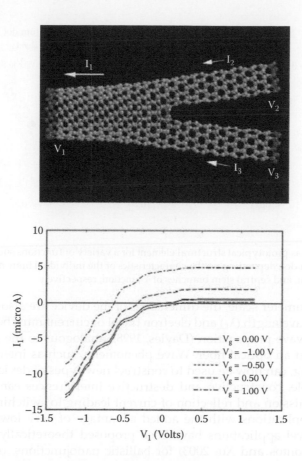

FIGURE 11.6

(See color insert.) Asymmetry and rectification-like behavior in current and voltage characteristics of single-walled Y-junction nanotube is indicated through quantum conductivity calculations. (From Andriotis, A.N. et al. 2001. *Physics Review Letters,* 87, 066802. With permission.)

11.6 Applications of Y-CNTs to Novel Electron Functionality

Generally, the motivation for use of new CNT morphologies such as Y-junctions based on SWNTs or MWNTs in addition to the miniaturization of electronic circuits is the possible exploration of new devices and technologies through new physical principles. The existence of negative curvature fullerene-based units (Scuseria, 1992) and branching in nanotubes necessitates the presence of topological defects in the form of pentagons, heptagons, and octagons at the junction regions for maintaining a low energy sp^2 configuration (Andriotis et al., 2002). These intrinsic defects are natural scattering centers that could affect or modulate the electrical transport characteristics of a nanotube.

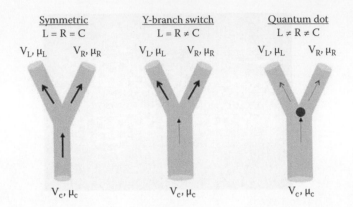

FIGURE 11.7
CNT Y-junction as prototypical structural element for a variety of functions such as switching or as a quantum dot, depending on the characteristics of the individual branches. L, R, and C refer to left, right, and central stem branches of Y-junction, respectively.

At the nanometer scale, the dimensions of the device are also comparable to electron wavelength (λ_F) and electron travel or current must be considered in terms of wave propagation (Davies, 1998), analogous to the propagation of light down an optical fiber. Wave phenomena, such as interference and phase shifting, can now be used to construct new types of devices.

For example, constructive and destructive interferences can be used to cause transmission and reflection of current leading to switching and transistor-like applications with the added advantage of very low power dissipation. Novel applications have been proposed theoretically (Xu, 2001a; Xu, 2002; Csontos and Xu, 2003) for ballistic nanojunctions, of which the Y-junction is only one example. Several of these applications have been demonstrated in preliminary experiments and will be elucidated later in this chapter. A brief overview follows.

11.6.1 Switching and Transistor Applications

In a basic Y-junction switch, an electric field can direct electrons into either of two branches, while the other branch is cut off (Wesstrom, 1999). It has been shown in computer simulations (Palm and Thylen, 1992) that a sufficient lateral field for electron deflection is created by applying a very small voltage of the order of millivolts (mV).

The specific advantage of a Y-junction switch is that it does not need single-mode electron waveguides for its operation and can operate over a wide range of electron velocities and energies; the reason is that the electrons are only deflected and not stopped by barriers. An operational advantage over a conventional FET could be that the current is switched between two outputs rather than completely turned on or off (Palm and Thylen, 1996), leading to higher efficiency of operation.

An electrical asymmetry can also be induced through structural or chemical means across the two branches in a nanostructured junction. The Y-junction region, for instance, can possess a positive charge (Andriotis et al., 2001) for two reasons: the presence of (1) topological defects due to the formation of nonhexagonal polygons at the junction to satisfy the local bond order (Crespi, 1998) in which delocalization of the electrons over an extended area leads to a net positive charge, and (2) catalyst particles that are inevitably present during synthesis (Gothard et al., 2004; Teo et al., 2004).

This positive charge and the induced asymmetry are analogous to a "gating" action that could be responsible for rectification. While the presence of defects at the junction seems to assist switching, there is also a possibility that such defects may not be needed as some instances of novel switching behavior in Y-junctions are observed in the noticeable absence of catalyst particles. Additional studies are necessary to elucidate this aspect, but such an observation is significant in that a three-dimensional array of Y-junction devices based on CNTs would be much easier to fabricate if a particle is not always required at the junction region.

11.6.2 Rectification and Logic Function

It is possible to design logic circuitry, based on electron wave guiding in Y-junction nanotubes (Xu, 2002), to perform operations similar to and exceeding the performance of conventional electronic devices (Palm and Thylen, 1996). When finite voltages are applied to the left and the right branches of a Y-junction, in a push–pull fashion (i.e., $V_{left} = -V_{right}$ or vice versa), the voltage output at the stem would have the same sign as the terminal with the lower voltage.

This dependence follows from the principle of continuity of electro-chemical potential ($\mu = -eV$) in electron transport through a Y-junction and forms the basis for the realization of an AND logic gate, i.e., when either of the branch voltages is negative (say, corresponding to a logic state of 0), the voltage at the stem is negative and positive voltage (logic state of 1) at the stem is obtained only when both the branches are at positive biases.

The change of μ is also not completely balanced out due to the scattering at the junction, and results in nonlinear interaction of the currents from the left and right sides (Shorubalko et al., 2003). To compensate, the resultant center branch voltage (V_S) is always negative and varies parabolically (as V^2) with the applied voltage (Xu, 2001a).

11.6.3 Harmonic Generation and Frequency Mixing

The nonlinear interaction of the currents and the V^2 dependence of the output voltage at the junction region also suggest the possibility of higher frequency/harmonic generation. When an AC signal of frequency ω, V_{L-R} = A cos [ωt], is applied between the left (L) and right (R) branches of the Y-junction, the output signal from the stem (V_S) would be of the form:

$$V_S = \mathbf{a} + \mathbf{b} \cos [2\omega t] + \mathbf{c} \cos [4\omega t]$$

where **a**, **b** and **c** are constants. The Y-junction can then be used for second and higher harmonic generation or for frequency mixing (Lewen et al., 2002). The second harmonic (2ω) output is orthogonal to the input voltage and can be easily separated out. These devices can also be used for an ultra-sensitive power meter, as the output is linearly proportional to V^2 to very small values of V.

A planar CNT Y-junction with contacts present only at the terminals suffers from fewer parasitic effects than a vertical transistor structure and high frequency operation, up to 50 GHz at room temperature (Song et al., 2001), is possible. It can be seen from the brief discussion above that several novel devices can be constructed on CNT Y-junction technology and may be the forerunners of a new paradigm in nanoelectronics.

11.7 Experimental Work on Electrical Characterization

Compared to the large body of work on electrical transport through linear nanotubes, the characterization of nonlinear nanotubes is still in its infancy. The samples for electrical measurements are typically prepared by suspending nanotube Y-junctions, say, in isopropanol and depositing them on a SiO_2/Si substrate with patterned Au pads.

Y-junctions, in proximity to the Au contact pads, are then located at low voltages (<5 kV) using a scanning electron microscope (SEM) and contacts patterned to each branch of the Y-junction, either through electron beam lithography (Kim et al., 2006) or focused ion beam-induced metal deposition (Gopal et al., 2004); see Figure 11.8. In the latter case, special care needs to be taken to not expose the nanotube to the ion beam to prevent radiation damage. The early measurements in Y-junction nanotubes explored the theoretical idea of rectification between any two branches of the Y-junction.

11.7.1 Rectification Characteristics

Initial work in measurement of current–voltage (I–V) characteristics of Y-junctions was accomplished through two-terminal measurements using the stem as one terminal and the two branches connected together as the other terminal. With a stem:branch diameter ratio approximately 60:40 nm, diode-like behavior was observed (Papadapoulos et al., 2000).

The authors ascribing SWNT-like p-type semiconductor characteristics modeled the behavior on a p–p isotype heterojunction where the concen-

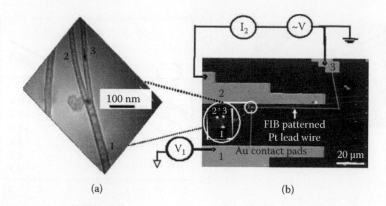

FIGURE 11.8
(a) MWNT Y-junction electrical measurement configuration imaged in (b) scanning electron micrograph.

tration (N) of carriers (holes) varied inversely as the fourth power of the diameter (D), i.e.,

$$\frac{N_{stem}}{N_{branch}} \sim \left(\frac{D_{branch}}{D_{stem}}\right)^4.$$

The doping mismatch would then presumably account for the rectifying behavior.

In the absence of further characterization, it is difficult to see how semiconducting characteristics could be assigned to >40-nm MWNTs. A subsequent publication from the same group (Perkins et al., 2005) ascribed the electrical characteristics to be dominated by activated conduction, presumably through carrier hopping.

More complex nonlinear, quasi-diode-like behavior has been seen (Bandaru et al., 2005) in MWNT Y-junctions (Figure 11.9) that corresponds to a saturation of current at positive bias polarities. This is predicted from theoretical considerations (Xu, 2001b) where the current cannot decrease beyond a certain value, but saturates at a value (~ 0.5 µA at positive V_{1-2}) corresponding to the intrinsic potential of the junction region itself. Such behavior has also been seen in Y-junctions fabricated from 2-DEG systems (Shorubalko et al., 2003; Wallin et al., 2006). The underlying rationale, in more detail, is as follows.

The necessity of maintaining a uniform electrochemical potential ($\mu = eV$) in the overall structure gives rise to nonuniform and nonlinear interactions (Xu, 2001a). For example, when V_2 (voltage on branch 2) decreases, μ_2 (= − eV_2) increases and an excess electron current flows toward the central junction. The balance between the incoming current and the outgoing currents, at the junction itself, is achieved by increasing μ_2 (decreasing V_1). On the other hand, when μ_2 decreases, μ_1 decreases also, but cannot decrease past

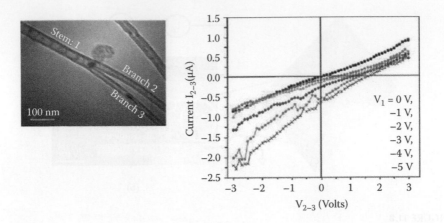

FIGURE 11.9

(See color insert.) Current (I) and voltage (V) characteristics of MWNT Y-junction. A constant direct current voltage is applied on the stem, while the I–V behavior across branches 2 and 3 is monitored. The gating action of the stem voltage (V_1) and the asymmetric response are to be noted. (From Bandaru, P.R. et al. 2005. *Nature Materials*, 4, 663. With permission.)

a certain critical point, i.e., the fixed electrochemical potential dictated by geometry or defects of the junction.

Work on SWNT-based Y-junctions has revealed the diode like behavior in richer detail. It was established through Raman spectroscopy analyses (Choi and Choi, 2005) that constituent Y-junction branches could be either metallic or semiconducting. This gives rise to the possibility of forming intrinsic metal (M)–semiconductor (S) junctions within or at the Y-junction region. Experiments carried out on SWNTs in the 2 to 5-nm diameter range (Choi and Choi, 2005) that could, depending on the chirality, correspond to metallic or semiconducting nanotubes with energy gaps in the range of 0.37 to 0.17 eV (Ding et al., 2002), reveal ambipolar behavior in which carrier transport due to both electrons and holes may be important.

Considering, for example, that the Fermi level (E_F) for the metallic branch of the Y-junction is situated around the middle of the semiconductor branch band gap, positive or negative bias on the semiconductor branch can induce electron or hole tunneling from the metal (Figure 11.10). Such temperature-independent tunneling behavior was invoked through modeling the I–V characteristics to be of the Fowler–Nordheim type. However, at higher temperatures (>100 K), thermionic emission corresponding to barrier heights of 0.11 eV could better explain the electrical transport results.

On an interesting note, it should be mentioned that the contact resistance at a branch–metal contact could also play a major role and contribute to the relatively low currents observed in experiments (Meunier et al., 2002). While MWNTs should theoretically have a resistance smaller than h/e^2 (~26 kΩ), ideal and reproducible ohmic contacts through metal evaporation have been difficult to achieve.

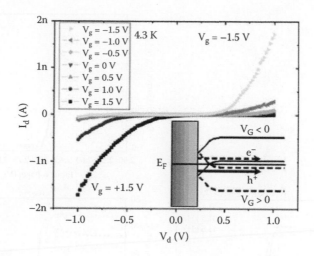

FIGURE 11.10

(See color insert.) I–V measurements on single walled Y-CNTs in which a metallic CNT interface with a semiconducting CNT (see band diagram in inset) indicates ambipolar behavior as a function of applied gate voltage on the semiconducting nanotube. (From Kim, D.H. et al. 2006. *Nanoletters*, 6, 2821. With permission.)

11.7.2 Electrical Switching Behavior

Intriguing experiments have brought forth the possibility of using CNT Y-junctions for switching applications as electrical inverters analogous to earlier Y-switch studies in 2-DEG systems (Palm and Thylen, 1992; Hieke and Ulfward, 2000; Shorubalko et al., 2003). In this measurement (Bandaru et al., 2005), a DC voltage was applied on one branch of the Y-junction while the current through the other two-branches was probed under a small AC bias voltage (<0.1 V). As the DC bias voltage is increased, at a certain point the Y-junction goes from nominally conducting to a "pinched-off" state. This switching behavior was observed for all the three-branches of the Y-junction at different DC bias voltages.

The absolute value of the voltage at which the channel is pinched off was similar for two branches (~2.7 V, as seen in Figures 11.11a and b) and is different for the third stem branch (~5.8 V, as in Figure 11.11c). The switching behavior was seen over a wide range of frequencies, up to 50 kHz, the upper limit set by the capacitive response of the Y-junction when the branch current tends to zero.

The detailed nature of the electrical switching behavior is not understood at present. The presence of catalyst nanoparticles (Figure 11.1) in the conduction paths may blockade current flow and their charging could account for the abrupt drop-off of the current. The exact magnitude of the switching voltage would then be related to the exact size of the nanoparticle, which suggests the possibility of nanoengineering the Y-junction to get a variety of switching behaviors. However, it was deduced (Andriotis and Menon, 2006),

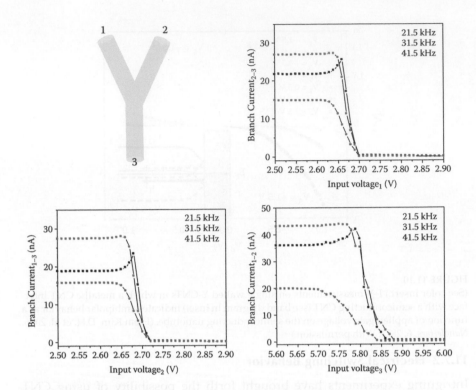

FIGURE 11.11

(See color insert.) Abrupt modulation of current through two branches of Y-junction, indicative of electrical switching, revealed by varying voltage on the third branch. The voltage at which the switching action occurs on the two branches (1 and 2) is similar and smaller (~2.7 V; see a and b) compared to the turn-off voltage (~5.8 V) on the stem (3) in c. Such abrupt switching characteristics are seen up to 50 kHz, the upper limit arising from the capacitive response of the Y-junction. (From Bandaru, P.R. et al. 2005. *Nature Materials*, 4, 663. With permission.)

through tight binding molecular dynamics simulations on SWNTs, that interference effects could be solely responsible for the switching behavior, even in the absence of catalyst particles.

An associated possibility is the intermixing of the currents in the Y-junction, where the electron transmission is abruptly cut off due to the compensation of currents, for example, the current through branches 2 and 3 is cancelled by current leakage through stem 1. The simultaneous presence of an AC voltage on the source–drain channel and a DC voltage on the control or gate terminal could also result in an abrupt turn-off, due to defect-mediated negative capacitance effects (Beale and Mackay, 1992). Further research is needed to clarify the exact mechanisms of these interesting phenomena.

A suggestion was also made that AND logic gate behavior could be observed (Bandaru et al., 2005) in a Y-junction geometry. The continuity of the electro-chemical potential from one branch of the CNT Y-junction to another is the basis for this behavior of the margins (see Section 11.5 for a more detailed explanation).

11.7.3 Current Blocking Behavior

Other interesting characteristics were seen when the CNT Y-morphologies were in situ annealed in the ambient, in a range of temperatures from 20 to 400°C, and I–V curves measured for various configurations of the Y-junction. The observations are summarized in Figure 11.12 and are fascinating from the view of tunability of electrical characteristics.

As the annealing is continued, the onset of nonlinearity in the I–V characteristics is observed (Figure 11.12b). With increasing times, the nonlinearity increases, but is limited to one polarity of the voltage. This is reminiscent of diode-like behavior and can be modeled as such (Figure 11.12c and inset).

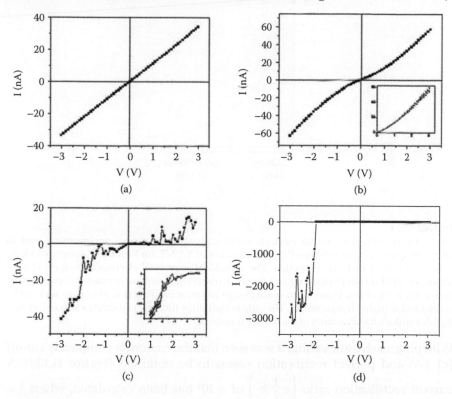

FIGURE 11.12
I–V characteristics for CNT Y-junction subject to high temperature annealing (~150°C) as a function of annealing time, after cooling down to room temperature. (a) Ohmic behavior is observed prior to annealing. At increasing times, nonlinearity (b) is introduced and can be modeled in terms of space charge currents (see inset) as $I = AV + BV^{3/2}$. Further annealing results in (c) diode-like behavior (inset shows a fit: $I = I_o (-1)$) and finally in (d) a current blocking or rectification behavior in which a 10^4-fold suppression in current is seen.

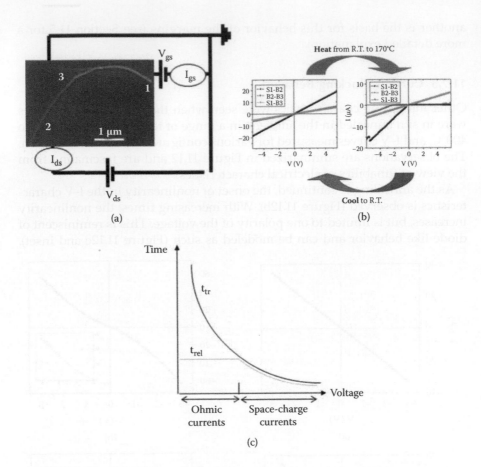

FIGURE 11.13
(See color insert.) (a) Scanning electron micrograph showing circuit arrangement used to probe the current (I) and voltage (V) characteristics of Y-CNT. (b) Reversible current blocking behavior induced in CNT Y-junction. The individual segments' electrical transport characteristics exhibit different blocking and linear characteristics and are geometry dependent. (c) Transition from ohmic behavior to space charge behavior is a function of voltage and can be accelerated at higher temperatures. The ratio of the transit time (t_{tr}) to dielectric relaxation time (t_{rel}) determines the dynamics of the carrier transport.

With progressive annealing, it was seen that the current is completely cut-off (<1 pA) and perfect rectification seems to be obtained (Figure 11.12d). A current rectification ratio $\left(= \dfrac{I_{ON}}{I_{OFF}} \right)$ of > 10^4 has been calculated, where I_{ON} denotes the current through the Y-device at –3 V while I_{OFF} is the current at +3 V. The transition to linear characteristics was more rapid at elevated temperatures of annealing (>200°C).

In yet another set of Y-junction samples, reversible behavior from ohmic to blocking type was observed (Figure 11.13). Interestingly, it was observed that the current through one set of branches (S1-B3 in Figure 11.13b) was not affected by the annealing, while the blocking voltages differed for the other two configurations (~0.1 V for S1-B2 and ~2.6 V for B2-B3).

The time-dependent behavior and rectification seem to be functions of both temperature and time of annealing. The sharp cut-off of the current at positive voltages is also remarkable [The cut-off of the currents at the positive polarity of the voltage seems to be indicative of hole transport in the Y-junctions through electrostatic doping, due to the work function of Pt, with Φ_{Pt} ~5.7 eV larger than Φ_{CNT}]. While it is well known that contacts to p-type CNTs through high work-function metals where $\Phi_M > \Phi_{CNT}$ (~4.9 eV) result in ohmic conduction (Javey et al., 2003; Yang et al., 2005), the transition from conducting to blocking behavior is generally more gradual.

A continuous change in the current was also seen when a gate voltage modulates the p-CNT channel conduction and in CNT based p–n junctions (Lee et al., 2004) with diode-like behavior (Lee et al., 2004; Manohara et al., 2005; Yang et al., 2005). Consequently, we think that the observed behavior is unique to the CNT branched topologies examined here.

The annealing-induced rectification behavior in CNT Y-junctions may be intrinsic to the Y-nanotube form or be related to the nature of the contacts. It was reported that SWNTs contacted by Pt (Javey et al., 2003) could yield non-metallic behavior arising from a discontinuous (Zhang et al., 2000) contact layer to the nanotubes. It is also possible that the outer contacting walls of the MWNT Y-junction are affected by the annealing procedure, resulting in a modification of the Schottky barrier (Collins et al., 2001a).

Several CNT device characteristics, such as transistors (Appenzeller et al., 2002) and photodetectors (Freitag et al., 2003) are Schottky barrier mediated. Exposure to oxygen is also known to affect the density of states of CNTs and the I–V characteristics (Collins et al., 2000). However, the low temperatures (<300°C) employed in the experiment preclude oxidation (Ajayan et al., 1993; Collins et al., 2001b) and the blocking behaviors that were observed in our experiment cannot be justified on the above principles.

A hint for explaining this intriguing I–V behavior is obtained through modeling the I–V characteristics of Figure 11.12b, an intermediate stage in the annealing process. A supralinear behavior, i.e., I proportional to AV + BV$^{3/2}$, can be fitted (Figure 11.12b inset) that may be indicative of space-charge limited currents (A and B are numerical constants). To further understand the transport behavior, it becomes necessary to examine the role of the contacts in detail. Prior to annealing, where we observe linear behavior (Figure 11.12a), the ohmic contact is a reservoir of free holes. Generally, the ratio of the hole transit time (t_{tr}) to the dielectric relaxation time (t_{rel}) in the CNT determines the carrier dynamics and currents.

A large t_{tr}/t_{rel} ratio, obtained, say, at smaller voltages would result in ohmic currents while a lower t_{tr}/t_{rel} ratio at increased voltages would imply space charge currents (Figure 11.13c). As t_{rel} is inversely proportional (Muller and Kamins, 1986) to conductivity (σ), such effects would be more important at

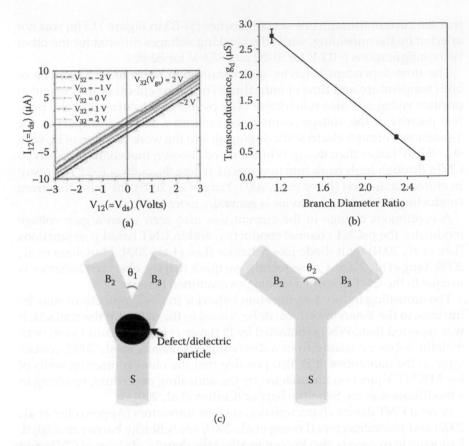

FIGURE 11.14
(See color insert.) (a) Current (I) and voltage (V) characteristics of Y-CNT consisting of metallic branches. The inset shows the circuit diagram. (b) Conductance (g_d) through two nanotubes of different diameters in the Y-CNT is inversely proportional to their ratio (= d_s/d_c). (c) Proposed Y-junction CNT based switching devices. (i) Smaller angle (θ_1) between the branches (B_2 and B_3) can result in a higher gating efficiency for the stem (S). (ii) Y-CNT with uniform gating and electrical switching characteristics can be fabricated by synthesizing all the constituent nanotubes to be of the same diameter and θ_1 = 120 degrees. (From Park, J. et al. 2006. *Applied Physics Letters*, 88, 243113. With permission.)

elevated temperatures for metallic Y-CNTs. We suggest then that the saturation of the current at positive voltages may be due to space charge effects. In other words, at sufficiently high voltages (>0.1 V for S1-B2 and >2.6 V for B2-B3 in Figure 11.13b), further hole generation is choked due to preponderance of space charge and negligible current is observed.

Note that such a scenario can also be predicted from the band structure diagrams at the metal contact–CNT interface (Figure 1.14). At (ii) forward bias, there is a significant barrier to hole transport and space charge processes that could be important. At (iii) negative voltages, however, no hole barriers exist and ohmic conduction is obtained. The space charge hypothesis is also supported by the observation that over a period of time, the blocking behavior reverts to ohmic conduction. In an earlier study (Antonov and Johnson, 1999) of a straight SWNT diode containing a charged impurity, similar space charge-based arguments were invoked to explain the observed current rectification. We are currently conducting experiments to probe the time constants and dynamics of current flow through the Y-CNTs.

We speculate that the different diameters of the branches of the Y-junction (Figure 11.1 or Figure 11.13a) may play a role in the current blocking behavior. The basis (Figure 11.13b) is that rectification was not observed for a particular (S1-B3) two-terminal configuration. It was found that the magnitude of the conductance (g_d) between any two terminals of the Y-junction is inversely proportional to the ratio of the diameters through which the current is flowing (see Figure 11,13b), i.e., a larger discrepancy between the branch diameters lowers the conductance. In analogy to fluid flow through a pipe, the hole current can similarly be modulated by the Y-junction geometry. It would be instructive to characterize the blocking behavior systematically as a function of branch diameters in the Y-junction geometry.

Shockley (1950) noted the phenomenon that current was low when voltage was high in terms of space charge distribution and remarked on its presence in the collector region of a junction transistor amplifier. The I–V characteristics observed in the CNT Y-forms here could be compared to those of a transit-time device such as an BARITT (barrier injection and transit time) diode (Streetman and Banerjee, 2000). Such devices consider space charge limited currents for creating negative resistance regions that can be exploited for switching, oscillation, amplification, and other functions in high-speed circuitry (Sze, 1981). The time delay associated with space charge dissipation is useful for generating microwaves and the BARITT devices are consequently used as local oscillators in microwave receivers.

11.7.4 Metallic Y-Junctions

It is also possible that the constituent nanotubes in the Y-junction can exhibit metallic conduction, in which case the behavior is analogous to that of three interconnected metal lines! Such behavior has indeed been observed (Park et al., 2006; Figure 11.14a). A control voltage (V) is applied on one of the three terminals (gate) of the Y-CNT while the current (I) through the other two terminals (source and drain) is monitored.

The I–V curves of the Y-CNTs, as a function of gate (V_{gs})/bias (V_{ds}) voltage are shown in Figure 11.14a. The notable features are (a) ohmic conduction, (b) a proportional displacement of the current with V_{gs} and V_{ds}, and (c) a geometry-dependent conductance. The devices are also seen to have both current and voltage gain.

In line with an earlier (Perkins et al., 2005) study, we define the differential current gain (g_{diff}) as the ratio of

$$\frac{\partial I_{gs}}{\partial V_{gs}} \text{ to } \frac{\partial I_{ds}}{\partial V_{gs}},$$

at a constant source–drain voltage (V_{ds}). In our Y-CNT devices a g_{diff} of up to 10 is obtained at room temperature. The voltage gain is calculated (Tans et al., 1998) by considering the voltage change (ΔV_{ds}) for a given increment in gate voltage (V_g), at a constant value of the current. The voltage gain when the stem (1 in Figure 11.13a) is used for electrical gating is ~6 while the value drops to ~0.3 when the branch (3 in Figure 11.13a) is used for a gate.

Other parameters of importance in quantifying electrical characteristics are the transconductance,

$$g_m = \left(\frac{\partial I_{ds}}{\partial V_{gs}} \right)_{V_{ds}},$$

the ratio of the output current (I_{ds}) to the modulating or gating voltage (V_{gs}) at constant bias voltage (V_{ds}), and the output conductance,

$$g_d = \left(\frac{\partial I_{ds}}{\partial V_{ds}} \right)_{V_{gs}},$$

the ratio of the output current to the bias voltage at a constant gate voltage. It was found that the magnitude of the g_d is inversely proportional to the ratio of the diameters of the branches through which the current is flowing (Figure 11.14b), i.e., a larger discrepancy between the branch diameters lowers the conductance. On the other hand, the g_m that denotes the gating efficiency for current modulation seems to be directly proportional to the diameter of the gating nanotube (Park et al., 2006).

From the Y-CNT electrical characteristics it is seen that the mode of operation of the device is quite different from that of a conventional FET and more akin to a Schottky barrier type FET (Muller and Kamins, 1986). The current modulation may be controlled by the electric field (= $\Delta_r \cdot V$) at the gating terminal, and not the carrier density in the channel. In this case, the device performance can be improved by increasing the diameter of the gate so as to have a greater modulating effect on the source–drain current.

The assumption of electric field control can also be tested (Park et al., 2006) by varying the angle θ between the branches, of the Y-CNT. From elementary electrostatics (Jackson, 1999), the surface charge density at a distance ρ from the intersection of the gate with the junction varies as $\rho^{(\pi/\theta - 1)}$ implies a greater modulating effect at a sharper angle. As an example, for the Y-CNT in Figure 11.4a, the stem would have a higher gating efficiency than in case 11.4b.

The manipulation of the junction region to be dielectric or insulating (say, by localized ion implantation) can also be used to change the electrochemical potential and vary device characteristics. It is interesting to note the similarity of Figure 11.14c to a metal oxide semiconductor field effect transistor (MOSFET) n which the stem is analogous to a gate, the dielectric particle plays the role of the gate oxide, and the channel is formed by conduction through the branches. On the other hand, uniform electrical gating or switching characteristics may be obtained by making all the nanotubes the same diameter and length at an angle of 120 degrees to each other (Figure 11.14c).

We also suggest that even more interesting applications are facilitated by our experimental results and observations. For example, a single Y-CNT switching device can be made to have diverse operating characteristics through the use of different gating terminals. Y-junction based technology can be used for devices with (1) multiple programmable characteristics, and as (2) components in field programmable gate arrays (FPGAs) for reconfigurable computing (Rabaey, 1996) where it would be possible to dynamically alter circuit paths. The large-scale integration density and high speed of signal propagation intrinsic to nanotubes would be advantageous in this regard.

11.8 Topics for Further Investigation

For the novel applications of Y-CNTs to be practicable, it is necessary to gain better control of the geometry through synthesis or processing of Y-CNTs. It is plausible that remarkable features of Y-junction-based transistors such as the abrupt switching (Bandaru et al., 2005) and differential gain (Perkins et al., 2005) observed in earlier studies could be related to the presence and location of defects. In situ engineering of CNT morphology, e.g., exposure to intense electron-beam radiation (Terrones et al., 2002) could also be utilized to tailor individual Y-CNT characteristics. Other aspects that merit further study are discussed below.

Characterization of the detailed morphologies of the Y-junctions and their effects on electrical properties—Catalyst particles have been found at the junction, along the length of the Y-junctions, and at the tips (Li et al., 2001; Figure 11.1). The effect of the location, type, and composition of these particles and their scattering characteristics on the electrical transport would yield insight into their influence.

Detailed study of the growth mechanisms of multi-junction carbon nanotubes vis-à-vis influence of catalyst particles—It has been contended (Teo et al., 2004) that a perturbation during growth may promote the formation of branches or junctions. For example, if the temperature is reduced during the growth process, a catalyst particle over-saturated with carbon can be induced to nucleate another branch. On the other hand, if the Ti catalyst does play a role, the importance of carbide formers in inducing side-branch growth is interesting from a basic thermodynamic view, i.e., is a high negative heat of formation (ΔH_f) of the carbide formers necessary? It is worth noting that Y-junctions and other morphologies, such as H-junctions and T-junctions (Ting and Chang, 2002), are also formed when methane (that may yield positive ΔH_f) is used as a precursor. The composition of the catalyst particle that nucleates the nanotube branches (Teo et al., 2004), and the effects of stress generated at the growing nanotube tips may also play important roles (Ting and Chang, 2002) in Y-junction growth.

Importance of electrical conductivity through nanotubes with bends, junctions, and catalyst particles for the future viability of nanotube-based electronics—While the MWNT has an intrinsic resistance, say, due to ballistic transport, a capacitive and inductive component will also have to be considered due to the presence of particles, intertube transport etc. Such a study will also yield insight into the speeds of operation of nanotube-based devices and affecting factors in terms of the RC delay.

It has been theoretically proposed (Burke, 2003) that straight SWNTs are capable of THz operation and can even be used as nanoantennae (Burke, 2002) for radiation and detection in these very high frequency ranges. Whether SWNT or MWNT Y-junctions are capable of being operated at such high frequencies, due to the presence of defects, and determining an ultimate limit (Guo et al., 2002) to their performance is worth investigating.

Creation of novel CNT-based circuits based on Y-junctions and branched morphologies (Xu, 2005)—These circuits may be fabricated so as to construct universal logic gates such as NAND and X-NOR. Such a demonstration together with the possibility of multi-functionality, e.g., where a catalyst particle could be used for photon detection (Kosaka et al., 2002) will be important. A variety of such novel circuits can be fabricated, leading to a potentially new paradigm for nanoelectronics that goes well beyond traditional FET architecture shape-controlled logic elements.

Paramount importance of assembly into a viable circuit topology and large scale fabrication—Currently, the Y-junctions are grown in mats or bundles and individually isolated and measured. A scheme by which each junction is assembled in this way is not practical. Ideas for self-assembly and controlled placement (Huang et al., 2005) of nanotubes will have to be addressed.

While success has been achieved in coordinating the placement of loose nanotubes, e.g., through chemically functionalized substrates (Liu et al., 1999), dip-pen lithography (Rao et al., 2003), an orienting electric field for exploiting the dipole character of nanotubes (Joselevich and Lieber, 2002), magnetic fields (Hone et al., 2000), and microfluidic arrangements (Huang et al., 2001), the assembly of loose nanotubes is difficult to scale up and remains a critical issue.

An array of Y-junctions can be prepared on the same nanotube stem, as found in preliminary studies, by exposing only periodic locations (Qin et al., 2005) along the length of a nanotube, sputter depositing catalyst, and growing parallel Y-junction branches from these linear arrays of catalysts using electric field-induced direction control during subsequent CVD processes, e.g., at any angle from the main stem. The above growth technique could be used to make multiple Y-junction devices in series or parallel. This proof of growth will go a long way in demonstrating the feasibility of large scale nanoelectronic device assembly—a highly important question at present.

11.9 Conclusions

The study of Y-junctions is still in its infancy. Their synthesis, while reasonably reproducible, is still challenging in terms of precise placement on a large scale. It is worth mentioning that this is an issue even with linear nanotubes and may well determine the feasibility of CNT applications in general.

Presently, both single-walled and multi-walled Y-CNTs, the constituent branches of which may have different diameters or semiconducting or metallic characters, have been synthesized. Such internal diversity gives rise to novel phenomena such as (1) rectification and current blocking behavior, (2) electrical switching, and (3) logic gate characteristics. It is also interesting to investigate whether the three terminals of a Y-CNT can be interfaced into a transistor-like paradigm. Future investigations should correlate detailed physical structures of the nanostructure morphologies to electrical measurements to gain a better understanding of the conduction processes vis-à-vis the role of defects and geometry in nonlinear structures. Such a comprehensive and correlated study would be useful to the nanotube and nanowire community and pave the way to the realization of shape-controlled nanoelectronic devices exclusive to the nanoscale.

Acknowledgments

I gratefully acknowledge support from the National Science Foundation (Grant ECS-05-08514) and the Office of Naval Research (Award N00014-06-1-0234). Discussions and interactions with Professors A. Rao (Clemson University), H.Q. Xu (Lund University), and Dr. K. Yang, are deeply appreciated.

References

Ajayan, P.M. 1999. Nanotubes from carbon. *Chemical Reviews*, 99, 1787–1799.

Ajayan, P.M., Ebbesen, T.W., Ichihashi, T. et al. 1993. Opening carbon nanotubes with oxygen and implications for filling. *Nature*, 362, 522–525.

Andriotis, A.N. and Menon, M. 2006. Are electrical switching and rectification inherent properties of carbon nanotube Y junctions? *Applied Physics Letters*, 89, 132116.

Andriotis, A.N., Menon, M., Srivastava, D. et al. 2001. Rectification properties of carbon nanotube Y-junctions. *Physical Review Letters*, 876, 066802.

Andriotis, A.N., Menon, M., Srivastava, D. et al. 2002. Transport properties of single-wall carbon nanotube Y-junctions. *Physical Review B*, 65, 65416–165411.

Andriotis, A.N., Srivastava, D., and Menon, M. 2003. Comment on intrinsic electron transport properties of carbon nanotube Y-junctions. *Applied Physics Letters*, 83, 1674–675.

Antonov, R.D. and Johnson, A.T. 1999. Subband population in a single-wall carbon nanotube diode. *Physical Review Letters*, 83, 3274–3276.

Appenzeller, J., Knoch, J., Martel, R. et al. 2002. Carbon nanotube electronics. *IEEE Transactions on Nanotechnology*, 1, 184–189.

Bachtold, A., Hadley, P., Nakanishi, T. et al. 2001. Logic circuits with carbon nanotube transistors. *Science*, 294, 1317–1320.

Bachtold, A., Strunk, C., Salvetat, J.P. et al. 1999. Aharonov-Bohm oscillations in carbon nanotubes. *Nature*, 397, 673–675.

Bandaru, P.R. 2007. Electrical properties and applications of carbon nanotube structures. *Journal of Nanoscience and Nanotechnology*, 7, 1239–1267.

Bandaru, P.R., Daraio, C., Yang, K. et al. 2007. A plausible mechanism for the evolution of helical forms in nanostructure growth. *Journal of Applied Physics*, 101, 094307.

Bandaru, P.R., Daraio, C., Jin, S. et al. 2005. Novel electrical switching behavior and logic in carbon nanotube Y-junctions. *Nature Materials*, 4, 663–666.

Baughman, R.H., Zakhidov, A.A., and de Heer, W.A. 2002. Carbon nanotubes: the route toward applications. *Science*, 2976, 787.

Beale, M. and Mackay, P. 1992. The origins and characteristics of negative capacitance in metal insulator–metal devices. *Philosophical Magazine B*, 65, 47–64.

Bockrath, M., Cobden, D.H., McEuen, P.L. et al. 1997. Single-electron transport in ropes of carbon nanotubes. *Science*, 275, 1922–1924.

Buitelaar, M., Bachtold, T.N., Iqbal, M. et al. 2002. Multiwall carbon nanotubes as quantum dots. *Physical Review Letters*, 88, 156801.

Burke, P.J. 2002. Luttinger liquid theory as a model of the gigahertz electrical properties of carbon nanotubes. *IEEE Transactions on Nanotechnology*, 1, 129–144.

Burke, P.J. 2003. An RF circuit model for carbon nanotubes. *IEEE Transactions on Nanotechnology*, 21, 55–58.

Castrucci, P., Scarselli, M., De Crescenzi, M. et al. 2004. Effect of coiling on the electronic properties along single-wall carbon nanotubes. *Applied Physics Letters*, 85, 3857–3859.

Choi, Y.C. and Choi, W. 2005. Synthesis of Y-junction single-wall carbon nanotubes. *Carbon*, 43, 2737–2741.

Collins, P.G. and Avouris, P. 2000. Nanotubes for electronics. *Scientific American*, December, 62–69.

Collins, P.G., Bradley, K., Ishigami, M. et al. 2000. Extreme oxygen sensitivity of electronic properties of carbon nanotubes. *Science,* 287, 1800–1804.

Collins, P.G., Arnold, M.S., and Avouris, P. 2001a. Engineering carbon nanotubes and nanotube circuits using electrical breakdown. *Science,* 292, 706–709.

Collins, P.G., Hersam, M., Arnold, M. et al. 2001b. Current saturation and electrical breakdown in multiwalled carbon nanotubes. *Physical Review Letters,* 86, 3128–3131.

Crespi, V.H. 1998. Relations between global and local topology in multiple nanotube junctions. *Physical Review B,* 58, 12671.

Csontos, D. and Xu, H.Q. 2003. Quantum effects in the transport properties of nanoelectronic three-terminal Y-junction devices. *Physical Review B,* 67, 235322–235321.

Dai, H. 2001. Nanotube growth and characterization. In Dresselhaus, M.S. et al., Eds., *Topics in Applied Physics* Vol. 80. Berlin: Springer, pp. 29–53.

Dai, H. 2002. Carbon nanotubes: from synthesis to integration and properties. *Accounts of Chemical Research,* 35, 1035–1044.

Davies, J.H. 1998. *The Physics of Low-Dimensional Semiconductors.* New York: Cambridge University Press.

Delaney, P., Di Ventra, M., and Pantelides, S. 1999. Quantized conductance of multi-walled carbon nanotubes. *Applied Physics Letters,* 75, 3787–3789.

Ding, J.W., Yan, X.H., and Cao, J.X. 2002. Analytical relation of band gaps to both chirality and diameter of single-wall carbon nanotubes. *Physical Review B,* 66, 073401.

Dresselhaus, M.S., Dresselhaus, G., and Jorio, A. 2004. Unusual properties and structure of carbon nanotubes. *Annual Review of Materials Research,* 34, 247–278.

Forro, L. and Schonenberger, C. 2001. Physical properties of multi-wall nanotubes. In Dresselhaus, M.S. et al., Eds., *Topics in Applied Physics* Vol. 80. Berlin: Springer.

Freitag, M., Martin, Y., Misewich, J.A. et al. 2003. Photoconductivity of single carbon nanotubes. *Nanoletters,* 3, 1067–1071.

Freitag, M., Radosavljevic, M., Zhou, Y. et al. 2001. Controlled creation of a carbon nanotube diode by a scanned gate. *Applied Physics Letters,* 79, 3326–3328.

Fuhrer, M.S., Nygard, J., Shih, L. et al. 2000. Crossed nanotube junctions. *Science,* 288, 494.

Gopal, V., Radmilovic, V.R., Daraio, C. et al. 2004. Rapid prototyping of site-specific nanocontacts by electron and ion beam assisted direct-write nanolithography. *Nanoletters,* 4, 2059–2063.

Gothard, N., Daraio, C., Gaillard, J. et al. 2004. Controlled growth of Y-junction nanotubes using Ti-doped vapor catalyst. *Nanoletters,* 42, 213–217.

Guo, J., Datta, S., Lundstrom, M. et al. 2002. Assessment of silicon MOS and carbon nanotube FET performance limits using a general theory of ballistic transistors. Paper presented at International Electron Devices Meeting, San Francisco.

Hieke, K. and Ulfward, M. 2000. Nonlinear operation of the Y-branch switch: ballistic switching mode at room temperature. *Physical Review B,* 62, 16727–16730.

Hone, J., Laguno, M.C., Nemes, N.M. et al. 2000. Electrical and thermal transport properties of magnetically aligned single wall carbon nanotube films. *Applied Physics Letters,* 77, 666.

Huang, X.M. H., Caldwell, R., Huang, L. et al. 2005. Controlled placement of individual carbon nanotubes. *Nanoletters,* 5, 1515–1518.

Huang, Y., Duan, X., Wei, Q. et al. 2001. Directed assembly of one-dimensional nanostructures into functional networks. *Science,* 291, 630–633.

Iijima, S., Ichihashi, T., and Ando, Y. 1992. Pentagons, heptagons and negative curvature in graphite microtubule growth. *Nature,* 356, 776–778.

Jackson, J.D. 1999. *Classical Electrodynamics*. New York: John Wiley.

Javey, A., Guo, J., Wang, Q. et al. 2003. Ballistic carbon nanotube field-effect transistors. *Nature*, 424, 654–657.

Joselevich, E. and Lieber, C.M. 2002. Vectorial growth of metallic and semiconducting single-wall carbon nanotubes. *Nanoletters*, 2, 1137–1141.

Kang, S.J., Kocabas, C., Ozel, T. et al. 2007. High-performance electronics using dense, perfectly aligned arrays of single-walled carbon nanotubes. *Nature Nanotechnology*, 2, 230–236.

Kaul, A.B. and Bandaru, P.R. (In press). Review Chapter, Electronic and Photonic Applications of One-Dimensional Carbon and Silicon Nanostructures, in A. Umar (Ed.), *Encyclopedia of Semiconductor Nanotechnology*, Valencia, CA: American Scientific Publishers.

Kim, D.H., Huang, J., Shin, H.K. et al. 2006. Transport phenomena and conduction mechanisms of single walled carbon nanotubes (SWNTs) at Y- and crossed junctions. *Nanoletters*, 6, 2821–2825.

Kim, P., Shi, L., Majumdar, A. et al. 2001. Thermal transport measurements of individual multiwalled nanotubes. *Physical Review Letters*, 87, 215502.

Kosaka, H., Rao, D.S., Robinson, H. et al. 2002. Photoconductance quantization in a single-photon detector. *Physical Review B*, 65, 201307.

Lee, J.U., Gipp, P.P., and Heller, C.M. 2004. Carbon nanotube p-n junction diodes. *Applied Physics Letters*, 85, 145–147.

Lewen, R., Maximov, I., Shorubalko, I. et al. 2002. High frequency characterization of GaInAs/InP electronic waveguide TBS switch. *Journal of Applied Physics*, 91, 2398–2402.

Li, W.Z., Wen, J.G., and Ren, Z.F. 2001. Straight carbon nanotube Y junctions. *Applied Physics Letters*, 79, 1879–1881.

Liu, J., Casavant, M.J., Cox, M. et al. 1999. Controlled deposition of individual single-walled carbon nanotubes on chemically functionalized templates. *Chemical Physics Letters*, 303, 125–129.

Manohara, HM., Wong, E.W., Schlecht, E. et al. 2005. Carbon nanotube Schottky diodes using Ti Schottky and Pt ohmic contacts for high frequency applications. *Nanoletters*, 5, 1469–1474.

Martel, R., Derycke, V., Appenzeller, J. et al. 2002. Carbon nanotube field-effect transistors and logic circuits. Paper presented at Design Automation Conference, New Orleans.

McEuen, P.L. 1998. Carbon-based electronics. *Nature*, 393, 15–16.

McEuen, P.L., Fuhrer, M.S., and Park, H. 2002. Single-walled carbon nanotube electronics. *IEEE Transactions on Nanotechnology*, 1, 78–85.

Meunier, V., Nardelli, M.B., Bernholc, J. et al. 2002. Intrinsic electron transport properties of carbon nanotube Y-junctions. *Applied Physics Letters*, 81, 5234–5236.

Muller, R.S. and Kamins, T.I. 1986. *Device Electronics for Integrated Circuits*, 2nd ed. New York: John Wiley.

Nichols, J.A., Saito, H., Deck, C. et al. 2007. Artificial introduction of defects into vertically aligned multiwall carbon nanotube ensembles: application to electrochemical sensors. *Journal of Applied Physics*, 102, 064306.

Palm, T. and Thylen, L. 1992. Analysis of an electron-wave Y-branch switch. *Applied Physics Letters*, 60, 237–239.

Palm, T. and Thylen, L. 1996. Designing logic functions using an electron waveguide Y-branch switch. *Journal of Applied Physics*, 79, 8076–8081.

Papadapoulos, C., Rakitin, A., Li, J. et al. 2000. Electronic transport in Y-junction carbon nanotubes. *Physical Review Letters*, 85, 3476–3479.

Park, J., Daraio, C., Jin, S. et al. 2006. Three-way electrical gating characteristics of metallic Y-junction carbon nanotubes. *Applied Physics Letters*, 88, 243113.

Perkins, B.R., Wang, D.P., Soltman, D. et al. 2005. Differential current amplification in three-terminal Y-junction carbon nanotube devices. *Applied Physics Letters*, 87, 123504.

Postma, H.W.C., Teepen, T., Yao, Z. et al. 2001. Carbon nanotube single-electron transistors at room temperature. *Science*, 293, 76–79.

Qin, L., Park, S., Huang, L. et al. 2005. On-wire lithography. *Science*, 309, 113–115.

Rabaey, J. 1996. *Digital Integrated Circuits: A Design Perspective*. New York: Prentice Hall.

Radosavljevic, M., Freitag, M., Thadani, K.V. et al. 2002. Non-volatile molecular memory elements based on ambipolar nanotube field effect transistors. arxiv:cond-mat/0206392.

Rao, S.G., Huang, L., Setyawan, W. et al. 2003. Large-scale assembly of carbon nanotubes. *Nature*, 425, 36–37.

Rueckes, T., Kim, K., Joselevich, E. et al. 2000. Carbon nanotube-based nonvolatile random access memory for molecular computing. *Science*, 289, 94–97.

Saito, R., Dresselhaus, G., and Dresselhaus, M.S. 1998. *Physical Properties of Carbon Nanotubes*. London: Imperial College Press.

Satishkumar, B.C., Thomas, P.J., Govindaraj, A. et al. 2000. Y-junction carbon nanotubes. *Applied Physics Letters*, 77, 2530–2532.

Scuseria, G.E. 1992. Negative curvature and hyperfullerenes. *Chemical Physics Letters*, 195, 534–536.

Shockley, W. 1950. *Electrons and Holes in Semiconductors*, Vol. D. Princeton, NJ: Van Nostrand.

Shorubalko, I., Xu, H.Q., Omling, P. et al. 2003. Tunable nonlinear current-voltage characteristics of three-terminal ballistic nanojunctions. *Applied Physics Letters*, 83, 2369–2371.

Song, A.M., Lorke, A., Kriele, A. et al. 1998. Nonlinear electron transport in an asymmetric microjunction: a ballistic rectifier. *Physical Review Letters*, 80, 3831–3834.

Song, A.M., Omling, P., Samuelson, L. et al. 2001. Room-temperature and 50 GHz operation of a functional nanomaterial. *Applied Physics Letters*, 79, 1357–1359.

Streetman, B.G. and Banerjee, S. 2000. *Solid State Electronic Devices*, 5th ed. New York: Prentice Hall.

Sze, S.M. 1981. *Physics of Semiconductor Devices*, 2nd ed. New York: John Wiley.

Tans, S.J., Devoret, M.H., Dai, H. et al. 1997. Individual single-wall carbon nanotubes as quantum wires. *Nature*, 386, 474–477.

Tans, S.J., Verschueren, A.R.M., and Dekker, C. 1998. Room-temperature transistor based on a single carbon nanotube. *Nature*, 393, 49–52.

Teo, K.B.K., Singh, C., Chhowalla, M. et al. 2004. Catalytic synthesis of carbon nanotubes and nanofibers. In Nalwa, H.S., Ed., *Encyclopedia of Nanoscience and Nanotechnology*. Stevenson Ranch, CA: American Scientific Publishers.

Terrones, M., Banhart, F., Grobert, N. et al. 2002. Molecular junctions by joining single-walled carbon nanotubes. *Physical Review Letters*, 89, 075505–075510.

Tian, W., Datta, S., Hong, S. et al. 1998. Conductance spectra of molecular wires. *Journal of Chemical Physics*, 109, 2874–2882.

Ting, J.M. and Chang, C.C. 2002. Multijunction carbon nanotube network. *Applied Physics Letters*, 80, 324–325.

Tsukagoshi, K., Watanabe, E., Yagi, I. et al. 2004. Multiple-layer conduction and scattering property in multi-walled carbon nanotubes. *New Journal of Physics*, 6, 1–13.

Wallin, D., Shorubalko, I., Xu, H.Q. et al. 2006. Nonlinear electrical properties of three-terminal junctions. *Applied Physics Letters*, 89, 092124.

Wesstrom, J.O. 1999. Self-gating effect in the electron Y-branch switch. *Physical Review Letters*, 82, 2564–2567.

White, C.T. and Todorov, T.N. 1998. Carbon nanotubes as long ballistic conductors. *Nature*, 393, 240.

Worschech, L., Xu, H.Q., Forchel, A. et al. 2001. Bias-voltage-induced asymmetry in nanoelectronic Y-branches. *Applied Physics Letters*, 79, 3287–3289.

Xu, H.Q. 2001a. Diode and transistor behaviors of three-terminal ballistic junctions. *Applied Physics Letters*, 80, 853–855.

Xu, H.Q. 2001b. Electrical properties of three-terminal ballistic junctions. *Applied Physics Letters*, 78, 2064–2066.

Xu, H.Q. 2002. A novel electrical property of three-terminal ballistic junctions and its applications in nanoelectronics. *Physica E*, 13, 942–945.

Xu, H.Q. 2005. The logical choice for electronics? *Nature Materials*, 4, 649–650.

Yang, M.H., Teo, K.B.K., Milne, W.I. et al. 2005. Carbon nanotube Schottky diode and directionally dependent field-effect transistor using asymmetrical contacts. *Applied Physics Letters*, 87, 253116.

Yao, Z., Postma, H.W.C., Balents, L. et al. 1999. Carbon nanotube intramolecular junctions. *Nature*, 402, 273–276.

Zhang, Y., Franklin, N.W., Chen, R.J. et al. 2000. Metal coating on suspended carbon nanotubes and its implication to metal–tube interaction. *Chemical Physics Letters*, 331, 35–41.

Zhou, D. and Seraphin, S. 1995. Complex branching phenomena in the growth of carbon nanotubes. *Chemical Physics Letters*, 238, 286–289.

12

Nanoscale Effects in Multiphase Flows and Heat Transfer

Navdeep Singh, Donghyun Shin, and Debjyoti Banerjee

CONTENTS

ABSTRACT   Phase change heat transfer is the most efficient mode for cooling due to the ability of these multiphase flows to transfer huge amounts of thermal energy across small temperature differences. Recent advances in multiphase flows have demonstrated that applying porous nanoparticle coatings to heat exchanging surfaces is a more efficient strategy for thermal management than using coolants that are doped with minute concentrations of nanoparticle-based emulsions. Both strategies are reported to enhance the thermal characteristics for cooling applications as well as energy conversion and thermal energy storage applications. Various efforts at analytical and numerical modelling of these nanoscale effects in multiphase flows have shown that different optimal size ranges exist for the different flow and heat transfer regimes.

12.1   Introduction

Increase in device density of electronic chips along with progressive enhancement in device performance (e.g., computation speed or clock speed) has resulted in enhanced density of power generation. Consequently, cooling loads for MEMS, micro-, nano, and opto-electronic devices in the future are projected to surpass ~1 to 10 kW/cm^2 within the next 10 years.

Conventional thermal management schemes using forced convection of air using fans or direct liquid cooling will not be adequate for meeting these enormous cooling needs within a short form factor. This technological need will significantly affect the growth of the chip interconnect and electronic packaging industries. The estimated annual sales for these industries are currently approaching ~$16 billion worldwide and are expected to increase exponentially in the next 10 years. This is an important technical bottleneck for these industries and in defense applications and therefore has a significant commercial impact.

Thermal management techniques involving phase change heat transfer (e.g., boiling and condensation or multiphase flows) are expected to meet the cooling requirements of these commercial systems. The performances of these commercial systems are limited by the requirements for removal of high heat fluxes within a narrow operating temperature range. For example, boiling of highly subcooled liquids can yield very high heat fluxes that are conducive for cooling applications. Boiling is a liquid-to-vapor phase change process that occurs at a solid–liquid interface when the solid surface temperature is sufficiently higher than the saturation temperature. It is an extremely efficient mode for heat transfer. Heat flux values as high as 17 kW/cm^2 have been achieved for forced flow boiling under sub-cooled conditions.

This is three times the heat flux at the surface of the Sun but at one-fifth of the temperature difference.

Other applications for high heat flux cooling include commercial systems for energy conversion, energy conservation, and biotechnology. Solar thermal energy systems often use concentrated solar power in which the surface temperatures of solar towers can exceed 600°C and the temperatures of solar furnaces can exceed 3000°C. Energy conservation in micro-climate cooling systems and building HVAC (heating, ventilation and air-conditioning) systems involves the design of more efficient condensers that can reject heat with small temperature differentials, and therefore require very high values of overall heat transfer coefficients.

In biotechnology applications often the prevention of evaporation along with very rapid temperature changes (e.g., rapid thermocycling for DNA amplification using the polymerase chain reaction or PCR) necessitates the development of devices with high heat transfer coefficients. Also, cooling of electronics in high temperature and pressure environments is needed for deep drilling (>15,000 feet) by the energy industry for oil and gas exploration.

All these applications involve thermal management systems that are governed by phase change heat transfer issues and multiphase flows. Contemporary advances in the science and technology of multiphase flows and heat transfer has focused primarily on microscale and nanoscale manipulation of the materials and transport phenomena as well as the associated flow configurations.

Bulk convections of heterogeneous multi-component mixtures are defined as multiphase flows. Such convective motion can be self-induced (due to temperature differences such as thermal gradients or chemical concentration difference from chemical potential gradients). Such convections can also be induced by external motive forces, for example pressure gradients or electromagnetic interactions (e.g., electrokinetic flows). The flows can be isothermal (no temperature change) or adiabatic (no heat transfer) or associated with heat transfer (heating and cooling). In general multiphase flows typically encompass two different classes:

Multiphase flows without phase change—This flow consists of a single fluid with an immiscible component dispersed fairly uniformly within the flow. Examples include:

1. Nanofluids that are colloidal suspensions of solid nanoparticles in liquid (e.g., organic or inorganic nanoparticles)

2. Slurries or noncolloidal suspensions of solid microparticles that usually settle or float due to gravity when the flow stops

3. Dispersion of gas bubbles in liquids with size ranges varying from miniscule with distinct interphase separation (i.e., meniscus formation) to gas slugs that cover the entire flow cross section to stratified flows in which the lighter gas phase separates under the influence

of external body forces (such as gravity) to form two distinct continuous streams of flow often with different flow speeds for both components (also characterized as slip velocity or drift flux velocity models)

4. Aerosols consisting of suspension of liquid mist or solid particles suspended in gas phase
5. Fluidized bed convection in which heavy solid particles are suspended by the pressure from the relative slip velocity of the flowing gas phase (in this case without chemical reaction)

Multiphase flows with phase change—One of the components undergoes a change into another phase or forms a different chemical species within the flow due to some form of chemical reaction or interaction. This class of multiphase flows can be further categorized into:

1. Pool or flow boiling in which the liquid phase is converted to vapor phase inside the flow conduits
2. Condensation in which the flowing vapor phase loses heat to the flow conduit walls and forms a distinct liquid phase, initially as a mist and subsequently as liquid slugs and a stratified liquid phase (similar to the flow morphologies mentioned in item 3 above)
3. Nanofluids with encapsulated nanoparticles in which the encapsulated phase undergoes phase change or the nanoparticles induce different forms of phase change in the solvent phase
4. Fluidized bed convection in which the flow phenomena are also associated with chemical reactions such as combustion
5. Nanostructured surfaces fabricated within flow conduits (or engineered nanostructures on flow surfaces) that enhance transport phenomena such as heating and cooling along with phase change (melting to solidification, boiling to condensation, or chemical reactions)

Various investigations of nanostructured surfaces, such as engineered nanofins and synthesized nanocoatings, have led to improvements in heat transfer. A few experimental studies have been reported for enhanced boiling heat transfer and condensation heat transfer on nanostructured surfaces. In many of these studies, the use of "small" heaters resulted in issues concerning repeatability of experiments.

In contrast, literature reports on enhancement of thermal properties of a wide range of nanofluids and their performance in augmenting heat transfer have been often contradictory and controversial. This is primarily due to the variability in the experimental procedures by the investigators and the lack of proper characterization of the nanomaterials. Often in these studies, the nanomaterials and the heat transfer surfaces or measurement probes were not characterized before and after performing the experiments. As a result, the efforts to model these flow configurations have also been controversial, resulting in a lack of consensus on the transport mechanisms responsible for the observed phenomena.

In this chapter we will discuss the experimental results and theoretical, analytical, and numerical models presented by many research groups involving nanoscale effects in multiphase flows and heat transfer. These studies are particularly relevant for a variety of engineering applications.

12.2 Surface Modification (Nanofins and Nanocoatings)

The simplest configuration of multiphase flow involving phase change is realized by pool boiling of liquids. This involves a container filled with a test liquid and a heater placed within the container. For a given heat transfer fluid and heater material at a given system pressure, the overall heat flux through the boiling surface can be enhanced by:

1. Augmenting the effective heat transfer surface area by increasing the surface roughness or by fabricating fins or enhanced heat transfer surfaces

2. Enhancing the area density of nucleating vapor bubbles by modifying the contact angle or surface modification

3. Disrupting the vapor film formation that degrades the heat transfer by augmenting the perimeters of the vapor films, since the peak in the profile of heat flux distribution occurs at the contact lines of the vapor films

Hence, the three approaches mentioned above for enhancing thermal performance in pool boiling involve some form of modification of the boiling surface. Nanofabrication techniques provide several attractive options for engineering nanostructured surfaces. In the next section, we will discuss several phase change experiments on nanostructured surfaces, such as engineered nanofins and synthesized nanocoatings.

12.2.1 Pool Boiling on Organic Nanostructured Surfaces

Ahn et al. (2006) were the first to report pool boiling performance in the nucleate and film boiling regimes on flat silicon substrates coated with vertically aligned multi-walled carbon nanotubes (MWCNTs) using a refrigerant liquid (PF-5060). These experiments were initially performed for the refrigerant liquid at the boiling temperature, i.e., at the saturation temperature. Subsequently

Sathyamurthi et al. (2009) refined the experimental procedure and performed experiments under subcooled conditions where the refrigerant liquid phase was maintained at temperatures lower than the saturation temperature.

Both studies used two different heights of MWCNTs: Type A (9 μm coating height) and Type B (25 μm height). The MWCNT sizes ranged from 8 to 16 nm in diameter with a random spacing of ~10 nm. The authors in the two studies reported that the MWCNT enhanced critical heat flux (CHF) by as much as 60% and augmentation of the CHF (maximum value of heat flux in pool boiling; see Zuber, 1959) was independent of the height of the MWCNT. Both heights of MWCNT provided about the same augmentation in CHF, within the limits of experimental measurement uncertainty.

Ahn et al. (2006) and Sathyamurthi et al. (2009) performed additional experiments using MWCNT coatings in the film boiling region. Film boiling occurs at very high heater temperatures; a "continuous" film of vapor blankets the heater surface which is disrupted chaotically, leading to the suspended liquid phase touching the heater surface at chaotic intervals. The chaotic disruption occurs due to myriad instability waves that are superposed at the liquid–vapor interface and within the vapor film as well as within the bulk liquid (due to departing vapor bubbles and enhanced mixing of fluid masses at different temperatures).

The results from these studies showed that the film boiling heat flux for the Type B MWCNT (25 μm in height) was augmented by as much as 150 to 300%, while almost no enhancement in heat flux for Type A (9 μm in height) was observed when compared to heat flux for a bare silicon surface. This implies that the height of the Type B MWCNT was higher than that of the vapor film at the point of minimum vapor film thickness and therefore disrupted the vapor film. The disruption of the vapor film resulted in enhanced transient contacts between the liquid phase and heater surface. The protrusions of the MWCNT from the heater surface also contributed to enhancing the heat transfer from the transient contacts by acting as "nanofins."

Conversely, the height of the Type A MWCNT was shorter than that of the vapor film; therefore, no enhancement of heat flux by the MWCNT occurred within the vapor film. In both studies, flow visualization of the vapor bubbles in nucleate and pool boiling also confirmed the disruption of the vapor films by the nanostructures (MWCNT coatings). The flow visualization images clearly showed disruption of the vapor film by Type B MWCNTs whereas in the case of Type A MWCNTs, the vapor film was found to have similar hydrodynamic structures as those of a bare silicon surface.

Figure 12.1 shows comparison of pool boiling curves for PF-5060 on silicon and MWCNT-coated surfaces. The wall superheat (the difference between the heater surface temperature and the boiling point of the refrigerant) is plotted on the x-axis and the measured heat flux from the heater is plotted on the y-axis. The boiling curves obtained at lower wall superheats are called the "nucleate boiling regime" that terminates at the CHF (maximum heat flux in pool boiling). At higher wall superheats, the film boiling occurs with a much

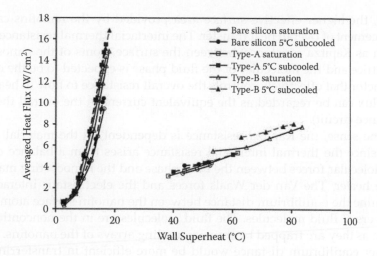

FIGURE 12.1
(See color insert.) Pool boiling curve of PF-5060 on silicon and on CNT-coated surfaces. (From Sathyamurthi, V. et al. 2009. *J. Heat Transfer*, 131, 071501. With permission.)

lower wall heat flux compared to the nucleate boiling regime, as shown in the figure. The plots (boiling curves) for bare surfaces and MWCNT-coated surfaces show the observed effects and peculiarities of MWCNT-coated surfaces described earlier in this section.

Pool boiling using MWCNT coated on hybrid microstructures was explored by Launay et al. (2006). The experiments were performed on plain and hybrid structures involving pin fins and three-dimensional (3D) structures coated with MWCNT using PF-5060 refrigerant and distilled water. Heat flux augmentation ranging from 60 to 900% was reported. Subsequently, Ujereh et al. (2007) also reported pool boiling experiments using CNT-coated silicon and copper surfaces in FC-72. However, this study was restricted for lower wall superheats to nucleate boiling and CHF. Significant augmentation of nucleate boiling heat transfer was reported and these observations are consistent with the measurements of Ahn et al. (2006) and Sathyamurthi et al. (2009).

Based on the experimental results, the transport mechanisms responsible for augmentation of pool boiling heat transfer on MWCNT-coated surfaces were classified into two factors that are coupled nonlinearly (Ahn et al., 2006; Sathyamurthi et al., 2009) and are discussed next.

12.2.1.1 Nanofin Effect

The "hair" shaped MWCNT nanostructures protruding from the heater surface caused augmentation of pool boiling heat flux by primarily modifying the effective surface area available for heat transfer to the boiling liquid. This can be attributed to a combination of mechanisms.

First, the higher effective surface area provided by the nanofins caused enhancement of overall heat transfer. The interfacial thermal resistance (also known as Kapitza resistance) between the surface atoms of the nanostructure lattice and the molecules of the fluid phase is expected to be the dominant factor that effectively controls the overall resistance to flow of heat flux (heat flux can be regarded as the equivalent current in the overall thermal resistance circuit).

In one sense, the Kapitza resistance is dependent on the material properties since the thermal interfacial resistance arises from a balance of the intermolecular forces between the fluid phase and the nanocoating material on the heater. The Van der Waals forces and the electrostatic interactions determine the equilibrium distance between the nanofin surface atoms and the layers of fluid molecules. The fluid molecules are in the noncontinuum regime as they are trapped by the surrounding arrays of the nanofins.

Closer equilibrium distance would be more efficient in transferring the thermal energy from the lattice atoms of the nanofins by means of low frequency thermal vibrational modes (low frequency phonons). Hence the interatomic potentials that are specific to each material determine the interfacial thermal resistance which in turn affects the nanofin effectiveness in transferring heat.

Molecular dynamics simulations have shown that silicon nanostructures have considerably lower thermal interfacial resistance ($\sim 10^{-11}$ m^2-K/W) for transferring heat to fluids (Murad and Puri, 2009) than organic nanostructures that have much higher thermal interfacial resistance ($\sim 10^{-8}$ m^2-K/W) for transferring heat to fluids (Shin and Banerjee, 2009). In addition, fluid molecules exist in a compressed phase around the lattice structures of the nanoparticles. The compressed phase is shown to acquire a semi-crystalline structure in the vicinity of the lattice atoms, where the fluid molecules are observed to acquire a periodic ordering in the vicinity of the lattice surface, thus behaving as if they are a part of the lattice structure.

The compressed phase was measured using electron-microscopy and shown to be ~1 nm thick (Oh et al., 1995). This phenomenon of the formation of a compressed phase was validated by a number of studies using molecular dynamics (Beskok, et al. 2005). The compressed phase also would modify the transient heat transfer characteristics from the nanofin surface to the bulk fluid by serving as a thermal capacitor, providing additional thermal inertance and modifying the interfacial thermal resistance that contributes to the overall thermal resistance network.

The second factor is that the nanofin protrusions disrupted the microlayer. During pool boiling, a thin film of liquid is trapped under the growing bubble at the liquid–vapor–solid contact line on the heater; this is the microlayer. Capillary wicking of the liquid layer within the nanostructures can modify the structure of the microlayer, leading to enhanced boiling heat flux.

Third, the nanofin protrusions were also found to disrupt the vapor film and augment the film boiling heat flux.

Finally, the densely packed MWCNT nanostructures are also expected to trap a layer of superheated liquid (which from a thermodynamic view is in a meta-stable state). The flow within this superheated liquid layer trapped within the MWCNT would be in the noncontinuum regime. The flow and thermodynamic stability of this regime is as yet unexplored in the literature.

The porous nanostructures would aid in replenishing the liquid on the boiling surface by capillary pumping of liquid from the periphery. The effect of such nanoscale-confined flows that operate in noncontinuum tranport regimes on the nucleation characteristics of vapor bubbles also begs further scientific investigation.

12.2.1.2 Thermo-Physical Property Effect

The enhanced thermo-physical properties of MWCNT would also provide a more efficient structure for transferring heat to the refriegrant liquid.

MWCNT-coated surfaces are expected to be more efficient in aiding the rapid formation of a thermal boundary layer due to the enhanced thermal conductivity of the MWCNT and the protrusion of the nanofin shaped structures into the fluid layer on the heated surface.

The liquid–vapor contact line on the heater surface provides a region of very high transient and oscillating heat flux (MW/m^2). Such high heat flux regions are also associated with inversion of the surface temperature profiles on the heater surface, i.e., formation of cold spots (Banerjee et al., 1996; Banerjee and Dhir, 2001a; Banerjee and Dhir, 2001b). These cold spots are ephemeral and are co-located on the heater surface at the perimeters of the vapor bubbles. The thickness of these annular shaped cold spots is dependent on the thermal diffusivity of the heater material.

MWCNT coatings by virtue of their high thermal diffusivity values are expected to augment the thickness of these cold spots, resulting in enhanced transient heat transfer from the heater to the boiling liquid.

However, the physical and chemical behavior of the vapor–liquid contact line on the MWCNT nanostructure is expected to be different from the behavior on a plain surface. The dynamic nature of such contact line length on nanostructures (e.g., dynamic contact angle) is as yet unknown and can be expected to have fractal behavior.

Based on these experiments (Ahn et al., 2006; Launay et al., 2006; Ujereh et al., 2007), the more dominant mechanism for pool boiling on nanostructured surfaces remains unclear, i.e., whether the shape effect (nanofin effect which is also an effect of length-scale) is more dominant than the effect of the thermophysical properties of the nanostructures. In other words, if two nanostructured surfaces composed of different materials have the same geometrical features, would the pool boiling heat flux be augmented by the same amount? To answer this question, the pool boiling on nanostructures fabricated from other materials needs to be explored.

12.2.2 Pool Boiling on Inorganic Nanostructured Materials

Sriraman (2007) performed pool boiling experiments using PF-5060. The silicon nanofin surface was fabricated by a step-and-flash nanoimprint lithography (SFIL) technique that enabled precise control of the height, diameter, and pitch of the nanofins. The nanofins were 200 nm in diameter with a pitch of ~900 nm. Their height varied from 10 to 600 nm by precise control over the dry etching process using reactive ion etching (RIE) and deep reactive ion etching (DRIE).

Surprisingly, the maximum heat flux (CHF) was enhanced by 120% (compared to 60% for MWCNT) and the enhancement of CHF was independent of the height of the nanofins as long as the height of the nanofins exceeded ~100 nm. This shows that the nanofin effect is the more dominant mechanism for augmentation of pool boiling on nanostructures, since nanofins with lower thermal conductivity augmented the pool boiling heat flux by higher margins.

This is direct experimental evidence that the microlayer of liquid trapped under a vapor bubble is ~100 nm thick since the nanofins exceeding a height of 100 nm exhibited the same CHF value. This study therefore enables the resolution of the controversy in the literature in which various studies debated whether the thickness of the microlayer is ~100 nm or if it is ~1μm or somewhere between those thicknesses (Mirzamoghadam and Catton, 1988).

Chen et al. (2009) studied pool boiling of water on nanowires made of silicon and copper and reported enhancement of the CHF by over 100%. Hendricks et al. (2010) fabricated nanostructured surfaces by deposition of ZnO on Al and Cu substrates. From their pool boiling experiments, they reported that the CHF for nanostructured surfaces was augmented by 400% when compared with pool boiling on a bare Al surface. Wu et al. (2010) investigated CHF phenomena during pool boiling of water and FC-72 on surfaces modified with TiO_2 nanoparticle coatings and reported that the CHF was enhanced by 50% for water and 40% for FC-72.

Im et al. (2010) fabricated copper nanowire on a silicon substrate by electrochemical deposition. The results showed that the copper nanowire increased the heat transfer coefficient by an order of magnitude (~10 times) while the CHF was enhanced only marginally. However, the CHF was found to occur at a much lower wall superheat. The results reported by Im et al. (2010) therefore contradict the results for pool boiling heat flux measurements by Chen et al. (2009). Such discrepancies probably arise due to poor design of experiments since the pool boiling experiments were performed for the small heater regime. This will be explained in the next section.

12.2.3 Flow Boiling on Nanostructures

Flow boiling studies on nanostructured surfaces have received less attention than pooling boiling due to complexity associated with the fabrication and assembly of the required experimental apparatus. Sunder and Banerjee (2009) characterized nanoscale temperature transients using surface micromachined temperature nanosensors (thin film thermocouples or TFTs) during flow boiling on a plain silicon wafer. The thin film layers constituting the K-type thermocouples of the TFT were ~200 nm thick and ~10 to 50 mm wide and composed of chromel and alumel layers that overlapped at the locations of temperature measurements.

The power spectrum of the surface temperature transients measured by the TFT was analyzed as the heater temperature was increased during flow boiling conditions. With increase in wall superheat there was a progressive increase in the number of frequency peaks and power levels of the peaks. This showed that coupled hydrodynamic and thermal features of microscale and nanoscale lengths were generated due to the various coupled and nonlinear transport mechanisms in flow boiling.

Subsequently, Singh et al. (2010) performed flow boiling experiments on MWCNT-coated surfaces using the same experimental apparatus. It was observed that the flow boiling heat flux was enhanced by 180% at boiling inception during forced convection. The results showed that the fully developed flow boiling heat flux was partitioned between pool boiling heat flux for similar wall superheat and forced convective heat flux on a flat plate heated to the same temperature. The boiling inception was found to match the heuristic models in the literature.

Also, it was observed that at higher mass flow rates and at higher mass velocities (mass flow rate divided by cross-sectional area), the augmentation in pool boiling heat flux was reduced. Khanikar et al. (2009a) performed flow boiling experiments with water on CNT-coated surfaces. The result showed enhancement of heat flux in the nucleate boiling regime as well as CHF at only low mass velocity.

The level of enhancement in flow boiling heat flux was eliminated at higher mass velocities—consistent with the measurements of Singh et al. (2009). Khanikar et al. (2009a) explained that high mass velocity causes CNT folding on the heated wall, resulting in decrease of wall roughness and turbulence. However, this explanation is not tenable since CNTs are quite stiff and the shear rates in flow boiling are estimated to be much lower than that required to cause considerable deformation of CNTs.

It is quite likely that the adhesion of the CNT to the substrate was an issue in these experiments. Khanikar et al. (2009b) performed scanning electron microscopy (SEM) and observed significant change in morphology of the CNT-coated surface after undergoing multiple tests at high mass velocity.

A major drawback of the studies by the various investigators in the pool boiling and flow boiling community (Launay et al., 2006; Ujereh et al., 2007; Chen et al., 2009; Khanikar et al., 2009a and b; Im et al., 2010; Hendricks et al., 2010) is that small heaters were used in their experiments. Consequently, the experiments suffered from lack of consistency and repeatability of the experimental data, especially when the various studies are compared.

It should be noted that for boiling experiments, the lateral dimension must be an order of magnitude larger than the most dominant Taylor hydrodynamic instability wavelength (λ_d), in order for a heater to be considered infinite. Otherwise, the pool boiling heat flux values on small heaters are effectively functions of the relative size of the heater. In the small heater regime, vapor bubble nucleation occurs in a confined space, leading to artificially higher values of heat flux that are strongly sensitive to the edge effects of the heater (Lienhard et al., 1973a; Lienhard and Dhir, 1973b). Hence, logical comparisons of various studies can be made only if the pool and flow boiling experiments involve infinite heaters; otherwise small heaters cause inconsistencies in the experimental measurements and prevent logical comparisons of experiments.

For the boiling experiments designed for an infinite heater, the lateral dimension of the heater must exceed ~5 to 10 times the "most dangerous" Taylor instability wavelength. This requires that the heater dimensions exceed ~5 cm on each side when multiphase flow experiments involving phase change are performed using water or organic refrigerants. Due to fabrication constraints, however, the studies mentioned in this paragraph used heaters with lateral dimensions of 1 cm or smaller. Hence, comparison of the results from these studies show varying levels of inconsistencies and require close examination before they can be used in practical engineering applications.

Taylor instability occurs when a heavier fluid is suspended on a lighter fluid (such as a vapor layer under a liquid layer). Pertubation of the vapor–liquid interface causes the growth and propagation of a range of instability waves known as Taylor instability. The Taylor instability wave with the smallest possible wavelength (critical wavelength = λ_c) is:

$$\lambda_c = \sqrt{\frac{\sigma}{g(\rho_l - \rho_v)}} \qquad (12.1)$$

where σ is the surface tension (N/m^2), g is the gravitational acceleration (m/s^2), ρ is the density (kg/m^3), and the subscripts l and v are for liquid and vapor properties, respectively. The most dominant perturbation with the maximum growth rate ("most dangerous" wavelength = λ_d) is:

$$\lambda_d = \lambda_c \sqrt{3} = \sqrt{\frac{3\sigma}{g(\rho_l - \rho_v)}} \qquad (12.2)$$

It should be noted that the heater dimension in pool and flow boiling experiments requires heaters whose lateral dimensions exceed ~5 to 10 times the value of λ_d for the particular test liquid to achieve the infinite heater configuration that ensures repeatability of the measurements.

It is expected that the vapor bubbles nucleate in the superheated liquid layer on top of the nanostructures (and not in the intervening spaces between the nanostructures) since the required wall superheat temperature for nucleation was observed experimentally to be quite low. If the vapor bubbles nucleate within the nanostructures, the required wall superheat would be quite high. However, the analysis for nucleation criteria is based on continuum models. The nucleation criteria need to be explored and validated for noncontinuum effects that dominate the fluid flow behavior within the nanostructures, which is as yet unexplored in the pool boiling literature.

12.2.4 Miscellaneous Topics

The formation of "nanobubbles" has been reported by various studies in the literature (Ishida et al., 2000). Huge capillary pressures within the nanobubbles should cause them to collapse (or dissolve into the liquid phase). However, the formation of these nanobubbles seems to defy the thermodynamic rules for bubble nucleation.

A potential reason for the nanobubbles to exist is that in reality the pressure inside a nanobubble is lower than the theoretical estimate. This is possibly due to the lower values of surface tension in the nanobubbles (or the Young–Laplace equation may not be applicable at this stage since the thickness and curvature of the meniscus are the same size). Charge build-up at the nanoscale can cause electrostatic gradients leading to reduced surface tension (electrowetting effects are potentially magnified at the nanoscale). Alternately, the extremely large values of shear stress gradients caused can also cause reduction of the surface tension on the nanobubbles.

These stress gradients can fluctuate rapidly due to fluctuating concentration of impurities on the liquid–vapor interface. In addition, the effects of the nanobubbles on heat transfer are as yet unknown. Hence this topic has not been explored in this chapter in much detail and probably is a subject of future review when sufficient data in the literature will afford a better understanding of this nanoscale phenomenon.

The enhancement of boiling and condensation in microchannels and nanochannels has also been excluded from this chapter. Interested readers can find additional information in relevant annual conference proceedings (e.g., *Proceedings of 8th International Conference on Nanochannels, Microchannels, and Minichannels* held in Montreal in 2010).

12.3 Nanofluids

The thermophysical properties of fluids mixed with suspended solid particles have been explored since the time of Maxwell (1873). It is expected that these mixtures will have enhanced thermal properties compared with those of the pure fluid. However, poor dispersion of the solid particles can cause degradation in the practical usability of these mixtures for engineering applications.

Agglomeration of suspended particles is the biggest hindrance to real applications of these solid–liquid mixtures. Earlier attempts at suspending millimeter-sized particles often lead to problems of clogging channels and valves. Subsequent attempts involving even micrometer-sized particles were found to cause problems due to agglomeration into bigger clusters of particles over time, eventually leading to precipitation of these agglomerated particles.

Nanofluids are colloidal systems consisting of minute concentration of nanometer-sized particles dispersed in a solvent medium. Since nanoparticles have exceptionally large surface areas, they are able to accumulate considerable levels of surface charge. The increased surface charge causes mutual repulsion of the nanoparticles and enables them to be well suspended in a solvent which in turn prevents agglomeration or clustering of the nanoparticles.

This effect is very sensitive for inorganic nanoparticles and especially to the pH (ionic concentration) of ionic solvents. For dielectric solvents and organic nanoparticles, surface functionalization by chemical synthesis or by use of surfactants is required to stabilize the nanoparticle suspensions.

Furthermore, the enhanced thermal properties of the solid nanoparticles due to the contribution from the high surface energy per unit mass (compared to mesoscale particles of the same material) may also contribute to improving the effective thermal properties of the nanofluids. In addition, nanoparticles may induce phase transitions within the solvent phase, for example, by forming a compressed phase of the solvent on the surfaces of the nanoparticles.

The compressed phase of the solvent is expected to have properties akin to the solid phase of the solvent. Hence, nanoparticles are expected to augment the resultant thermophysical properties of the mixture only if the induced phase transformations in the solvent provide a material that has enhanced properties. The nanomaterials (added or induced) also provide a percolation network for preferential transport compared to the bulk phase.

Many research groups have reported conflicting measurements on the enhancement of thermal properties of a wide range of nanofluids. In the next section, experimental results showing thermal properties of various nanofluids will be listed. The thermophysical properties and behaviors that will be explored or summarized in this chapter include thermal conductivity, specific heat, viscosity, and pool boiling. For a more detailed and comprehensive review, we suggest the following references: Taylor and Phelan (2009), Wen et al. (2009); Wang and Majumdar (2006), and Keblisnski et al. (2005).

12.3.1 Thermal Conductivity

Studies on effective thermal conductivity of nanofluids have been reported since nanofluids were introduced by Choi (1995). However, inconsistency in the experimental results from different studies has not been resolved yet. The properties of the nanofluids are strongly sensitive to the synthesis and testing conditions. The inconsistencies among studies are primarily due to lack of proper monitoring and control of the relevant experimental parameters. In general, nanoparticles improve the effective thermal conductivities of nanofluids. Table 12.1 lists the effective thermal conductivity values of a wide range of nanofluids.

Many of these studies relied on hot wire measurement techniques that are likely to produce uncertainties due to the precipitation of the nanoparticles on the heated probes, resulting in changes of the calibration constants for the probes. In general, nonintrusive techniques such as those based on optical methods (e.g., laser flash apparatus or LFA) are preferable for these measurements since the techniques are more reliable.

TABLE 12.1

Effective Thermal Conductivities of Nanofluids

First Author and Year	Nanofluid	Concentration (vol. %)	Particle Size (nm)	Enhancement (%)
Lee et al., 1999	Al_2O_3/water	4.0	23.6	23
Wang et al., 1999	CuO/ethylene glycol	14.8	23	54
Xie et al., 2002	SiC/water	4.18	26	17
Das et al., 2003a	CuO/water	4.0	28.6	36
Wen and Ding, 2004	Al_2O_3/water	1.59	42	10
Chon et al., 2005	Al_2O_3/water	4.0	47	29
Hong et al., 2005	Fe/ethylene glycol	0.55	10	18
Xuan and Li, 2000	Cu/water	7.5	100	75
Eastman et al., 2001	Cu/ethylene glycol	0.28	<10	10
Patel et al., 2003	Ag/water	0.00026	10 to 20	8
Choi et al., 2001	MWCNT/poly-α-olefin	1.0	25 nm (D) × 50 μm (L)	157
Xie et al., 2003	MWCNT/ethylene glycol	1.0	15 nm (D) × 30 μm (L)	13
Assael et al., 2004	MWCNT/water	0.6	100 nm (D) × 50 μm (L)	38
Assael et al., 2005	MWCNT/water	0.6	130 nm (D) × 10 μm (L)	28

12.3.2 Specific Heat

Specific heat of nanofluids has not been investigated as extensively as thermal conductivity. However, this topic is also rife with counter-intuitive results. While aqueous nanofluids have shown reductions in specific heat capacity values when compared to pure water, in literature reports, the specific heat capacities of other nanofluids have been shown to increase anomalously. For example, when molten salts and poly-α-olefin (PAO) were doped with nanoparticles at 1 and 0.6% mass concentrations, the specific heat of the resultant nanofluid was enhanced by ~20 to 100% and ~20 to 50%, respectively.

As noted earlier, in these instances, the nanoparticles were found to induce phase transitions in the solvent phase due to the high specific surface energy of the nanoparticles. On performing electron microscopy of the solid phase of these nanocomposite materials, higher density phase transitions were found to nucleate around the nanoparticles (compressed layer). The thickness of the compressed layer was estimated to be ~1 nm from electron microscopy studies.

A significant part of the bulk solvent phase also showed the formation of a percolation network composed of the phase transition materials induced by the presence of the nanoparticles. The percolation network was found to branch out at considerable distances from the nanoparticles forming a "web" nanostructure that also "trapped" the rest of the nanoparticles in the percolation network. Table 12.2 lists experimental results for measurement of effective specific heat capacity values of a wide range of nanofluids.

It is possible that the nanoparticles formed intermetallic complexes with the molten salts. Additionally, it is possible that the molten salt materials initiated chemical reactions (reversible or irreversible) on the surfaces of the nanoparticles that may appear as enhancements of the effective specific heat capacities of the nanocomposite materials. Such surface chemical reactions (similar to

TABLE 12.2

Effective Specific Heats of Nanofluids

First Author and Year	Nanofluid	Concentration	Particle Size (nm)	Enhancement (%)
Nelson et al., 2009	Graphite/ Poly-α-olefin	0.6 wt.%	20 μm (D) × 100 nm (L)	50
Shin and Banerjee, 2011a	SiO_2/Li_2CO_3-K_2CO_3	1.0 wt.%	30	20
Shin and Banerjee, 2011b	CNT/Li_2CO_3-K_2CO_3	0.5 wt.%	30 nm (D) × 1.5 μm (L)	18
Shin and Banerjee, 2010	SiO_2/Li_2CO_3-K_2CO_3	1.5 wt.%	10 to 30	110 to 125
Zhou et al., 2008	Al_2O_3/water	21.7 vol.%	45	−40
Namburu et al., 2007	SiO_2/water	10 vol.%	20	−12
Vajjha et al., 2009	ZnO/water/ethylene glycol	7 vol.%	77	−20

absorption and adsorption type interactions) can be reversible or irreversible at the various temperature ranges at which the experiments were performed, often exceeding the melting points of the high temperature materials.

Such results have huge implications for practical engineering applications. For example, PAO is widely used for cooling of electronics components in avionics. Molten salts are used for energy storage in solar thermal energy applications. Such dramatic enhancements in specific heat capacities of molten salts are expected to reduce the cost of solar energy by as much as 40%, potentially making it competitive with the cost of conventional (fossil fuel-based) energy generation.

12.3.3 Viscosity

Viscosity is an important property for determining the practical efficacy of coolants in engineering applications. Viscosity values provide an estimate for pump power that must be expended for coolant and lubricants. The pump power requirement must be balanced with any perceived enhancement in heat flux. Various figures of merit are usually derived for specific situations to determine the efficacy of various coolants and additives. Table 12.3 summarizes the viscosity measurements of nanofluids reported by several studies.

While the solvents may be Newtonian (linear stress versus strain rate relationships), the addition of minute concentrations of nanoparticles can convert the nanofluids into non-Newtonian compounds exhibiting nonlinear behavior. This behavior contradicts predictions from Einstein's equation for the viscosity of spherical additives in very dilute concentrations. Hence, the viscosity of nanofluids can be a complex function of temperature, concentration, shape, size, additives (e.g., surfactants), and pH. In general, the viscosities of nanofluids increase with concentrations of nanoparticles and can increase or decrease with temperature (for various pure solvents the viscosity decreases as temperature increases). The effect of particle size on viscosity is still controversial and further systematic studies are needed to resolve these issues.

TABLE 12.3

Viscosities of Nanofluids

First Author and Year	Nanofluid	Viscosity
Li et al., 2002	CuO/water	Decrease with temperature; increase with concentration
Ding et al., 2006	CNT/water	
Chen et al., 2007	TiO_2/EG	
Nguyen et al., 2007	Al_2O_3/water	
Namburu et al., 2007	SiO_2/water	
Nelson et al., 2009	Graphite/PAO	Increase with temperature

EG = ethylene glycol. PAO = poly-α-olefin.

12.3.4 Pool and Flow Boiling of Nanofluids

In Section 12.2, we discussed pool boiling experiments on nanostructured surfaces. Similarly, nanofluids have been reported for anomalously enhanced boiling heat transfer. Table 12.4 lists results of boiling experiments on a wide range of nanofluids. The results imply that the presence of nanoparticles improves boiling heat transfer. Increased boiling surface area due to precipitation of nanoparticles also contributes to the enhanced heat transfer. The average enhancement of the heat transfer of nanofluids ranges from 30 to 200% in comparison with those of the base fluids.

Hence, various literature reports on the boiling performances of the same nanofluid are contradictory. The controversies also arise because the experimental parameters were not rigorously monitored. In addition, the synthesis protocols for the nanofluids also affect the uniformity of dispersion and agglomeration or precipitation of the nanoparticles on the heater surface, leading to another pathway for variability in the experiments.

TABLE 12.4

Nanofluid Boiling Experiments

First Author and Year	Nanofluid	Boiling Surface	Enhancement (%)
Liu et al., 2007	CuO/water	Copper block	25 to 50
Wen and Ding, 2005	Al_2O_3/water	Stainless steel disc	40
Tu et al., 2004	Al_2O_3/water	Ti heater	64
You et al., 2003	Al_2O_3/water	Copper block	200
Shi et al., 2007	Al_2O_3, Fe/water	Copper block	60
Witharana, 2003	Au, SiO_2/water/ethylene glycol	Copper block	20
Truong, 2007	$Al2O_3$, SiO_2/water	Stainless steel wire	68
Ahn et al., 2009	MWCNT/PF-560	Nanostructured copper block	33
Coursey and Kim, 2008	Al_2O_3/water/ethane	Oxidized copper block	50
Bang and Chang, 2005	Al_2O_3/water	N.A.	–20
Das et al., 2003b	Al_2O_3/water	N.A.	–40
Jackson and Bryan, 2007	Au/water	Copper block	–25
Kim et al., 2007	Al_2O_3,ZrO_2,SiO_2/water	Stainless steel wire	Decrease
Milanova and Kumar, 2005	$Al2O_3$,SiO_2,CeO_2/water	NiCr wire	Decrease
Sajith et al., 2008	Al_2O_3, Cu/ water	NiCr wire	Decrease
Trisaksri and Wongwises, 2009	TiO_2, Cu/HCFC 141b	Copper block	Decrease

N.A. = not available.

Excessive precipitation of the nanoparticles, especially for highly concentrated nanofluids, can cause fouling of the heater surfaces, leading to degradation of the pool boiling heat flux. In contrast, at very dilute concentrations of nanoparticles, a dispersed precipitation of the nanoparticles on the heater surfaces leads to the formation of nanofins that can enhance the nucleation density of bubbles and lead to dramatic enhancement of pool boiling heat flux.

To complicate the scenario, often the investigators would reduce the experimental data by assuming the thermophysical property values of the nanofluids using standard correlations from the literature (e.g., for viscosity and specific heat) without measuring them. Such assumptions often lead to erroneous deductions. These studies require a thorough and systematic investigation of heater surface characteristics for each experimental iteration and for the thermophysical property values of the nanofluids.

A comprehensive review of the pool boiling of nanofluids is provided by Taylor and Phelan (2009). Flow boiling experiments using a variety of nanofluids were reported by Henderson et al. (2009). The nanofluids were synthesized by doping R-134a refrigerant with nanoparticles of silica and CuO (stabilized in polyester oil). The silica nanoparticles did not yield a stable suspension in the refrigerants. This led to precipitation and caused degradation in the heat transfer by 55%. However, the CuO nanoparticles formed a stable suspension and yielded ~100% enhancement in the flow boiling heat transfer coefficients.

12.4 Nanofluid Applications

12.4.1 Thermal Management

Employing nanofluids for cooling applications has been one of the most widely researched areas in nanotechnology. Enhanced heat transfer capability of nanofluids, whose thermal properties are highly improved over conventional coolants, can significantly reduce the size of a thermal management system. This can dramatically reduce system and material costs.

However, adding nanoparticles in a fluid can increase thermal properties and viscosity. Higher viscosity of a nanofluid indicates resistance due to internal friction between nanoparticles and liquid molecules. This may considerably alter pumping efficiency. Therefore, a key for enhanced cooling capability using nanofluid may depend on achieving an ideal balance between improvement due to enhanced thermal properties and higher pumping power requirements due to a nonlinear increase in viscosity with shear rates (or flow rates).

Cooling applications of nanofluids can range from use in power plants to automotive cooling to microelectronics cooling. Buongiorno et al. (2009b) explored the potential use of nanofluids as coolants for nuclear reactors. Singh et al. (2006) explored their use in automobile radiators and reported that nanofluids can reduce the heat exchanger size by ~10%, with an associated saving in fuel use by ~5%. Ma et al. (2006) studied the use of nanofluids in oscillating heat pipes and found radical improvements in their performance. This study was performed to explore the future applications of oscillating heat pipes in cooling of high heat flux electronics and optoelectronics devices.

Nanofluids composed of encapsulated nanoparticles involving phase change of the encapsulated material were explored by Hong et al. (2010). Silica-encapsulated indium nanoparticles and polymer-encapsulated wax nanoparticles were dispersed in poly-α-olefin (PAO) and water, respectively. The heat transfer coefficient was reported to be enhanced by the nanofluids by 60 and 75% in this study, compared to that of the base fluid (pure PAO and water).

12.4.2 Solar and Thermal Energy Storage Applications

In concentrating solar power (CSP) systems, solar thermal energy is concentrated by arrays of solar collectors. The thermal energy is collected and transferred by a heat transfer fluid (HTF) to thermal energy storage (TES). The heat stored by the material used in TES is extracted subsequently, especially during periods of peak demand, for power generation by using Rankine or Stirling thermodynamic cycles. (The Rankine cycle is used to generate electrical power by driving a steam turbine using the heat from the CSP-TES; the Stirling cycle generates electrical power by driving a piston rotary unit using alternate expansion and compression of air or a gas using the heat from the CSP-TES. In both cases, the waste heat is rejected using cooling water from a large reservoir or the heat is rejected to the ambient air.)

TES also enables reliability for continuous power generation by storing surplus thermal energy in the daytime and utilizing it during the off-peak hours (at nights and during rainy days).

Conventional TES media are typically organic materials such as synthetic oils such as Therminol® whose operating temperatures are less than 500°C (since the C-H bonds are disrupted at this temperature). Increasing the operating temperature to 500 to 600°C can significantly enhance the overall system efficiency by enhancing the Carnot efficiency. However, very few materials are compatible at these high temperatures.

To obtain highly effective TES, candidate materials are also required to have high specific heat, appropriate thermal conductivity, and stability at very high temperatures. Molten salts such as alkali–carbonate, alkali–nitrate, or their eutectics have high melting points and are stable up to 700°C (Janz et al., 1979). However, specific heats or thermal conductivities of the molten salts are usually less than those of conventional TES materials, i.e., less than 2J/gK and 1W/mK, respectively (Araki et al., 1988; Wicaksono et al., 2001).

As discussed in Section 12.3, nanofluids have been reported for their highly enhanced thermal properties. Similarly, the low thermal properties of the molten salts can be improved by dispersing nanoparticles. Research investigations focusing on enhancing thermal properties of molten salt-based nanofluids for TES applications are of current interest (Shin and Banerjee, 2010a and c; Shin et al., 2010b).

12.4.3 Other Applications of Nanofluids

Since the initial studies on the efficacy of nanofluids (Choi, 1995), further investigations of nanofluids focused on the anomalous enhancement of their thermal properties. However, a few studies have explored other valuable aspects of nanofluids. Li et al. (2004) explored IrO_2 and ZrO_2 nanoparticles suspended in a lubricant and observed improved lubricant performance and a remarkable decrease in friction when applied on steel surfaces. Kao et al. (2008) explored the use of aluminum nanofluids to improve the combustion of diesel fuels and observed reductions in fuel consumption and in exhaust emissions of smoke particulates and nitrous oxide. Nanofluids were also explored in healthcare applications such as in nanofluid drug delivery systems (Kaparissides et al., 2006; Kleinstreuer, 2008).

12.5 Modeling of Nanoscale Transport in Multiphase Flows

Keblinski et al. (2002) noted that four possible transport mechanisms were responsible for the anomalous increases of the thermal conductivity of nanofluids: (1) Brownian motions of nanoparticles, (2) molecular level layering of liquid at the liquid–particle interface (formation of a compressed phase), (3) the nature of heat transport in nanoparticles, and (4) the effect of nanoparticle clustering. To investigate these mechanisms, several numerical methods were explored in the literature and discussed next.

12.5.1 Lattice Boltzmann Model

Xuan et al. (2005) proposed a lattice Boltzmann model to simulate flow and energy transport processes within nanofluids. The lattice Boltzmann equations derived from spatial and temporal discretization of the discrete Boltzmann equation for multi-component system is given as:

$$f_i^\sigma (x + e_i\Delta t, t + \Delta t) - f_i^\sigma (x,t) = -\frac{1}{\tau^\sigma}\left[f_i^\sigma (x,t) - f_i^{\sigma,eq} (x,t)\right] \qquad (12.3)$$

where $\sigma = 1,2....n$ and $\tau^\sigma = \lambda^\sigma/\Delta t$ are the dimensionless collision relaxation time constants of the rth component of the fluid, e_i is the lattice velocity vector, the i subscript represents the lattice velocity direction, and $f_i^\sigma (x,t)$ is the population of the particles of component σ with velocity e_i (along with the direction i) at lattice x and time t.

This numerical model comprises a lattice, an equilibrium distribution, and a kinetic equation called the lattice Boltzmann equation (LBE). The lattice is selected based on the hydrodynamic problem. The equilibrium distribution is selected so each component obeys the macroscale Navier–Stokes equations.

Based on the model, the authors concluded that the Brownian force is the dominant factor affecting random displacement and aggregation of the nanoparticles. Also, distribution of nanoparticles can be improved by forcing bulk convection and raising the temperature of the medium, which in turn enhances the energy transport within the nanofluid.

Zhou et al. (2010) developed a multi-scale analytical approach based on the lattice Boltzmann model to study the fluid flow and thermal transport of nanofluids. Using this model on the mesoscale, the nanofluid is approximated as a binary component fluid, while at the macroscale it is treated as a single component system. The method consists of dividing the computational domain into two regions with fine and coarse mesh. In the fine mesh regions, the multi-component lattice Boltzmann model is used to study the binary components of nanofluid; in the coarse mesh region, the single component lattice Boltzmann model is adopted.

The continuity at the boundary is ensured by the mass, momentum, and energy conservation across it. The model is validated by studying the sudden start Couette flow and convective heat transfer in parallel plate channels. The nanofluid consists of water as base fluid with 10-nm Cu particles at 1% concentration. The model successfully simulated the microscale characteristics of the nanofluid and has much better efficacy in predicting the behaviors of nanofluids when compared with pure multi-component lattice Boltzmann models.

12.5.2 Linear Analytical Models

The effective thermal conductivity for solid–liquid mixtures of microscale and nanoscale size at low concentrations can be defined by the Maxwell equation (Maxwell, 1873) as:

$$k_{eff} = \frac{k_p + 2k_b + 2\left(k_p - k_b\right)\varphi}{k_p + 2k_b - \left(k_p - k_b\right)\varphi} k_b \qquad (12.4)$$

where k_b is the thermal conductivity of the base fluid and k_p is the thermal conductivity of the particle. Later Hamilton and Crosser (1962) modified the Maxwell formula to calculate the effective thermal conductivity for nonspherical particles. They introduced a shape factor given by n = 3/Ψ, where Ψ is the particle sphericity and is obtained as the ratio of surface area of a sphere to its volume, where the sphere has a volume corresponding to that of the particle. Hence the previous equation is modified as follows:

$$k_{eff} = \frac{k_p + (n-1)k_b - (n-1)(k_b - k_p)\varphi}{k_p + (n-1)k_b + (k_b - k_p)\varphi} k_b \tag{12.5}$$

This equation is applicable for materials in which the ratio of the thermal conductivity of the solid particles to the base fluid is 100 times or more, i.e., k_p/k_b > 100. According to these equations, the thermal conductivity of a nanofluid depends on the thermal conductivity of the base fluid, the nanoparticles, and the volume ratio of the nanoparticles (Φ).

Experimental studies show that the effective thermal conductivity of the nanofluids also depends on the volume fraction, surface area, shapes of the nanoparticles, and temperature of the medium. To account for the effects of nanolayers (i.e., the compressed phase) on thermal conductivity, Yu and Choi (2003) modified the Maxwell formula as:

$$k_{eff} = \frac{k_{pe} + 2k_b + 2(k_{pe} - k_b)(1-\beta)^3 \varphi}{k_{pe} + 2k_b - (k_{pe} - k_b)(1+\beta)^3 \varphi} k_b \tag{12.6}$$

where k_{pe} is the modified thermal conductivity of the nanoparticle given as:

$$k_{pe} = \frac{\left[2(1-\gamma) + (1+\beta)^3 (1+2\gamma)\gamma\right]}{-(1-\gamma) + (1+\beta)^3 (1+2\gamma)} k_p \tag{12.7}$$

where γ is the ratio of the thermal conductivity of the nanolayer to the nanoparticle and β is the ratio of nanolayer thickness to the radius of nanoparticle. Later these authors modified the Hamilton and Cross model to account for the thermal conductivity of the nanolayers in nonspherical particles (Yu and Choi, 2004). Xue (2003) derived a relationship based on Maxwell theory and average polarization theory by considering the interface effect between solid particles and base fluid. This relationship is expressed as.

$$9\left(1 - \frac{\varphi}{\lambda}\right)\frac{k_{eff} - k_b}{2k_{eff} + k_b} +$$
$$\frac{\varphi}{\lambda}\left[\frac{k_{eff} - k_{c,x}}{k_{eff} + B_{2,x}(k_{c,x} - k_{eff})} + 4\frac{k_{eff} - k_{c,y}}{2k_{eff} + (1 - B_{2,x})(k_{c,y} - k_{eff})}\right] = 0 \tag{12.8}$$

Here the nanoparticles are assumed to be elliptical with interfacial shells with half radius (a, b, c) where λ is expressed as:

$$\lambda = \frac{abc}{\left[(a+t)(b+t)(c+t)\right]} \tag{12.9}$$

$B_{2,x}$ is the depolarization factor derived from the polarization theory and $k_{c,x}$ and $k_{c,y}$ are the effective dielectric constants. Xue and Xu (2005) estimated the effective thermal conductivity of the nanofluids by modeling the nanoparticles to be surrounded by interfacial shells and derived the following relationship:

$$\left(1-\frac{\varphi}{\alpha}\right)\frac{k_{eff}-k_b}{2k_{eff}+k_b}+\frac{\varphi}{\alpha}\frac{\left(k_{eff}-k_2\right)\left(2k_2+k_1\right)-\alpha\left(k_1-k_2\right)\left(2k_2+k_{eff}\right)}{\left(2k_{eff}+k_2\right)\left(2k_2+k_1\right)+2\alpha\left(k_1-k_2\right)\left(k_2-k_{eff}\right)}=0 \tag{12.10}$$

where α is the ratio of the volume of spherical nanoparticles to complex structures of the nanoparticles and k_1 and k_2 are the thermal conductivity of nanoparticle and interfacial shell (i.e., compressed phase), respectively.

Xie et al. (2005) proposed that the interfacial structures formed by liquid molecule layering (i.e., formation of the compressed phase) play a dominant role in determining the effective thermal conductivities of nanofluids. Based on the heat conduction equations and hard sphere models, they estimated the effective thermal conductivity of the nanofluids as:

$$\frac{k_{eff}-k_f}{k_f}=3\Theta\varphi_T+\frac{3\Theta^2\varphi_T^2}{1-\Theta\varphi_T} \tag{12.11}$$

where

$$\Theta=\frac{\beta_{lf}\left[\left(1+\gamma\right)^3-\dfrac{\beta_{pl}}{\beta_{fl}}\right]}{\left(1+\gamma\right)^3+2\beta_{lf}\beta_{pl}},\beta_{lf}=\frac{k_l-k_f}{k_l+2k_f},\beta_{pl}=\frac{k_p-k_l}{k_p+2k_l},\beta_{fl}=\frac{k_f-k_l}{k_f+2k_l} \tag{12.12}$$

φ_T is the total volume fraction of the original nanoparticle and nanolayer (compressed phase) and is given as:

$$\varphi_T=\frac{4}{3}\pi\left(r_p+\delta\right)^3 n=\varphi\left(1+\gamma\right)^3 \tag{12.13}$$

where φ is the original volume fraction of nanoparticles and $\gamma=\delta/r_p$ is the ratio of the nanolayer thickness to the original particle radius.

Xuan et al. (2003) studied nanofluids formed by suspended Cu nanoparticles of mean radius 10 nm in water. Using the Brownian motion model and diffusion-limited aggregation model, authors studied the irregular motion and aggregation processes of nanoparticles. A theoretical model was also developed that took into account the physical properties of the nanoparticles, the solvent phase, and the structures of the nanoparticles. According to this model the effective thermal conductivity is given by:

$$\frac{k_{eff}}{k_f} = \frac{k_p + 2k_f - 2\varphi(k_f - k_p)}{k_p + 2k_f + \varphi(k_f - k_p)} + \frac{\rho_p \varphi c_p}{2k_f}\sqrt{\frac{k_B T}{3\pi r_c \eta}} \tag{12.14}$$

The predictions from these simulations and analytical models were in good agreement with the experimental results. Nan et al. (2004) developed a model to study the effective thermal conductivity of composites filled with carbon nanotubes. The effective thermal conductivity is obtained as:

$$\frac{K_e}{K_m} = \frac{3 + f(\beta_x + \beta_z)}{3 - f\beta_x} \tag{12.15}$$

with

$$\beta_x = \frac{2(K_{11}^c - K_m)}{K_{11}^c + K_m}, \beta_z = \frac{K_{33}^c}{K_m} - 1, K_{11}^c = \frac{K_c}{1 + \dfrac{2a_K}{d}\dfrac{K_c}{K_m}}, K_{33}^c = \frac{K_c}{1 + \dfrac{2a_K}{L}\dfrac{K_c}{K_m}} \tag{12.16}$$

Zheng et al. (2007) developed a modified model to predict the effective thermal conductivity in nanofluids. Their model combined the possible effects of liquid layering and interfacial resistance into one effective interfacial resistance. They used average length efficiency (ALE) to denote the strength of entanglement of carbon nanotubes. Based on this analysis, they predicted the effective thermal conductivity of the nanofluid to be:

$$\frac{k_{eff}}{k_b} = \frac{\dfrac{k_{p,m}}{k_b} + \alpha - \alpha\varphi_N\left(1 - \dfrac{k_{p,m}}{k_b}\right)}{\dfrac{k_{p,m}}{k_b} + \alpha + \varphi_N\left(1 - \dfrac{k_{p,m}}{k_b}\right)} \tag{12.17}$$

where

$$k_{p,m} = \frac{k_{p,N}}{1 + \dfrac{2k_{p,N}}{L_N G}}, \alpha = 2\varphi_N^{0.2}\frac{L_N}{d}, \varphi_N = \eta\varphi, L_N = \eta L, k_{p,N} = \eta k_p \tag{12.18}$$

The myriad analytical models proposed since Maxwell's equations have been marginally successful in explaining the observed anomalous enhancement of the thermal conductivity of nanofluids. To complicate the situation, a recent study refutes the observed enhancements in the thermal conductivity of nanofluids and proposes that Maxwell's equations are valid for nanofluids after all (Buongiorno et al., 2009a). However, these studies were performed by multiple investigators at multiple locations using intrusive metrology techniques and were primarily restricted to aqueous nanofluids. The experimental techniques employed by Buongiorno et al. in are circumspect for their repeatability, and the applicability of the studies to other classes of nanofluid materials is also not clear.

12.5.3 Fractal Models

Wang et al. (2003) developed a model to predict the effective thermal conductivities of nanofluids by employing a combination of the effective medium approximation and a fractal model for the nonmetallic nanoparticles. The method takes into account the size effect and surface adsorption of the nanoparticles. The model predictions were found to be in marginal agreement with the experimental results for 50-nm CuO nanoparticles suspended in deionized water for a volume fraction less than 0.5%.

Xu et al. (2006) developed a new model to predict the thermal conductivities of nanofluids using a fractal model. The method accounts for nanoparticle size based on the fractal distribution and heat convection between nanoparticles and the solvent phase due to the Brownian motion of nanoparticles. The model takes into account the dependence of thermal conductivity on sizes of nanoparticles, fractal dimension, temperature, and properties of the fluid.

The authors found that the effective thermal conductivity decreases with increases in the sizes of the nanoparticles; the smaller the nanoparticle size, the higher their velocity due to Brownian motion. This leads to enhancement in the heat transferred by convection. Smaller nanoparticle size means more contribution from the convective transport mechanisms. Also the contribution to thermal conductivity by the Brownian motion of nanoparticles was reported to increase as the concentration of nanoparticles increased until a critical concentration of 12.6% was reached, after which it was reported to start decreasing.

12.5.4 Monte Carlo Simulations

Evans et al. (2008) and Prasher et al. (2006) studied the effects of the aggregation of nanoparticles and the effects on interfacial thermal resistance and thermal conductivities of nanocomposites and colloidal nanofluids. For this study, they developed a three level homogenization model based on the fractal morphology of nanoparticle clusters in colloids (de Rooji et al., 1993). They used random walk Monte Carlo simulations to study effective thermal conductivity of fractal aggregate nanocomposites (Van Siclen, 1999).

The authors reported that the enhancement in the effective thermal conductivities of nanofluids and nanocomposites can occur by the formation of clusters of nanoparticles due to aggregation. They also demonstrated the dependence of thermal conductivity on cluster morphology, filler conductivity, and interfacial thermal resistance.

Feng et al. (2008) studied the thermal conductivities of nanofluids using Monte Carlo simulations. They developed a numerical model for the effective thermal conductivity of nanofluids based on the fractal characteristics of the nanoparticles. The model takes into account the effects of microconvection due to the Brownian motions of nanoparticles, thermal conductivities of the base fluid and the nanoparticles, volume fraction, fractal dimension for the nanoparticles, sizes of nanoparticles, and the temperatures of the fluids. The solutions of the model using Monte Carlo simulations were found to be in marginal agreement with the experimental data in the literature.

12.5.5 Molecular Dynamics

Nie et al. (2004) developed a hybrid multi-scale method for simulating microscale and nanoscale fluid flows. The hybrid method consists of solving Navier–Stokes equations for the continuum regions and molecular dynamic (MD) simulations for the interfacial regions where noncontinuum flow regimes exist. The coupling is achieved by imposing continuity of fluxes at the boundaries of a defined overlap region. The authors studied the sudden start Couette flow and channel flow with nanoscale rough walls, and the results were found to be in agreement with the analytical solutions and full molecular dynamics simulations.

Subsequently Eapen et al. (2006) performed equilibrium molecular dynamic simulations to study the thermal conductivities of nanofluids with Xe as the base fluid and Pt nanoparticles. The thermal conductivity was calculated based on the Green-Kubo theory. The authors reported an enhancement in the thermal conductivities of the nanofluids from their calculations. Also the level of enhancement was found to be many times larger than those predicted by Maxwell's model.

Sarkar and Selvam (2007) used equilibrium molecular dynamic (EMD) simulations to explore the transport mechanisms responsible for the observed enhancements in the thermal conductivities of nanofluids. The EMD simulations employed the Green-Kubo approach in which thermal conductivity was estimated using the fluctuation–dissipation theorem. The authors considered Ar as the base fluid and used Cu nanoparticles to simulate the nanofluid system. They observed a linear increase in the thermal conductivity of the nanofluid with increased concentration of the Cu nanoparticles.

Also the rate of increase in the thermal conductivity was found to be steeper at lower concentrations (up to 0.4%). The authors also observed significant movement of liquid atoms in the nanofluid as compared to the movement of atoms in the neat solvents. The movement of the nanoparticles was significantly slower, by as much as 28 times, compared to the solvent molecules for a mass concentration of 1%. The authors deduced that the enhanced movement of the liquid molecules is responsible for rapid transport of thermal energy within the nanofluid. This is reflected as a significant enhancement of the effective thermal conductivities of the nanofluids. The conclusion from this study was that the enhancement is not due to the slow Brownian motion of the nanoparticles but due to the rapid movement of the solvent molecules.

Galliero and Volz (2008) developed a new algorithm using nonequilibrium molecular dynamic simulations to study the thermodiffusion of a single nanoparticle through the estimation of the thermophoretic forces acting on the solute nanoparticle. They observed that the thermodiffusion amplitude and thermal conductivity decreased with nanoparticle concentration. The decrease was more pronounced for the smallest nanoparticles. The authors attributed this trend to the concentration of the thermal flux lines induced by the presence of nanoparticles that are more conductive than the fluid (volume effect) and due to the presence of Kapitza resistance at the nanoparticle–fluid interface (surface effect).

They also observed that for the systems under consideration, the surface effects dominated the volumetric effects. The authors studied the effects of particle size, mass, internal stiffness, solvent quality, and viscosity on the thermodiffusion of a single nanoparticle. They found that under all conditions the thermodiffusion was positive, i.e., the nanoparticle tended to migrate toward the colder zone in the simulation domain. Also the thermal diffusion was observed to be independent of the size and mass of the nanoparticle, which was found to increase with the increase in the relative stiffness of the internal bonds within the nanoparticle and inversely proportional to the viscosity of the fluid. The results were found to be consistent with the Stokes–Einstein laws for viscosity.

Sachdeva and Kumar (2009) studied water-based nanofluids using molecular dynamics simulations. They observed that a hydration layer formed at the particle–fluid interface where the attraction or cohesive potential between the liquid molecules dominates the thermal characteristics of the nanofluid. The primary factors responsible for the enhancement of thermal properties in nanofluids were reported to be the collision modes occurring within the hydration layer. Collision modes lead to more enhanced interactions than the kinetic or potential modes in the hydration layer. Also the thermal conductivity was reported to be higher for the wetting particles than for the neutral or nonwetting particles.

12.5.6 Brownian Dynamics Models

Bhattacharya et al. (2004) developed a simulation technique to study the thermal conductivities of nanofluids using Brownian dynamics models. This technique does not account for the motions of fluid molecules and therefore the transport mechanisms for short time-scales are not computed deterministically. The solvent particles are omitted from the simulation and their effects on the solute are represented by a combination of random forces and frictional terms.

This technique has the advantage of being less expensive than molecular dynamic simulations and can be coupled with the equilibrium Green-Kubo approach to estimate the thermal conductivities of nanofluids. The authors studied the copper–ethylene glycol and aluminum oxide–ethylene glycol nanofluid systems with the model. The results of the simulations were in good agreement with the experimental results.

Jain et al. (2009) studied the effective thermal conductivities of alumina–ethylene glycol nanofluids. Brownian dynamic simulations coupled with the Green-Kubo method were performed to study the effects of particle size, volume fraction, and temperature. The results showed a linear dependence of thermal conductivity on particle concentration. The thermal conductivity was reported to decrease with increases in particle size. The results were found to be in good agreement with those of Bhattacharya et al. (2004) and other results cited in the literature. The authors also affirmed that the Brownian motion of the nanoparticles is the most dominant factor for the anomalous increase in the thermal conductivity of the nanoparticles.

12.5.7 Nanofin Models

As noted earlier, the dominant factor responsible for enhancing the pool boiling heat flux is the increase in the surface area due to the presence of the nanostructures (i.e., nanofin effect). Order of magnitude analysis of carbon and silicon nanofins revealed that the interfacial thermal resistance to heat flow from nanofins to the surrounding molecules was three orders of magnitude for carbon as compared to silicon nanofins. In order to study the thermal performance of carbon nanotubes and silicon nanostructures in multiphase flows, interfacial thermal resistance must be studied.

Interfacial thermal resistance represents a barrier to heat flow at the boundary between two phases or two dissimilar materials. Thermal resistance at the interface was first reported by Kapitza, with his measurement of the temperature drop at the interface between helium and a solid (Kapitza, 1941). Hence, the interface thermal resistance is also known as Kapitza resistance.

For a temperature drop of ΔT at the interface, the heat flux J_Q is related to the Kapitza resistance R_k as $R_k = \Delta T / J_Q$. Interfacial thermal resistance has been typically estimated using nonequilibrium molecular dynamic (NEMD) simulations. To simulate this resistance using NEMD simulations, a thermal current is created in the system. There are two approaches to create a thermal current in a system. The first is the constant heat flux method in which an equal amount of energy is added and removed from two plates at the boundary of a computational domain to create a thermal current in the system. In the second approach to create thermal current, the two plates are maintained at different temperatures. The difference in the temperatures leads to a thermal current in the system as heat flows from hot plate to cold plate.

In both the approaches, the thermal current creates a temperature difference at the interface. Using the ratio of the temperature difference and heat flux in the system, Kapitza resistance can be computed.

Xiang et al. (2008) studied the role of the interface on interface resistance in microscale and nanoscale heat transfer using MD simulations. They reported that the thermal contact resistance increased exponentially with decreasing area of contact. They also reported that the structural defects within the materials increased the thermal interfacial resistance values.

Wang and Liang (2008) studied cross-plane thermal conductivity and interfacial thermal resistance between bilayer films using nonequilibrium MD simulations. The bilayer films were composed of argon atoms and other materials identical to argon but with different atomic masses. The results showed large temperature jump at the interface suggesting important role of interfacial resistance to heat transfer. They found interfacial resistance to be dependent on the mass ratio of the atoms but independent of the film thickness.

Murad and Puri (2008) studied the effects of pressure and liquid properties on interfacial thermal resistance. They found that high fluid pressure and hydrophilic surfaces facilitate better acoustic matching and thus decrease interface resistance.

Maruyama et al. (2004) performed the first molecular dynamic simulations to study the interfacial thermal resistance between carbon nanotubes in a bundle. They calculated the resistance to be $6.46 \times 10^{-8} \, \text{m}^2\text{K}/\text{W}$. Zhong and Lukes (2006) studied the dependence of interfacial thermal resistance between two carbon nanotubes as a function of nanotube spacing, overlap, and length. From the simulations, they found a fourth order of magnitude reduction in the interfacial thermal resistance if the nanotubes are bought into intimate contact. The resistance values were also reduced for longer nanotubes and nanotubes with more overlap area. They attributed the increase to the increase in area available for heat transfer between the nanotubes.

Huxtable et al. (2003) provided the experimental validation for these numerical simulations by using transient absorption measurements on individual single-walled carbon nanotubes. The nanotubes were encased in cylindrical micelles of sodium dodecyl sulfate (SDS) surfactant that were then dispersed in D_2O to the study the interfacial thermal resistance between the nanotube and the solvent phase. A Ti:sapphire mode-locked laser was used to produce a series of subpicosecond pulses with wavelengths in the range of 740 to 840 nm. A decay constant of 45 ps was calculated for the heat flow from the carbon nanotube to SDS, which was deduced to −correspond to an interfacial resistance of $8 \times 10^{-8} \, \text{m}^2\text{K}/\text{W}$.

Unnikrishnan et al. (2008) studied the effects of chemical additives like CuO on carbon nanotubes suspended in water. A marginal increase in the Kapitza resistance was reported for additives mixed with CNT dispersed in water. Recently, Singh and Banerjee (2010) studied the effects of polymer chains on interfacial thermal resistance using MD simulations. They found that the interfacial thermal resistance depends on the ability of the polymer molecules to wrap around the carbon nanotubes.

12.6 Conclusions and Future Directions

A review of the nanoscale transport mechanisms in multiphase flows reveals a discipline that is ripe for additional research and technological development to meet the needs of a variety of practical engineering applications. The heat transfer and transport mechanisms in multiphase flows can be enhanced by (1) the addition of nanoparticles for synthesizing nanofluids that have enhanced thermophysical properties or (2) creating nanostructures that act as nanofins on heat exchanging surfaces.

The properties of the nanofluids result from competition between Brownian diffusion and interfacial thermal (Kapitza) resistance caused by the formation of a compressed layer of solvent molecules on the surfaces of nanoparticles. Smaller sizes of nanoparticles enhanced Brownian diffusion but also caused a nonlinear increase in the Kapitza resistance. However, increasing the sizes of the nanoparticles led to a reduction in the Brownian diffusion. Hence, for aqueous nanofluids, the optimum diameter of nanoparticles is in the range of 10 to 50 nm.

Optimum sizes of nanoparticles can be explored in other solvents using a similar approach. The controversies associated with the study of nanofluids can be traced to the lack of rigor in characterizing the experimental parameters and synthesis conditions in various studies. In particular, the dispersion of the nanoparticles and the agglomeration and precipitation of the nanoparticles must be characterized rigorously for each experimental step to ensure repeatability and consistency.

Similarly, phase change heat transfer experiments performed using nanocoatings seem to be affected by the lack of consistency in ensuring the heater size corresponds to that of an infinite heater. However, in general, the nanostructured surfaces consistently yield enhancements in heat flux to varying degrees, depending on the material of the nanostructures and the fabrication methods used.

The experiments demonstrate that the nanofin effect is the more dominant mode for augmenting heat flux than the material properties themselves (in other words, materials with lower values of thermal conductivity are found to yield higher levels of enhancements in critical heat flux). Also, the greater surface area of the nanofins is the primary mode for augmenting the overall heat flux. In general, inorganic nanostructures are found to demonstrate higher levels of enhancement in heat transfer for multiphase flows involving phase change. This is due to the lower values of Kapitza resistance on these inorganic nanofins than on the organic nanofins.

Review of the literature data shows that the agglomeration and precipitation of nanoparticles on heat exchanging surfaces is the most dominant mechanism for enhancing heat flux in nanofluids. The concomitant disadvantage for nanofluids arises from the nonlinear enhancement of the viscosity of these nanofluids, thus increasing the pumping power requirements. Hence, application of surface nanocoatings on heat exchanging surfaces appears to be the most economical approach for augmenting heat transfer in engineering applications rather than using nanoparticle doping in the coolants used for thermal management. The primary advantage of nanofluids appears to be in emerging technologies such as thermal energy storage for solar thermal power applications and healthcare applications such as drug delivery.

Acknowledgments

During the preparation of this book chapter, the authors acknowledge the support of the Department of Energy (DOE) Solar Energy Program (Golden, CO) under Grant DE-FG36-08GO18154 ("Molten Salt–Carbon Nanotube Thermal Energy Storage for Concentrating Solar Power Systems"). The authors also acknowledge support from the following agencies and the program managers and lead personnel who enabled some of the results reported in this study:

1. Defense Advanced Projects Agency/Micro-Nano Fluidics Fundamental Focus Center (DARPA-MF[3], Dr. Dennis Polla, Dr. Abe Lee), 2006–2013.

2. DARPA Micro Technology Office (MTO), Seedling Program, Dr. Amit Lal, 2006–2008.

3. National Science Foundation (NSF), Small Grants for Exploratory Research, Dr. Al Ortega and Dr. Patrick Phelan, 2007–2008.

4. National Science Foundation (NSF); Small Business Innovation Research (SBIR) through NanoMEMS Research LLC, Dr. H.J. De Los Santos, 2006–2007.

5. Office of Naval Research (ONR), Thermal Management Program, Dr. Mark Spector, 2006–2010.

6. ONR-SBIR Phases I and II (through Aspen Thermal Systems), Dr. Mark Spector, 2009–2011.

7. Army Research Office (ARO), SBIR Phase II (through Lynntech, Inc.), 2007.

8. American Society for Engineering Education (ASEE)/ONR, Summer Faculty Fellowship Program (SFFP) at the Space and Naval Warfare Center Systems Command (SPAWAR- SSC), Dr. Ryan Lu, 2009.

9. ASEE/Air Force Office of Scientific Research (AFOSR), Summer Faculty Fellowship Program (SFFP) at Air Force Research Laboratories (AFRL) at Wright Patterson Air Force Base (WPAFB) Thermal Laboratory, Propulsion Division, Dr. R. Ponnappan, 2006–2007.

10. AFOSR/Small Business Technology Transfer and Research (STTR), Phase I (through NanoMEMS Research LLC), 2009.

11. AFRL/SBIR, Phase II (through Irvine Sensors Corp.), 2008.

12. AFRL Materials Directorate Bio/Microfluidics Program (through General Dynamics' Anteon Corporation), Dr. Rajesh Naik, 2006–2007.

13. National Aeronautics & Space Administration (NASA; through Texas Institute for Intelligent Bio/NanoMaterials and Structures), Dr. D. Lagoudas, 2008.

14. Pew Charitable Trust (through Center for Teaching Excellence (CTE) Peer Review of Teaching Project, Texas A&M University), 2007.

15. Texas Space Grants Consortium (TSGC), New Investigator Program, 2005–2007.

16. Mary Kay O'Connor Process Safety Center, Texas A&M University.

17. Texas Engineering Experimentation Station (TEES), Select Young Faculty Award, Dr. K. Bennett, 2008–2009.

18. Morris Foster Faculty Fellowship, Mechanical Engineering Department, Texas A&M University, Dr. Dennis O'Neal, 2007–2009.

19. 3M Corp. Innovation Center, Non-Tenured Faculty Award.

20. Qatar National Research Foundation (QNRF), 2009–2012.

References

Ahn, H.S., Sinha, N., Banerjee, D. et al. 2006. Pool boiling experiments on multi-walled carbon nanotube (MWCNT) forests. *ASME J. Heat Transfer*, 128, 1335–1342.

Ahn, H.S., Sathyamurthi, V., and Banerjee, D. 2009. Pool boiling experiments on a nanostructured surface. *IEEE Compon. Pkg. Tech.*, 32, 156–165.

Araki, N., Matsuura, M., Makino, A. et al. 1988. Measurement of thermophysical properties of molten salts: mixtures of alkaline carbonate salts. *Intl. J. Thermophys*, 9, 1071–1080.

Assael, M.J., Chen, C.F., Metaxa, I.N. et al. 2004. Thermal conductivity of suspensions of carbon nanotubes in water. *Intl. J. Thermophys.*, 25, 971–985.

Assael, M.J., Metaxa, I.N., Arvanitidis, J. et al. 2005. Thermal conductivity enhancement in aqueous suspensions of carbon multi-walled and double-walled nanotubes in the presence of two different dispersants. *Intl. J. Thermophys.*, 26, 647–664.

Banerjee, D. and Dhir, V.K. 2001a. Study of subcooled film boiling on a horizontal disc: part I: analysis. *J. Heat Transfer*, 123, 271–284.

Banerjee, D. and Dhir, V.K. 2001b. Study of subcooled film boiling on a horizontal disc: part II: experiments. *J. Heat Transfer*, 123, 285–293.

Banerjee, D., Son, G., and Dhir, V.K. 1996. Conjugate thermal and hydrodynamic analysis of saturated film boiling from a horizontal surface. *ASME HTD*, 334, 57–64.

Bang, I.C. and Chang, S.H. 2005. Boiling heat transfer performance and phenomena of Al_2O_3–water nanofluids from a plain surface in a pool. *Intl. J. Heat Mass Transfer*, 48, 2420–2428.

Beskok, A., Karniadakis, G.E., and Aluru, N. 2005. *Microflows and Nanoflows: Fundamentals and Simulation*, (Chapter 1). New York: Springer.

Bhattacharya, P., Saha, S.K., Yadav, A. et al. 2004. Brownian dynamics simulation to determine the effective thermal conductivity of nanofluids. *J. Appl. Phys.* 95, 6492–6494.

Buongiorno, J., Venerus, D.C., Prabhat, N. et al. 2009a. A benchmark study on the thermal conductivity of nanofluids. *J. Appl. Phys.*, 106, 094312.

Buongiorno, J., Hu, L.W., Apostolakis, G. et al. 2009b. A feasibility assessment of the use of nanofluids to enhance the in-vessel retention capability in light-water reactors. *Nucl. Eng. Des.*, 239, 941–948.

Chen, H., Ding, Y., He, Y. et al. 2007. Rheological behaviour of ethylene glycol based titania nanofluids. *Chem. Phys. Lett.*, 444, 333–337.

Chen, R., Lu, M.C., Srinivasan, V. et al. 2009. Nanowires for enhanced boiling heat transfer. *Nano Lett.*, 9, 548–553.

Choi, S.U.S. 1995. Enhancing thermal conductivity of fluids with nanoparticles. In Siginer, D.A. and Wang, H.P., Eds., *Developments and Applications of Non-Newtonian Flows*. New York: American Society of Mechanical Engineers, pp. 99–105.

Choi, S.U.S., Zhang, Z.G., Yu, W. et al. 2001. Anomalously thermal conductivity enhancement in nanotube suspensions. *Appl. Phys. Lett.*, 79, 2252–2254.

Chon, C.H., Kihm, K.D., Lee, S.P. et al. 2005. Empirical correlation finding the role of temperature and particle size for nanofluid Al_2O_3 thermal conductivity enhancement. *Appl. Phys. Lett.*, 87, 153107.

Coursey, J.S. and Kim, J. 2008. Nanofluid boiling: the effect of surface wettability. *Int. J. Heat Fluid Flow*, 29, 1577–1585.

Das, S.K., Putra, N., Thiesen, P. et al. 2003a. Temperature dependence of thermal conductivity enhancement for nanofluids. *J. Heat Transfer*, 125, 567–574.

Das, S.K., Putra, N., and Roetzel, W. 2003b. Pool boiling of nanofluids on horizontal narrow tubes. *Int. J. Multiphase Flow*, 29, 1237–1247.

Ding, Y., Alias, H., Wen, D. et al. 2006. Heat transfer of aqueous suspensions of carbon nanotubes CNT nanofluids. *Int. J. Heat Mass Transfer*, 49, 240–250.

Eapen, J., Li, J., and Yip, S. 2006. Probing transport mechanisms in nanofluids by molecular dynamics simulations. Paper presented at 18th National and 7th ISHMT-ASME Heat and Mass Transfer Conference, IIT, Guwahati, India.

Eastman, J.A., Choi, S.U.S., Li, S. 2001. Anomalously increased effective thermal conductivity of ethylene glycol-based nanofluids containing copper nanoparticles. *Appl. Phys. Lett.*, 78, 718–720.

Evans, W., Prasher, R., Fish, J. et al. 2008. Effect of aggregation and interfacial thermal resistance on thermal conductivity of nanocomposites and colloidal nanofluids. *Intl. J. Heat Mass Transfer*, 51, 1431–1438.

Feng, Y., Yu, B., Feng, K. et al. 2008. Thermal conductivity of nanofluids and size distribution of nanoparticles by Monte Carlo simulations. *J. Nanoparticle Res.*, 10, 1319–1328.

Galliero, G. and Volz, S. 2008. Thermodiffusion in model nanofluids by molecular dynamics simulations. *J. Chem. Phys.*, 128, 064505.

Hamilton, R.L. and Crosser, O.K. 1962. Thermal conductivity of heterogeneous two component systems. *IEC Fundam.*, 1, 182–191.

Henderson, K., Park, Y.G., Liu, L. et al. 2009. Flow-boiling heat transfer of R-134a-based nanofluids in a horizontal tube. *Intl. J. Heat Mass Transfer*, 53, 944–951.

Hendricks, T., Krishnan, S., Choi, C. et al. 2010. Enhancement of pool-boiling heat transfer using nanostructured surfaces on aluminum and copper. *Intl. J. Heat Mass Transfer*, 53, 3357–3365.

Hong, T.K., Yang, H.S., and Choi, C.J. 2005. Study of the enhanced thermal conductivity of Fe nanofluids. *J. Appl. Phys.*, 97, 1–4.

Hong, Y., Ding, S., Wu, W. et al. 2010. Enhancing heat capacity of colloidal suspension using nanoscale encapsulated phase-change materials for heat transfer. *J. Appl. Mater. Interfaces*, 2, 1685–1691.

Huxtable, S.T., Cahill, D.G., Shenogin, S. et al. 2003. Interfacial heat flow in carbon nanotube suspensions. *Nat. Mater.*, 2, 731–734.

Im, Y., Joshi, Y., Dietz, C. et al. 2010. Enhanced boiling of a dielectric liquid on copper nanowire surfaces. *Intl. J. Micro Nano Scale Transport*, 1, 79–96.

Ishida, N., Inoue, T., Miyahara, M. et al. 2000. Nano bubbles on a hydrophobic surface in water observed by tapping-mode atomic force microscopy. *Langmuir*, 16, 6377–6380.

Jackson, J. and Bryan, J. 2007. Investigation into the Pool Boiling Characteristics of Gold Nanofluids Master's Thesis, University of Missouri, Columbia.

Jain, S., Patel, H.E., and Das, S.K. 2009. Brownian dynamic simulation for the prediction of effective thermal conductivity of nanofluid. *J. Nanoparticle Res.*, 11, 767–773.

Janz, G., Allen, C., Bansal, N. et al. 1979. *Physical Properties Data Compilations Relevant to Energy Storage*. Washington, D.C.: U.S. Dept. of Commerce, National Bureau of Standards.

Kao, M., Ting, C., Lin, B. et al. 2008. Aqueous aluminum nanofluid combustion in diesel fuel. *J. Testing Eval.*, 36, 186–190.

Kaparissides, C., Alexandridou, S., Kotti, K. et al. 2006. Recent advances in novel drug delivery systems. *J. Nanotechnol. Online*, 2, 1–11.

Kapitza, P.L. 1941. The study of heat transfer in helium II. *J. Phys. USSR*, 4, 181–210.

Keblinski, P., Eastman, J.A., and Cahill, D.G. 2005. Nanofluids for thermal transport. *Mater. Today*, 8, 36–44.

Keblinski, P., Phillpot, S., Choi, S. et al. 2002. Mechanisms of heat flow in suspensions of nanosized particles nanofluids. *Intl. J. Heat Mass Transfer*, 45, 855–863.

Khanikar, V., Mudawar, I., and Fisher, T.S. 2009a. Flow boiling in a micro-channel coated with carbon nanotubes. *IEEE Trans. Comp. Pkg. Technol.*, 32, 639–649.

Khanikar, V., Mudawar, I., and Fisher, T. 2009b. Effects of carbon nanotube coating on flow boiling in a micro-channel. *Intl. J. Heat Mass Transfer*, 52, 3805–3817.

Kim, S.J., Bang, I.C., Buongiorno, J. et al. 2007. Surface wettability change during pool boiling of nanofluids and its effect on critical heat flux. *Intl. J. Heat Mass Transfer,* 12, 4105–4116.

Kleinstreuer, C., Li, J., and Koo, J. 2008. Microfluidics of nanodrug delivery. *Intl. J. Heat Mass Transfer,* 51, 5590–5597.

Launay, S., Fedorov, A.G., Joshi, Y., Cao, A., Ajayan, P.M. 2006. Hybrid micro-nano structured thermal interfaces for pool boiling heat transfer enhancement. *Microelectronics Journal,* 37, 11 (November), 1158–1164.

Lee, S., Choi, S.U.S., Li, S. et al. 1999. Measuring thermal conductivity of fluids containing oxide nanoparticles. *ASME J. Heat Transfer,* 121, 280–289.

Li, J.F., Liao, H., Wang, X.Y. et al. 2004. Improvement in wear resistance of plasma sprayed Yttria stabilized Zirconia coating using nanostructured powder. *Tribology Int.,* 37, 77–84.

Li, J.M., Li, Z.L., and Wang, B.X. 2002. Experimental viscosity measurements for copper oxide nanoparticle suspensions. *Tsinghua Sci. Technol.,* 7, 198–201.

Lienhard, J.H., Dhir, V.K., and Riherd, D.M. 1973a. Peak pool boiling heat-flux measurements on finite horizontal flat plates. *J. Heat Transfer,* 95, 477–482.

Lienhard, J.H. and Dhir, V.K. 1973b. Extended hydrodynamic theory of the peak and minimum pool boiling heat fluxes Contract NGL 18-001-035. NASA CR 2270.

Liu, Z., Xiong, J., and Bao, R. 2007. Boiling heat transfer characteristics of nanofluids in a flat heat pipe evaporator with micro-grooved heating surface. *Int. J. Multiphase Flow,* 33, 1284–1295.

Ma, H.B., Wilson, C., and Borgmeyer, B. 2006. Effect of nanofluid on the heat transport capability in an oscillating heat pipe. *Appl. Phys. Lett.,* 88, 143116.

Maruyama, S., Igarashi, Y., Taniguchi, Y. et al. 2004. Molecular dynamics simulations of heat transfer issues in carbon nanotubes. Paper presented at First International Symposium on Micro and Nano Technology, Honolulu.

Maxwell, J.C. 1873. *Treatise on Electricity and Magnetism.* Oxford: Clarendon Press.

Milanova, D. and Kumar, R. 2005. Role of ions in pool boiling heat transfer of pure and silica nanofluids. *Appl. Phys. Lett.,* 87, 185–194.

Mirzamoghadam, A. and Catton, I. 1988. A physical model of the evaporating meniscus. *J. Heat Transfer,* 110, 201–207.

Murad, S. and Puri, I.K. 2008. Thermal transport across nanoscale solid-fluid interfaces. *Appl. Phys. Lett.,* 92, 133105.

Murad, S. and Puri, I.K. 2009. Thermal transport through a fluid-solid interface. *Chem. Phys. Lett.,* 476, 267–270.

Namburu, P., Kulkarni, D., Dandekar, A. et al. 2007. Experimental investigation of viscosity and specific heat of silicon dioxide nanofluids. *Micro and Nano Lett.,* 2, 67–71.

Nan, C.W., Liu, G., Lin, Y. 2004. Interface effect on thermal conductivity of carbon nanotube composites. *Appl. Phys. Lett.,* 85, 3549–3551.

Nelson, I.C., Banerjee, D., and Ponnappan, R. 2009. Flow loop experiments using polyalphaolefin. *J. Thermophys. Heat Transfer,* 23, 752–761.

Nguyen, C. ., Desgranges, F., Roy, G. et al. 2007. Temperature and particle size-dependent viscosity data for water-based nanofluid-hysteresis phenomenon. *Intl. J. Heat Fluid Flow,* 28, 1492–1506.

Nie, X.B., Chen, S.Y., and Robbins, M.O. 2004. A continuum and molecular dynamics hybrid method for micro- and nanofluid flow. *J. Fluid Mech.,* 50, 55–64.

Oh, S.H., Kauffmann, Y., Scheu, C. et al. 2005. Ordered liquid aluminum at the interface with sapphire. *Science,* 310, 661–663.

Patel, H.E., Das, S.K., Sundararagan, T. et al. 2003. Thermal conductivities of naked and monolayer protected metal nanoparticle based nanofluids: manifestation of anomalous enhancement and chemical effects. *Appl. Phys. Lett.,* 83, 2931–2933.

Prasher, R., Evans, W., Meakin, P. et al. 2006. Effect of aggregation on thermal conduction in colloidal nanofluids. *Appl. Phys. Lett.,* 89, 143119.

Rooji, R. de, Potanin, A.A., Van den Ende, D. et al. 1993. Steady shear viscosity of weakly aggregating polystyrene latex dispersions. *J. Chem. Phys.,* 99, 9213–9223.

Sachdeva, P. and Kumar, R. 2009. Effect of hydration layer and surface wettability in enhancing thermal conductivity of nanofluids. *Appl. Phys Lett.,* 95, 223105.

Sajith, V., Madhusoodanan, R.M., and Sobhan, C.B. 2008. An experimental investigation of the boiling performance of water-based nanofluids. Paper presented at Micro/Nanoscale Heat Transfer International Conference, Tainan, Taiwan.

Sarkar, S. and Selvam, R.P. 2007. Molecular dynamics simulation of effective thermal conductivity and study of enhanced thermal transport mechanism in nanofluids. *J. Appl. Phys.,* 102, 074302.

Sathyamurthi, V., Ahn, H.S., Banerjee, D. et al. 2009. Subcooled pool boiling experiments on horizontal heaters coated with carbon nanotubes. *J. Heat Transfer,* 131, 071501.

Shi, M.H., Shuai, M.Q., Chen, Z.Q. et al. 2007. Study on pool boiling heat transfer of nanoparticle suspensions on plate surface. *J. Enhanced Heat Transfer,* 14, 223–231.

Shin, D. and Banerjee, D. 2009. Investigation of nanofluids for solar thermal storage applications. Paper presented at ASME Third International Conference on Energy Sustainability, San Francisco.

Shin, D., and Banerjee, D. 2010. Effects of silica nanoparticles on enhancing the specific heat capacity of carbonate salt eutectic (work in progress). *International Journal of Structural Change in Solids—Mechanics and Applications,* 2, 2, 25–31.

Shin, D., Jo, B., Kwak, H. et al. 2010. Investigation of high temperature nanofluids for solar thermal power conversion and storage applications. Paper presented at International Heat Transfer Conference, IHTC14, Washington.

Shin, D. and Banerjee, D. 2011a. Enhanced specific heat of silica nanofluid. *J. Heat Transfer,* 133, 024501.

Shin, D. and Banerjee, D. 2011b. Enhanced specific heat capacity of nanofluid synthesized by dispersing silica nanoparticles in molten alkali salt eutectic. *Intl. J. Heat Mass Transfer,* 54, 1064–1070.

Singh, D., Toutbort, J., and Chen, G. 2006. *Heavy Vehicle Systems Optimization Merit Review and Peer Evaluation: Annual Report.* Argonne National Laboratory.

Singh, N. and Banerjee, D. 2010. Investigation of interfacial thermal resistance on nanostructures using molecular dynamics simulations. Paper presented at 14th International Heat Transfer Conference, IHTC-14, Washington.

Singh, N., Sathyamurthi, V., Patterson, W., Arendt, J.L., and Banerjee, D. 2010. Flow boiling enhancement on a horizontal heater using carbon nanotube coatings. *International Journal of Heat and Fluid Flow,* 31, 201–207. 2010doi:10.1016/j.ijheatfluidflow.2009.11.002.

Sriraman, S.R. 2007. Pool Boiling on Nanofinned Surfaces Master's Thesis, Texas A&M University.

Sunder, M., and Banerjee, D. 2009. Experimental investigation of micro-scale temperature transients in sub-cooled flow boiling on a horizontal heater. *International Journal of Heat and Fluid Flow*, 30, 1, 140-149. doi:10.1016/j.ijheatfluidflow.2008.08.003, February.

Taylor, R.A. and Phelan, P.E. 2009. Pool boiling of nanofluids: comprehensive review of existing data and limited new data. *Intl. J. Heat Mass Transfer*, 52, 5339–5347.

Trisaksri, V. and Wongwises, S. 2009. Nucleate pool boiling heat transfer of TiO_2-R141b nanofluids. *Int. J. Heat Mass Transfer*, 52, 1582–1588.

Truong, B.H. 2007. Determination of Pool Boiling Critical Heat Flux Enhancement in Nanofluids Undergraduate Rhesis, Massachusetts Institute of Technology.

Tu, J.P., Dinh, N., and Theofanous, T. 2004. An experimental study of nanofluid boiling heat transfer. Paper presented at Sixth International Symposium on Heat Transfer, Beijing.

Ujereh, S., Fisher, T., and Mudawar, I. 2007. Effects of carbon nanotube arrays on nucleate pool boiling. *Intl. J. Heat Mass Transfer*, 50, 4023–4038.

Unnikrishnan, V.U., Banerjee, D., and Reddy, J.N. 2008. Atomistic meso-scale interfacial resistance-based thermal analysis of carbon nanotube systems. *Intl. J. Thermal Sci.*, 47, 1602–1609.

Vajjha, R. and Das, D.K. 2009. Specific heat measurement of three nanofluids and development of new correlations. *J. Heat Transfer*, 131, 071601.

Van Siclen, C. DeW. 1999. Walker diffusion method for calculation of properties of composite materials. *Phys. Rev. E*, 59, 2804–2807.

Wang, B.X., Zhou, L.P., and Peng, X.F. 2003. A fractal model for predicting the effective thermal conductivity of liquid with suspension of nanoparticles. *Intl. J. Heat Mass Transfer*, 46, 2665–2672.

Wang, S. and Liang, X. 2008. Thermal conductivity and interfacial thermal resistance in bilayered nanofilms by nonequilibrium molecular dynamics simulations. *Intl. J. Thermophysics*, 31, 1935–1944.

Wang, X., Xu, X., and Choi, S.U.S. 1999. Thermal conductivity of nanoparticle-fluid mixture. *J. Thermophysics Heat Transfer*, 13, 474–480.

Wang, X.Q. and Majumdar, A.S. 2006. Heat transfer characteristics of nanofluids: a review. *Intl. J Thermal Sci.*, 46, 1–19.

Wen, D. and Ding, Y. 2004. Effective thermal conductivity of aqueous suspensions of carbon nanotubes carbon nanotube nanofluids. *J. Thermophysics Heat Transfer*, 18, 481–485.

Wen, D. and Ding, Y. 2005. Experimental investigation into the pool boiling heat transfer of aqueous based γ-alumina nanofluids. *J. Nanoparticle Res.*, 7, 265–274.

Wen, D., Lin, G., Vafaei, S. et al. 2009. Review of nanofluids for heat transfer applications. *Particology*, 7, 141–150.

Wicaksono, H., Zhang, X., and Fujii, M. 2001. Measurements of thermal conductivity and thermal diffusivity of molten carbonates. *Thermophysics Prop.*, 22, 413–415.

Witharana, S. 2003. Boiling of Refrigerants on Enhanced Surfaces and Boiling of Nanofluids Doctoral Dissertation, Royal Institute of Technology.

Wu, W., Bostanci, H., Chow, L.C. et al. 2010. Nucleate boiling heat transfer enhancement for water and FC-72 on titanium oxide and silicon oxide surfaces. *Intl. J. Heat Mass Transfer*, 53, 1773–1777.

Xiang, H., Jiang, P.X., and Liu, Q.X. 2008. Non-equilibrium molecular dynamics study of nanoscale thermal contact resistance. *Mol. Simulation*, 34, 679–687.

Xie, H., Fujii, M., and Zhang, X. 2005. Effect of interfacial nanolayer on the effective thermal conductivity of nanoparticle–fluid mixture. *Intl. J. Heat Mass Transfer*, 48, 2926–2932.

Xie, H., Lee, H., Youn, W. et al. 2003. Nanofluids containing multiwalled carbon nanotubes and their enhanced thermal conductivities. *J. Appl. Phys.*, 94, 4967–4971.

Xie, H., Wang, J., Xi, T. et al. 2002. Thermal conductivity of suspensions containing nanosized SiC paraticles. *Intl. J. Thermophysics*, 23, 571–580.

Xu, J., Yu, B.M., Zou, M.Q. et al. 2006. A new model for heat conduction of nanofluids based on fractal distributions of nanoparticles. *J. Phys. D*, 39, 4486–4490.

Xuan, Y. and Li, Q. 2000. Heat transfer enhancement of nanofluids. *Intl. J. Heat Fluid Flow*, 21, 5864.

Xuan, Y., Li, Q., and Hu, W. 2003. Aggregation structure and thermal conductivity of nanofluids. *AIChE J.*, 49, 1038–1043.

Xuan, Y. and Yao, Z. 2005. Lattice Boltzmann model for nanofluids. *Heat Mass Transfer*, 41, 199–205.

Xue, Q., and Xu, W.M. 2005. A model of thermal conductivity of nanofluids with interfacial shells. *Mater. Chem. Phys.*, 90, 298–301.

Xue, Q.Z. 2003. Model for effective thermal conductivity of nanofluids. *Phys. Lett.*, 307, 313–317.

You, S.M., Kim, J.H., and Kim, K.H. 2003. Effect of nanoparticles on critical heat flux of water in pool boiling heat transfer. *Appl. Phys. Lett.*, 83, 3374–3376.

Yu, W. and Choi, S.U.S. 2003. The role of interfacial layers in the enhanced thermal of nanofluids: a renovated Maxwell model. *J. Nanoparticle Res.*, 5, 167–171.

Yu, W. and Choi, S.U S. 2004. The role of interfacial layers in the enhanced thermal conductivity of nanofluids: a renovated Hamilton-Cross model. *J. Nanoparticle Res.*, 6, 355–361.

Zheng, Y. and Hong, H. 2007. Modified model for effective thermal conductivity of nanofluids containing carbon nanotubes. *J. Thermophysics Heat Transfer*, 21, 658–660.

Zhong, H. and Lukes, J.R. 2006. Interfacial thermal resistance between carbon nanotubes: molecular dynamics simulations and analytical thermal modeling. *Phys. Rev. B*, 74, 125403.

Zhou, L., Xuan, Y., and Li, Q. 2010. Multiscale simulation of flow and heat transfer of nanofluid with lattice Boltzmann method. *Intl. J. Multiphase Flow*, 36, 364–374.

Zhou, S.Q. and Ni, R. 2008. Measurement of the specific heat capacity of water-based Al_2O_3 nanofluid. *Appl. Phys. Lett.*, 92, 1–3.

Zuber, N. 1959. Hydrodynamic aspects of boiling heat transfer Doctoral Dissertation, University of California, Los Angeles.

13

Nanoengineered Material Applications in Electronics, Biology, and Energy Harnessing

Daniel S. Choi, Zhikan Zhang, and Naresh Pachauri

CONTENTS

ABSTRACT Nanoengineered material systems are rapidly evolving areas of science and engineering that hold the promise of creating new techniques to manufacture devices and develop advanced technology: synthesis of nanomaterials, properties of nanomaterials, and engineered nanomaterials. Nanoengineering is fundamentally changing the way materials and devices will be produced and it will be central to the next era of the technology. The ability to develop and engineer materials at the nanoscale level and apply their unique properties to device applications will have great impacts on nanotechnology, industry, and society.

13.1 Introduction

One-dimensional (1D) nanomaterials such as wires and tubes have received much attention from the scientific community in recent years. Particularly, their novel electrical and mechanical properties are the subjects of intense research.

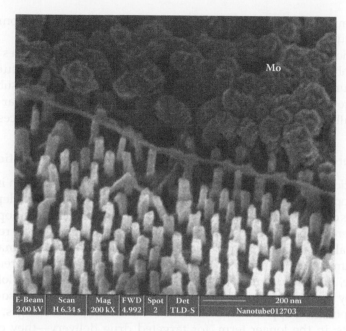

FIGURE 13.1
(See color insert.) Scanning electron micrograph of vertical array of carbon nanotubes on silicon substrate. (From Choi, D. et al. 2003. *Proceedings of IEEE Nano 2003 Conference*, Seoul. With permission.)

13.1.1 One-Dimensional Carbon Nanotubes and Their Applications

Carbon nanotubes (CNTs) were first observed in a scanning electron microscope (SEM) by Sumio Iijima in 1991. CNTs are extended tubes of rolled graphene sheets. Two types of CNT have been synthesized: single-walled (one tube) or multi-walled (several concentric tubes). Both types have diameters of only a few nanometers and are several micrometers to centimeters long.

CNTs have an important role in the context of nanomaterials because of their novel chemical and physical properties. These remarkable properties give CNTs a range of potential applications that range from reinforced composite sensors to nanoelectronics and display devices. Figure 13.1 is an SEM image of vertically grown CNTs on a silicon substrate.

13.1.2 One-Dimensional Nanowires and Their Applications

Nanowires are ultra fine wires or linear arrays of dots formed by self-assembly methods. They can be made from a wide range of materials. Semiconductor nanowires made of silicon, gallium nitride, and indium phosphides have revealed remarkable optical, electronic, and magnetic characteristics. Nanowires also have potential applications for high-density data storage, either as magnetic read heads or patterned storage media and as electronic and opto-electronic nanodevices for metallic interconnects of quantum devices and nanodevices.

Sophisticated growth techniques used to prepare these nanowires include self-assembly processes in which atoms arrange themselves naturally on stepped surfaces, chemical vapor deposition (CVD) onto patterned substrates, and electroplating or molecular beam epitaxy (MBE). The "molecular beams" are typically generated from thermally evaporated elemental sources.

13.1.3 Zero Dimensional (0D) Nanoparticles and Their Applications

Nanoparticles are always defined as particles smaller than 100 nm in diameter. Nanoparticles under this scale exhibit new or enhanced size-dependent properties compared with larger particles of the same material. Nanoparticles are of interest since they exhibit new properties (such as chemical reactivity and optical behavior) compared with larger particles of the same materials.

For example, titanium dioxide and zinc oxide become transparent at the nanoscale, but because they are able to absorb and reflect ultraviolet (UV) light, they have been applied in sunscreens. Nanoparticles also have a range of potential applications: in the short-term in new cosmetics, textiles, and paints; in the longer term, for targeted drug delivery—they could be used to deliver drugs to a specific site in the body without any side effects. Nanoparticles can also be arranged to increase the surface area and thus enhance activity relevant to a range of potential applications such as catalysts. Figure 13.2 shows silver nanoparticles prepared by colloidal chemistry.

In particular, nanoparticles of semiconductors (quantum dots) were theorized in the 1970s and initially created in the early 1980s. If semiconductor particles are made small enough, quantum effects will play a significant role so that the electrons and holes (absences of electrons) can exist in the particles. Recently, quantum dots have found applications in composites, solar cells (Gratzel cells), and fluorescent biological labels (to trace biological molecules). Recent advances in chemistry have resulted in the preparation of monolayer-protected, high-quality, monodispersed, crystalline quantum dots as small as 2 nm in diameter that can be treated conveniently and processed as typical chemical reagents.

Because of the nanometer size, significant differences exist between the mechanical properties of the nanomaterials and the bulk materials including hardness, elastic modulus, fracture toughness, scratch resistance, and fatigue strength. An enhancement of mechanical properties of nanomaterials can result from small size modifications that are generally perfect for the structures of the materials. The small size offers an internal structure free of imperfections such as dislocations, microtwins, and impurity precipitates, or the few defects or impurities present do not multiply sufficiently to cause mechanical failure.

The imperfections at nanoscale are not only highly energetic but also have the ability of migration to the surface to relax themselves during annealing,

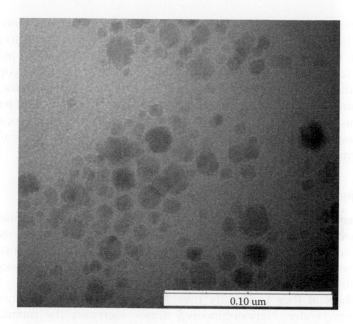

FIGURE 13.2
Transmission electron micrograph of silver nanocrystals prepared by colloidal chemistry.

purifying the material, and leaving perfect material structures inside the nanomaterials. Moreover, the external surfaces of nanomaterials also few or no defects compared to bulk materials, the function of which is to enhance the mechanical properties of nanomaterials.

The mechanical properties of nanomaterials enhanced by nanomaterials present many potential applications both in nanoscale (mechanical nanoresonators, mass sensors, microscope probe tips, and nanotweezers for nanoscale object manipulation) and macroscale applications (structural reinforcement of polymer materials, light-weight high-strength materials, flexible conductive coatings, wear-resistant coatings, and tougher and harder cutting tools).

The advance of nanotechnologies in the past decade has resulted in a burst of promising synthesis, processing, and characterization technologies that generate a variety of nanomaterials with highly controlled structures and related properties. More importantly, as the dimensions go down into the nanoscale, the temperature is in question. In non-metallic material systems, phonons are the main thermal energy carriers and they vary widely in frequency and mean free paths. The heat-carrying phonons often have large wave vectors and mean free paths in nanometer range at room temperature. As a result, the dimensions of the nanostructures are of the same magnitude as the mean free paths and wavelengths of photons.

However the general definition of temperature is attributed to the average energy of a material system in equilibrium. For macroscopic systems, the dimension is large enough to define a local temperature in each region within the materials and this local temperature will vary from region to region, which can help us study the thermal transport properties of the materials based on certain temperature distributions. Compared to macroscopic systems, the dimensions in nanomaterial systems are too small to define a local temperature. Moreover, it is also problematic to use the concept of temperature defined in equilibrium conditions for the nonequilibrium processes of thermal transport in nanomaterials, posing difficulties for theoretical analysis of thermal transport in nanoscales.

13.1.4 Applications of Nanoengineered Materials

Applications of nanoengineered materials include efficient production of energy, medical devices, cleaning environments, and many more. Researchers have devised materials like nanoscale dressings for wounds, nanopowders to treat chemical spills, nanosizing of sunscreen lotions, and nanocoatings for eye glasses and car mirrors.

Researchers at Rice University invented nanoparticles that can remove arsenic from water. Nanoengineered armor increases the mobility and endurance of military personnel and law enforcement officers. A soldier with a higher strength-to-weight ratio can pack more weapons and armor in the same volume. CNTs in combination with plastics can make composite "ropes" held together by van der Waals forces that are far lighter and stronger than steel.

Other potential applications of CNTs are energy storage and energy conversion devices; sensors; field emission displays and radiation sources; hydrogen storage media; and nanometer-sized semiconductor devices, probes, and interconnects. A single-walled nanotube when cut open laterally gives rise to a flat carbon sheet (graphene). The carbon–carbon bond length is about 0.142 nm.

Graphene is an excellent sensor due to its two-dimensional (2D) structure. A gas molecule adsorbed to the surface of graphene experiences a local change in electrical resistance. High electrical conductivity and low noise for graphene makes this change in resistance easily detectable over other materials.

Graphene is also a strong candidate for fabricating ballistic transistors. Graphene-based field effect transistors (FETs) may operate at much higher speeds (100 GHz) than Si FETs. Graphene can act as a single electron transistor (SET) by holding a single electron, i.e., the transistor can be turned on or off simply by shunting a single electron into and out of this restriction, which requires only a tiny amount of gate voltage. Chemically modified graphenes have good electrical conductivity and are extremely promising candidates for ultracapacitors. Figure 13.3 shows the conversion of a graphene sheet to a single-walled nanotube.

Nanoscale gold absorbs light and converts to heat that can kill unwanted cancer cells in the body. The nanostructures used to kill cancer cells are

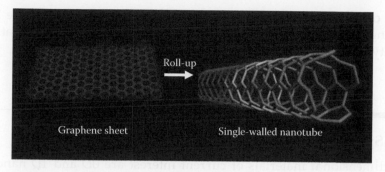

FIGURE 13.3
(See color insert.) Conversion of graphene sheet to single-walled nanotube.

called dendrimers. One of the materials used in the field of nanomedicine for drug delivery is the "buckyball" or "fullerene" shown in Figure 13.4. The most common buckyball is a hollow cluster of 60 carbon atoms shaped like a soccer ball. The other buckyballs reported are C_{70}, C_{84}, C_{28} (smallest), and C_{240} (largest).

Encapsulation of potassium or cesium into empty spaces of a fullerene can make it behave like an organic superconductor. Fullerenes have rugged structures and are highly reactive. Fullerenes incorporating alkali metals possess catalytic properties resembling those of platinum. The C_{60} molecule can also absorb large numbers of hydrogen atoms without changing the buckyball structure. This makes fullerenes better storage media for hydrogen than metal hydrides. Other possible applications for fullerenes are optical devices; chemical sensors, chemical separation devices, and batteries and other electrochemical applications.

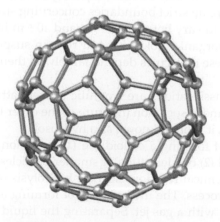

FIGURE 13.4
Atomic structure of buckyball (C_{60}).

The other applications of quantum dots are next-generation white light emitting diodes (LEDs), thermoelectrics, photonics, telecommunications, security inks, and solar cells and photovoltaics.

13.2 Synthesis of 0D and 1D Nanomaterials

Low-dimensional materials of current interest are 0D and 1D nanomaterials that exhibit anisotropic and a high surface-to-volume ratio properties. Both properties are important in terms of material sciences, new energy storage, semiconducting devices, and environmental protection. Thus, the ability to synthesize these materials in large quantities becomes a critical issue for industrial applications. With the development of nanotechnology, many types of low-dimensional nanomaterials have been synthesized and structurally characterized. Examples of 0D materials are nanodots, and nanoshells; 1D examples are nanotubes, nanowires, and nanorods.

13.2.1 Synthesis of 0D Nanoparticles and Nano (Quantum) Dots

13.2.1.1 Colloidal Chemistry

A colloid is a type of mixture in which one substance (dispersed or internal phase) is dispersed evenly throughout another (continuous phase or dispersion medium). According to this definition, a colloidal system is divided into minute particles (colloidal particles) and dispersed throughout a second substance. A colloidal system may be solid, liquid, gaseous, or a combination of these states. There are no strict boundaries concerning sizes of colloidal particles, but they tend to vary between 10^{-9} m and 10^{-6} m in size. Colloids are bigger than most inorganic molecules and remain suspended indefinitely. The properties of these molecules depend largely on their large specific surfaces (Table 13.1).

Further colloid classifications are lyophilic (solvent-attracting), lyophobic (solvent-repelling), and association (mixture of the other two). If water is the dispersing medium, it is often known as a hydrosol.

Two main ways of forming a colloid are (1) reduction of larger particles to colloidal size and (2) condensation of smaller particles into colloidal particles. Generally, chemical reactions such as hydrolysis or displacement are used in the latter process. The first method of forming an aerosol is to tear away a liquid spray with a gas jet. Separating the liquid into droplets with electrostatic repulsions by applying a charge to the liquid can be used to help the process. An emulsion is usually prepared by vigorously shaking

TABLE 13.1

Various Types of Colloids

Dispersing Medium	Dispersed Phase	Name
Solid	Solid	Solid sol
Solid	Liquid	Gel
Solid	Gas	Solid foam
Liquid	Solid	Sol
Liquid	Liquid	Emulation
Liquid	Gas	Foam
Gas	Solid	Solid aerosol
Gas	Liquid	Aerosol

Note: Dispersing medium (external phase) is the constituent found in greater extent in a colloid. Dispersed (internal) phase is the constituent found in a lesser extent.

and mixing the two constituents together, often with the addition of an emulsifying agent to stabilize the product formed.

For the second method, semi-solid colloids, known as gels, may be formed from the cooling of lyophilic sols that contain large linear molecules and have a much greater viscosity than the solvent. Usually, dialysis is employed to purify the colloids—a very slow process in which the aim is to remove most ionic materials that may have accompanied the formation.

13.2.1.2 Sol–Gel Process

A sol is a stable suspension of colloidal solid particles within a liquid. Based on the dispersion medium, a sol can be a solid, a gas dispersion medium, or a liquid dispersion medium. Based on the nature of the medium liozol called hydrosol (water), sol can be organic medium or, more specifically, spirits, esters, etc. Sols occupy an intermediate position between true solutions and coarse systems (suspensions, emulsions). A gel is a porous, three-dimensionally interconnected, solid network that expands in a stable fashion throughout a liquid medium and is limited only by the size of container.

The sol–gel process, also known as chemical solution deposition, is used primarily to fabricate nanoparticles by gelation and precipitation. It starts from a chemical solution that acts as a precursor for an integrated network (or gel) of discrete particles or network polymers. The first step of any sol–gel process is to select the precursors of the wanted materials.

The properties of a precursor can lead a reaction toward the formation of colloidal particles or polymeric gels. Then the colloidal particles obtained can be precipitated and treated by cold pressing, hot processing, and sintering, that can help produce desired ceramic in nanoscale. Before transformation into a gel, the colloidal particles can be dispersed into a stable sol to ensure that they will not be aggregated together during the following procedures. Generally, the particles in a colloidal sol usually are sized between 2 and 200 nm.

Metal alkoxides and metal chlorides undergoing various forms of hydrolysis and polycondensation reactions are typical precursors. The formation of a metal oxide is achieved by connecting the metal centers with oxo (M-O-M) or hydroxo (M-OH-M) bridges, therefore generating metal–oxo or metal–hydroxo polymers in solution. Thus, the sol evolves toward the formation of a gel-like diphasic system that contains both liquid and solid phases ranging from discrete particles to continuous polymer networks.

The process of sol–gel not only allows materials to have precise compositional control, but it also permits the production of new hybrid organic–inorganic materials that do not exist naturally. Also, the chemical processes are always carried out at low temperature to minimize the chemical interactions between the material and the container walls. The nucleation and growth of the primary colloidal particles can be controlled by the low processing temperature to control particle size, shape, and size distribution. Moreover, the easy setup makes sol–gel processes even more attractive.

13.2.2 Synthesis of 1D Nanowires and Nanotubes

13.2.2.1 Anisotropic Growth Mechanisms

Anisotropy is a material's directional dependence of a physical property. Anisotropy in polycrystalline materials can also be due to certain texture patterns that are often produced during manufacturing of the material. The mechanisms of anisotropic growth are that (1) only one facet has significantly higher growth rate than growth rates of the others and (2) different facets have different surface energies and thus different solubilities. One surface is supersaturated; the others are not. In addition, different surfaces, forming different defects exhibit different chemical potentials. Thus, preferential adsorption of impurities and growth can direct agents to block growth. Moreover, defects such as screw dislocation, stacking faults, and twins in anisotropic growth serve as continuous nucleation centers.

13.2.2.2 Chemical Vapor Deposition

Chemical vapor deposition (CVD) is a chemical process used to produce high-purity, high-performance solid materials. The process is often used to produce thin films and nanowires as well. In a typical CVD process, the substrate is exposed to one or several volatile precursors that react and/or decompose on the substrate surface to produce the desired deposit. Volatile by-products are produced at the same time and are removed by gas flow through the reaction chamber.

Microfabrication processes widely use CVD to deposit materials in various forms (monocrystalline, polycrystalline, amorphous, and epitaxial). The materials deposited are silicon, silicon carbide, carbon fibers, carbon nanofibers, filaments, carbon nanotubes, SiO_2, silicon–germanium, tungsten, silicon nitride, silicon oxynitride, titanium nitride, and various high-k dielectrics.

13.2.2.3 Vapor–Liquid–Solid (VLS) Growth

The vapor–liquid–solid method (VLS), a type of CVD, is a mechanism for the growth of 1D structures such as nanowires and nanotubes. Generally, crystal growth is very slow by direct adsorption of a gas phase onto a solid surface. The VLS mechanism circumvents this by introducing a catalytic liquid alloy phase that can adsorb a vapor to supersaturation levels in a short time, and from which crystal growth can subsequently occur from nucleated seeds at the liquid–solid interface. The physical characteristics of nanowires grown can be controlled based on the size and physical properties of the liquid alloy. The mechanism of VLS is typically described in three stages:

1. Preparation of a liquid alloy droplet on the substrate from which a wire is to be grown
2. Introduction of the substance to be grown as a vapor that adsorbs onto the liquid surface and diffuses in to the droplet
3. Supersaturation and nucleation at the liquid–solid interface, leading to axial crystal growth

The concentration gradient in VLS should be maintained below the equilibrium vapor pressure for the solid surface to prevent the side facets from growing.

First, a thin (~1 to 10 nm) gold (Au) film is deposited onto a silicon (Si) wafer substrate by sputtering or thermal evaporation. The temperatures used to anneal the wafer should be higher than the Au–Si eutectic point, creating Au–Si alloy droplets on the wafer surface (the thicker the Au film, the larger the droplets). The alloy formed by Au and Si can greatly reduce the melting temperature compared to the alloy constituents. The melting temperature of the Au–Si alloy reaches a minimum (~363°C) when the ratio of its constituents of Au to Si is 4 to 1.

During the procedure, lithography techniques can also be used to manipulate the diameters and positions of the droplets. One-dimensional nanowires can be grown by a liquid metal alloy droplet-catalyzed chemical or physical vapor deposition process that takes place in a vacuum deposition system. The function of Au–Si droplets on the surface of the substrate is to lower the activation energy of normal vapor–solid growth.

Au particles can form Au-Si eutectic droplets at temperatures above 363°C and adsorb Si from the vapor state until reaching a supersaturated state of Si in Au. The reason Au can adsorb Si from a vapor state is the ability of Au to form a solid–solution with Si. Furthermore, nanosized Au–Si droplets have much lower melting points because the surface area-to-volume ratio is increased and they become energetically unfavorable and unstable. Thus, the nanometer-scale particles act to minimize their surface energies by forming droplets (spheres or half-spheres).

Another point is that Si has a much higher melting point (~1414°C) than the eutectic Au–Si alloy. Therefore, Si atoms precipitate out of the supersaturated liquid alloy droplet at the liquid alloy–solid Si interface and the droplet rises from the surface.

13.2.2.4 Template-Based Growth

The template approach to produce free-standing nanowires has been developed extensively. The most common templates are anodized aluminum oxide (AAO) membranes (pore diameters of 20 to 200 nm and pore density of 10^{12}/cm^2) and radiation track-etched polycarbonate (PC) membranes (pore sizes of 20 nm to 1 μm and pore density of 10^9/cm^2).

Other membranes such as phase segregation-etched boron glass, porous silicon obtained via electrochemical etching of silicon wafers, zeolites, and mesoporous materials have also been used. The commonly used AAO membranes with uniform and parallel pores are manufactured by the anodic oxidation of aluminum sheets in acid solutions. Figure 13.5 is an SEM image of a free-standing AAO membrane.

Another membrane of PC is produced by bombarding a nonporous polycarbonate sheet with fission fragments to create damage tracks, and then using a chemical to etch tracks into pores. The distribution of these uniform pores in a PC membrane is random. All the template materials must be compatible with the specific processing conditions. During the synthesis via the following processing steps, the template materials should remain thermally and chemically inert. Nevertheless template-directed synthesis is an exception.

Another requirement is that the material to be deposited must wet the internal walls of pores. In the deposition process, the material deposited on the template must start from the bottom or from one end of the wall of the template channel and proceed from one side to the other. The advantages of using PC as template are easy handling and removal by using pyrolysis. The largest limitation is that PC's lack of flexibility, especially during heating and removal of the template, may damage the nanowires. The advantage of using an AAO membrane as a template is its strong resistance to high temperatures that allows the nanowires to densify completely before removal.

10.0 kV   100 nm

FIGURE 13.5
Scanning electron micrograph of free-standing AAO porous template.

13.2.2.5 Electrospinning

Electrospinning is the process of using an electrical charge to draw very fine (typically at microscale or nanoscale) fibers from a liquid. Electrospinning shares characteristics of both electrospraying and conventional solution dry spinning of fibers. The process is noninvasive and does not require coagulation chemistry or high temperatures to produce solid threads from solution. This makes the technique particularly suited to the production of fibers using large and complex molecules.

Electrospinning from molten precursors ensures that no solvent can be carried over into the final product. When a sufficiently high voltage is applied to a liquid droplet, the droplet becomes charged and electrostatic repulsion counteracts the surface tension Thus, the droplet is stretched at a critical point where a stream of liquid erupts from the surface (known as the Taylor cone).

If the molecular cohesion of the liquid is sufficiently high, stream break-up does not occur. Instead, a charged liquid jet is formed. As the jet dries in flight, the mode of current flow changes from ohmic to convective because charge migrates to the surface of the fiber. The jet is then elongated by means of whipping process caused by electrostatic repulsion initiated at small bends in the fiber until it is finally deposited on the grounded collector. The elongation and thinning of the fiber resulting from this bending instability leads to the formation of uniform fibers with nanometer-scale diameters. The parameters that can control the electrospinning process are:

1. Molecular weight, molecular weight distribution, and architecture (branched, linear, etc.) of polymer
2. Solution properties (viscosity, conductivity, and surface tension)
3. Electric potential, flow rate, and concentration
4. Distance between capillary and collection screen
5. Ambient parameters (temperature, humidity, air velocity in chamber)
6. Motion of target screen (collector)

13.3 Properties of Nanomaterials

13.3.1 Quantum Confinement

The quantum confinement effect occurs when the diameter of a particle is of the same magnitude as the wavelength of the electron wave function. When materials are in nanoscale, their electronic properties deviate substantially from those of bulk materials. When the confining dimension is large compared to the wavelength of a particle, the particle behaves like a free electron.

The band gap remains at its original energy because of its continuous energy state. Nevertheless, as the confining dimension decreases and reaches a certain limit such as nanoscale, the energy spectrum turns from a continuous to a discrete level. The result is that the band gap becomes size dependent, which ultimately results in a blue shift in optical illumination as the size of the particles decreases.

Specifically, the effect illustrates the phenomenon resulting from the squeezing of electrons and electron holes into a dimension that approaches a critical quantum measurement.

13.3.2 Optical Properties of Metal Nanoparticles (NPs)

The optical properties of metallic NPs are determined by the surface plasmon resonance (SPR) phenomenon. The SPR is due to the coherent motion of the conduction band electrons from one surface of the particle to the other upon interaction with an electromagnetic field. The distance the electron travels between scattering collisions with the lattice centers gives rise to intense absorption in the UV-visible range.

Generally, optical excitation of the SPR gives rise to surface plasmon absorption. This absorption relies on the sizes and the shapes of the particles and on the dielectric constant of the matrix. In comparison to noble metals that exhibit extremely weak photoluminescence in the bulk assigned to recombination of the excited electrons with the holes in the energy band, the luminescence yield of a metal NP increases by as much as six orders of magnitude because of the so-called lightning rod effect. The incoming (exciting) and outgoing (emitted) electric fields are amplified by the plasmon resonances around the particles. Hence, the luminescence energy and efficiency are subjected to the size effect.

In semiconductors, optical excitation leads to the promotion of an electron from the valence band (VB) into the conduction band (CB), leaving behind a hole in the valence band. After promotion occurs, the CB electron quickly jumps into the lowest energy CB state, while the hole moves to the top of the VB.

Photoluminescence is a phenomenon from recombination of the CB electron with the VB hole and is called band edge luminescence. The characteristic of luminescence is determined by a very small Stokes shift relative to band edge absorption. Since the promoted electron has opposite charges from the VB hole, a strong electrostatic attraction causes them to remain relatively localized within a nanometer-sized region. For a semiconductor particle, quantum confinement occurs when the nanocrystal radius becomes comparable to the Bohr radius that is critical for quantum effect.

The presence of dangling orbitals at the surface of the NP or defect states in the bulk provide traps for the excited electrons and holes prior to recombination. Hence, an electron excited to the CB or a VB hole can transfer itself nonradiatively to one of these surface or defect states before undergoing a (radiative or nonradiative) transition to a lower energy state. The net result is that either emission is shifted to much lower energies compared to those of band edge recombination or band edge luminescence is quenched.

Since it is difficult to fine tune defect emission for many applications, preservation band edge emission becomes the most important factor. Two methods have been widely used. The first is passivating the surface of the NP by chemically binding a capping agent. Another approach involves overcoating a nanocrystal of a given semiconductor with a thin layer of a semiconductor with a larger band gap.

13.3.3 Physical Properties of Nanomaterials

Among several novel mechanical properties of nanomaterials, high hardness has been the focus of many nanomaterials systems. A variety of superhard nanocomposites can be made of nitrides, borides, and carbides by plasma-induced chemical and physical vapor deposition. In appropriately synthesized binary systems, the hardness of a nanocomposite increases significantly over that given by the rule of mixtures in bulk.

For example, the hardness of $M_nN/a–Si_3N_4$ (M = Ti, W, V, ...) nanocomposites with the optimum content of Si_3N_4 close to the percolation threshold reaches 50 GPa although that of the individual nitrides does not exceed

21 GPa. These superhard nanocomposites show promise for hard protective coatings. Superhardness also comes from pure nanoparticles as well. For instance, Gerberich reports superhardness from nearly spherical, defect-free silicon nanospheres with diameters from 20 to 50 nm of up to 50 GPa—fully four times greater than the hardness of bulk silicon.

CNTs have motivated research since their discovery. They are the smallest carbon fibers discovered and carbon nanotubes exhibit excellent mechanical properties. The strength of the carbon fibers is enhanced by graphitization along the fiber axis. CNTs made of seamless cylindrical graphene layers are representative of ideal carbon fibers and should display the best mechanical properties of all carbon fiber species—high Young's modulus and high tensile strength.

Research theories predicted high Young's moduli of CNTs before direct experimental measurements were made. Young's moduli of single-walled CNTs have been calculated to range from 0.5 to 5.5 TPa, much higher than high-strength steel of ~200 GPa. The first experimental measurement of Young's moduli of multi-walled CNTs analyzed thermal vibrations of CNTs via transmission electron microscopy (TEM), yielding Young's modulus of 1.8 ± 0.9 TPa. The values of Young's moduli prove the high elastic modulus values of CNTs. The excellent mechanical properties of nanomaterials may also lead to many potential applications at all the nano-, micro-, and macroscales.

Another active application of nanomaterials is the enhancement of mechanical properties of polymeric materials by nanofillers. Micrometer size fillers have been used in traditional polymer composites to improve their mechanical properties (modulus, yield strength, and glass transition temperature). However, the ductility and toughness of the materials will be sacrificed for enhancement performances and large amounts of filler were needed to achieve the desired properties.

Comparably, polymer nanocomposites from nanosize fillers could lead to unique mechanical properties at very low filler weight fractions. Dramatic improvements in the yield stress (30%) and Young's modulus (170%) have been found in polypropylene filled with ultra-fine SiO_2, compared to micrometer-filled polypropylene. Significant increases in tensile strength (15%), strain-to-failure (150%), Young's modulus (23%), and impact strength (78%) with only 5 wt% nanoparticle content were reported. Elongation at break increased six times and a three times increase in modulus was found in a rubbery polyurethane elastomer filled with 40 wt% of 12-nm silica, compared to a micrometer-sized filler-reinforced polymer.

Vinylacetate, acrylic ester, synthetic rubber, and other polymer latexes have been used in coatings and adhesives; adhesion, durability, and abrasion resistance have been improved by combinations of colloidal silica and these polymer emulsions. The silica also serves as a stickiness preventive to improve the washing resistance of the coatings. In addition to nanoparticles, 1D nanomaterials such as carbon nanotubes are also superior candidates for nanofillers due to their high aspect ratios and excellent mechanical properties. Outstanding mechanical properties, such as high Young's modulus, stiffness, and flexibility are expected of nanocomposite materials from CNTs.

13.3.4 Thermal Properties of Nanomaterials

Many aspects of nanoscale materials have been developed and studied including optical electrical, magnetic, and mechanical properties. However, the thermal properties of nanomaterials have seen slower progress due to an extent to the difficulties of experimentally measuring and controlling thermal transport at nanoscale. Many difficulties surround experimental and theoretical characterization of the thermal properties of nanomaterials in comparison with their macroscopic counterparts. Recent advances have shown that certain nanomaterials have extraordinary thermal properties.

The thermal properties of nanomaterials have been modified by several factors such as small sizes, special shapes, and large interfaces that render behaviors very different from those of macroscopic materials. As described above, when a dimension is reduced to nanoscale, the sizes of the nanomaterials are almost the same magnitude as the wavelength and mean free path of the photons. As a result, the transport of photons within the materials will be changed significantly due to photon confinement and quantization of photon transport that lead to modification of thermal properties.

For example, nanowires from silicon have much smaller thermal conductivities than bulk silicon. The special structures of nanomaterials also affect their thermal properties. A direct example illustrating this view is that the tubular structures of CNTs have extremely high thermal conductivities in axial directions, leaving high anisotropy in the heat transport in the materials.

Also the interfaces are important factors for determining the thermal properties of nanomaterials. Generally, the internal interfaces impede the flow of heat because of photon scattering. At interfaces or grain boundaries between similar materials, phonons will be scattered by the interface disorder, while the differences between elastic properties and densities of vibrational states affect the transfer of vibrational energy across interfaces of dissimilar materials. As a result, the nanomaterial structures with high interface densities would reduce the thermal conductivities of the materials.

Combinations of these interconnected factors determine the special thermal properties of nanomaterials. CNTs are suitable examples. As we all know, CNTs are carbon nanostructures relating to diamond and graphite, both of which are known for their high thermal conductivities. The high stiffness of sp^3 bonds in diamond structures result in high phonon speed and consequently high thermal conductivities of the material.

The even stronger sp^2 bonds instead of sp^3 bonds in nanotube structures hold the carbon atoms together. As a result, the structures consisting of seamlessly joined graphitic cylinders have extraordinarily high thermal conductivities. The rigidity of these nanotubes, combined with virtual absence of atomic defects and coupling to soft photon modes of the embedding medium should make isolated nanotubes very efficient thermal conductors.

Conversely, 1D nanowires may offer ultra-low thermal conductivities very different from those of CNTs. In nanowires, because of the quantum confinement in 1D structures, phonons behave differently from those in the corresponding bulk materials. The surface phonon modes can be introduced to nanowire surfaces, resulting in many phonon polarizations other than the two transverse and one longitudinal acoustic branch found in bulk semiconductors. Those changes in the dispersion relations can modify the group velocity and density of state of each branch.

The phonon lifetime also changes due to the strong phonon–phonon interactions and the boundary scattering within nanostructures. Thus there is a great difference between the phonon transports and the thermal properties of the nanowires and those of the bulk materials. Although the nanowires have well defined crystalline orders like bulk materials, the observed thermal conductivity was more than two orders of magnitude smaller than that of bulk silicon, which also showed a strong dependence on nanowire size.

In addition to 1D nanowires, multilayers and superlattices are other types of nanostructures showing low thermal conductance. Multilayers and superlattices are thin film structures containing alternating layers of two different materials stacked on each other. In multilayer structures, the films can be amorphous or polycrystalline compared the single crystal films in superlattices.

Many aspects of multilayer or superlattice structures exert great effects on the phonon transport properties of the materials. When alternating layers of materials are stacked together, many collective modes of phonon transport appear in addition to the phonon modes in each single layer. When the coherence phonon length scales are much larger than those of single layers, the stacking effect will be more apparent. It is also a coupled behavior caused by interference of phonon waves reflected from multiple interfaces.

When the mean free paths of phonons span multiple interfaces, the phonon dispersion relation is modified and zone folding occurs, producing multiple phonon band gaps. Moreover, the phonon group velocities will be reduced significantly and the scattering rate will also be increased because of the modification of the phonon dispersion.

Another important factor is the interface that can determine phonon transport properties based on the high density of interfaces in the multilayer or superlattice structures. For instance, if a large mismatch between the two materials in the superlattice in the phonon dispersion relations occurs, phonons in a certain frequency range cannot propagate to the neighboring layer unless mode conversions appear at the interface. Also, dislocations and defects can be included in the interfaces of two different materials with different lattice constants. This can scatter phonons and reduce thermal conductivity. Physical roughness and alloying will also exist at the interface, depending on the processing as well, and affect phonon transport. Thus, the overall effect of these factors on the phonon transport in the multilayer and superlattice structures is a general decrease of thermal conductivities.

One other promising application of the thermal properties of nanomaterials is using nanofluids to enhance thermal transport. Nanofluids are generally defined as the solid–liquid composite materials made of nanomaterials in a size range of 1 to 100 nm suspended in a liquid. Both research and practical applications pay attention to the nanofluids because their thermal properties are greatly enhanced in comparison to those of their base fluids.

Many types of nanomaterials can be used in nanofluids including nanoparticles of oxides, nitrides, metals, metal carbides, and nanofibers such as single-walled and multi-wall CNTs that can be dispersed in a variety of base liquids specific to possible applications (e.g., water, ethylene glycol, and oils). The important feature of nanofluids is the significant increase of thermal conductivity compared with liquids without nanomaterials, as proven in many experimental works.

13.4 Nanoengineered Materials: Functionalized and Core–Shell Nanomaterials

13.4.1 Functionalized Fullerene

Fullerene (C_{60}) is a third allotrope of carbon that has poor solubility and forms aggregates in aqueous solvents. Functionalization of fullerenes increases solubility in polar solvents, making them suitable candidates for drug delivery. A supramolecular organization of dendritic amphiphiles containing fullerenes is called an amphifullerene. The structure has a 60-carbon core that contains both hydrophobic and hydrophilic moieties.

Amphiphilic fullerene-1 (AF-1) self-assembles into a hydrophilic vesicle with a hollow interior. Self-assembly of these forms of supramolecular structures are called "buckysomes." They are novel "nanovectors" that target specific disease sites in the human body. Paclitaxel-embedded buckysomes (PEBs) are spherical nanostructures of AF-1 embedding paclitaxel (an anticancer drug) inside its hydrophobic pockets. PEB's nanoscale dimensions (200 nm) allow it to pass easily through leaky tumor tissues and accumulate at tumor sites. Its advantages include (1) reduction of infusion times; (2) increased tumor uptake, resulting in better anti-cancer efficacy; and (3) no need for nonaqueous solvents that cause discomfort to patients.

Carboxyfullerenes are also reported as neuroprotective agents. Carboxyfullerene is efficient against excitotoxic necrosis and provides protection against two forms of neuronal apoptosis. Carboxyfullerenes can act as antioxidants because they are capable of suppressing iron-induced lipid peroxidation.

Fullerenes can quench ROS (reactive oxygen species) and generate ROS under specific conditions. Fullerenes exhibit extended π-conjugation, allowing them to absorb visible light and generate ROS. This may create a role for fullerenes in photodynamic therapy (PDT) which involves a combination of nontoxic photosensitizers and harmless visible light to generate ROS and kill cells. Depending on the functionalized fullerene, the technique can effectively photo-inactivate pathogenic microbial cells, malignant cancer cells, or both.

13.4.2 Functionalized Carbon Nanotubes

CNTs are hollow graphitic nanomaterials with diameters of 0.4 to 2 nm for single-walled types (SWCNTs) and 2 to 100 nm for multiple-walled (MWCNTs) types. As with fullerenes, CNT nonsolubility in aqueous media is a major issue in biomedical applications. Functionalization enables conjugation of proteins, carbohydrates, and nucleic acids with CNTs. A nanoscale bioelectronics system based on carbon nanotubes would exhibit better reliability, less power consumption, smaller space requirements, and long life.

The field of nanobioelectronics includes active field effect transistors, and electrochemical biosensors such as enzyme electrodes and immunosensors. These nanoelectrodes can be plugged into locations such as inside proteins where electrochemistry cannot perform. SWCNTs are one molecular layer thick; every atom is exposed to the adsorption of any molecule. The transfer of electrons is easier and hence the systems have high sensitivity over a broad range of gaseous and liquid analytes. This is the principle of nanoscale molecular sensors used to detect gas molecules with fast response time and high sensitivity at room temperature.

13.4.3 Functionalization of Carbon Nanotubes

13.4.3.1 Chemical Bond Formation

The functional groups are covalently linked to CNTs and the CNTs are oxidized with strong acids to develop carboxylic acid (–COOH) groups mainly at the open ends or defect sides. For side wall functionalization, nitrene cycloaddition and arylation using diazonium salts are usually employed. In some cases, functionalized CNTs obtained are reacted directly with biomolecules. In another technique, an additional bifunctional spacer is first reacted with a functionalized CNT and then linked to the biomolecule of interest.

A multi-step approach to functionalize CNT side walls involves (1) the reaction of a nitrobenzene diazonium salt, (2) reduction and reaction with a heterofunctional spacer to introduce the maleimide group, (3) reaction with a 5-thiol-modified single-stranded DNA.

The covalent method for functionalization is the simplest way of chemically modifying CNTs. Excessive oxidation can significantly change the structures and properties of the CNTs, making this method the least specific.

The side wall modification method is much less damaging to the structure and the reactive group has high specificity for attachment with biomolecules.

13.4.3.2 Physio-Absorption

This method is based on noncovalent interactions of CNTs and aromatic and/or hydrophobic molecules for further functionalization with biomolecules. The π-electrons on hydrophobic surfaces of CNTs interact noncovalently with complementary molecules and macro size biomolecules. These interactions can take place both inside and outside CNTs.

Polyethylene glycol (PEG) is one material used to noncovalently coat CNTs for specific protein absorption. PEG-coated CNTs are chemically modified to provide sites for chemical or affinity-based linking to proteins. Polyaromatic compounds such as pyrenes are functionalized with the appropriate functional group, e.g., NHS-activated acids, for attachment to proteins.

The advantage of the non-covalent approach is that the carbon nanotube structure is not changed significantly, making it easier to compare properties such as conductivity before and after biomodification. The disadvantages are lack of specificity and denaturing of the target biomolecule upon adsorption. The biological molecule can sometimes totally encapsulate a CNT that may be used the subsequently to spin hybrid CNT biofibers with greatly enhanced mechanical properties compared to other methods for spinning CNT fibers.

13.4.4 Core–Shell Nanomaterials

Core–shell nanoparticles have cores made of a material coated with another material. The shell material helps prevent the agglomeration of particles, prevents oxidation, and improves the monodispersity of nanoparticles. Core–shell nanoparticles range in size from 20 to 200 nm. The core–shell structure increases thermal and chemical stability, improves solubility, makes the particles less cytotoxic, and allows conjugation of other molecules. Figure 13.6 is a diagram of the atomic structure of a gold core–silver shell bimetallic nanoparticle. Classification is based on the material with which the core and shell of the nanocomposite are made.

13.4.4.1 Inorganic Core–Shell Nanoparticles

The core, shell, or both are made of inorganic materials like metals, semiconductors, or lanthanides. Metallic nanoparticles have cores and shells made of metal, metal oxide, or inorganic materials like silicas. The most widely used inorganic core and shell particles have gold or silver cores with silica shells. The silica coat makes the gold nanoparticles biocompatible.

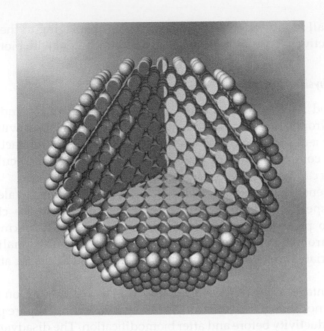

FIGURE 13.6
(See color insert.) Gold core–silver shell bimetallic nanoparticle.

The gold–silica nanoparticles are used for optical sensing. The thickness of the silica coat alters the optical properties of gold nanoparticles. The silver–silica nanoparticles are used in fluorescence imaging. Copper–copper oxide nanoparticles are used in optical bioimaging. A gold coating on iron oxide makes a particle microchip-compatible and the silica coat prevents agglomeration and also enables conjugation. Tin–tin oxide nanoparticles are used in food processing and humidity sensors. Gold core and palladium shell structures are used as catalysts.

Semiconductor nanoparticles have cores and shells made of semi conductor materials, semiconductor alloys, or metal oxides. The structures can be binary (core and shell) or tertiary (core and two shells). Quantum dots are binary structures that are alloys of group 3 and group 5 metals or group 4 and group 6 metals. Examples are CdSe/Cds, CdSe/ZnS, ZnSe/ZnS, and CdTe/CdS nanoparticles (for fluorescent bioimaging). The cores of lanthanide nanoparticles are made of lanthanide group elements with shells of inorganic materials like silica or a lanthanide material, for example, Ce- and Tb-doped core particles with $LnPO_4/H_2O$ shells. These materials find applications in electronics and bioimaging.

13.4.4.2 Organic–Inorganic Hybrid Core–Shell Nanoparticles

The cores of these particles consist of organic compounds and organic polymers. The shell is inorganic, made of metal, silica, or silicone. Examples are polymer–metal, polyethylene–silver, and polylactide–gold combinations. These are used in joint replacements due to their high resistance to corrosion and abrasion. The core is made of metal, metal oxide, or silica with a polymer of organic material or organic shell. Some of the particles in this category are SiO_2/PAPBA [poly(3-aminophenylboronic acid)], Ag_2S/PVA (polyvinyl alcohol), CuS/PVA, Ag_2S/PANI (polyaniline), and TiO_2/cellulose. These nanoparticles are used in optical devices, sensors, electrical devices, and dentistry.

13.4.4.3 Polymeric Core–Shell Nanoparticles

These nanoparticles have polymeric cores and shells. One example is polymethylmethacrylate (PMMA)-coated antimony trioxide compounded with polyvinyl chloride (PVC)/antimony trioxide composites. The interaction of PMMA and the PVC along with antimony trioxide enhance the toughness and strength of the PVC. In the field of electronics, the sensitivity to voltage change is improved by using junctions such as PDCTh [poly-(3,4-dicyano-thiophene)]/methoxy-5-(2′-ethyl-hexyloxy)-1,4-phenylene vinylene (MEH-PPV). Some examples are polystyrene (PSt)/polyvinyl acetate (PVAc), PSt/Ppy (polypyrrole), and PSt/PBA (polybutyl alcohol).

13.4.5 Synthesis of Core–Shell Nanoparticles

13.4.5.1 Polymerization

The surface of a silica particle is modified with a suitable initiator (generally bromine) and added to a solution containing the monomer of the shell polymer. The polymerization is revealed by a change in the optical clarity of the solution. In chemical oxidation polymerization, the monomer is polymerized by adding oxidizing agents. Metallic particles are given a coating of polyaromatic compounds. Most metallic nanoparticles are formed by chemical reaction followed by reduction. This method is used for silver/polyaniline and silver/polypyrrole and PbS/polypyrrole core–shell nanocomposite synthesis.

13.4.5.2 Sol–Gel Method

This technique is used for synthesis of metal–polymer core metal oxide shell nanoparticles and semiconductor nanoparticles. A solution containing a metal salt and silica-based compound is formed. The solution is heated; gelated and metal salt is reduced in the hydrogen atmosphere to metal nanoparticles. The metal particles are heated to form an oxide shell on the top. For polymer core and metal oxide shells, the polymer nanoparticles can be added to metal salt solution and then oxidized. Iron–iron oxide, tin–tin oxide, and copper–copper oxide in silica matrices are synthesized by this method.

13.4.5.3 Reverse Micelle Method

Micelles are the centers for nucleation and epitaxial growth of nanoparticles. Micelles are formed by mixing an aqueous reactant with a suitable surfactant. Synthesizing in emulsions forming micelles controls the sizes and morphologies of nanoparticles. The molar ratio of the surfactant to water affects size and morphology of the resultant particles. These particles are oxidized and polymerized to obtain core–shell structures. Examples include the synthesis of iron nanoparticles with gold shells and cobalt–platinum core–shell structures. Polymeric particles can be synthesized if the emulsion of the monomer is thermodynamically stable.

13.4.5.4 Sonochemical Synthesis

The process involves chemical reaction to form nanoparticles and sonication (frequency range is 20 kHz to 1 MHz) to improve the kinetics of reaction, break down the particles, and enhance the dispersion of particles in the solvent. In sonication, localized cavities (acting as microreactors) are formed for a short time. Temperature and pH are maintained at specific values to obtain the nanoparticles dispersed in matrix material. Reduction reactions are carried out for metallic shells and in situ polymerization for polymeric and nonmetallic shells. There are a number of composite nanoparticles such as iron oxide as a core material and gold as a shell material; cobalt or cobalt alloy as a core material and metals other than gold as a shell material, for example, silica (core)/poly(3-acrylamidophenylboronic acid (shell), Ag2S (core)/polyvinyl alcohol (shell), and CuS (core)/polyvinyl alcohol (shell).

13.4.5.5 Electrodeposition

Shells of charged polymers or inorganic materials are formed by this method. The matrix material for nanoparticles can be the electrode or electrolytic medium. The electric pulse is varied and the deposition occurs on one of the electrodes. Metal deposits during the negative cycle and polymer deposits during the positive cycle. Electrodeposition is used to synthesize polypyrrole–iron nanoparticles.

13.5 Applications of Nanoengineered Materials

13.5.1 Potential for Nanobiotechnology Innovations

There is a growing worldwide effort by the research and industrial communities to combine the advantages of nanotechnology and the life sciences into a new field called nanobiotechnology. The idea is to combine the nanometer scale engineering and fabrication capabilities of the microelectronics and nanotechnology industries with biotechnology and medicine to produce new classes of biomedical devices that utilize the advantages of both fields to improve the quality of life.

Nanobiotechnology is a rapidly emerging technology area that combines biology with nanometer-scale structures to design and fabricate biodevices. The objective of this research is to expand our knowledge of the fundamentals of nanotechnology, for example, to enable biological fluids to detect microbial food pathogens and detect proteins as biomarkers for a number of diseases. The focus is on understanding the behaviors of proteins in response to changes in environment and find ways to construct fluidic devices.

The *nano* term is derived from *nanometer* (10^{-9} meters); 1 micron (μ) corresponds to 10^3 nanometers. People should know how big a millimeter is. A micrometer is a thousand times smaller than a millimeter. A human hair measures ~80 to 100 micrometers. A nanometer is a thousand times smaller than a micrometer. The nanometer scale is at the levels of atoms and molecules.

During the twenty-first century, nanotechnology opened a door to possibilities in industries and academia. Nanobiotechnology discoveries initiated a "Second Industrial Revolution" around the middle of the twentieth century. The technology enables mass production of inexpensive biodevices and improved genetically engineered devices. To achieve the objectives nanobiotechnology research, a fair amount of effort must be focused in the following areas:

1. Applying the technologies to materials and structures at molecular level and building blocks of single cells to reveal structure information for medical device applications.
2. Developing devices to detect and manipulate biomolecules.
3. Understanding behavior of biomaterials in integrated systems.

What about fluidic devices to be integrated into biomaterials for nanobiotechnology? Microfluidic devices consists of microchannels and electrodes that have novel functions; their dimensions range from one to a few hundred micrometers. A very small volume of fluids can be carried in micrometer-scale channels. The surfaces of channels can be functionalized so proteins or DNA can be characterized or processed.

Nanobiotechnology is a multi-disciplinary area utilizing materials and methods whose critical dimensions lie between molecular levels and wavelengths of visible light. Another interest may be in exploring how to interface micro- or nanosystems and biological entities.

13.5.1.1 Major Areas of Nanobiotechnology

One example is the development of new drug delivery systems that will deliver exact amounts of drugs to precise locations in the body. For most of the pharmaceutical industry's history, drugs and medicines have consisted of simple, fast-acting chemical compounds that are dispensed orally (as pills and liquids) or as injectables. During the past three decades, however, formulations that control the rate and timing of drug delivery (time-release medications) and target specific areas of the body for treatment have become increasingly common and complex.

Current methods of drug delivery exhibit specific problems that scientists are attempting to address. For example, the potencies and therapeutic effects of many drugs are limited or reduced because of the partial degradation that occurs before they reach a desired target in the body. After ingestion, time-release medications deliver treatment continuously, rather than providing relief of symptoms and protection from adverse events only when necessary. Injectable medications could be made less expensively and administered more easily if they could simply be dosed orally or dispensed internally as needed. However, these improvements are not possible until methods are developed to safely transport drugs through specific areas of the body, such as the stomach, where low pH can destroy a medication, or through an area where healthy bone and tissue may be adversely affected.

The goal of all sophisticated drug delivery systems, therefore, is to deploy medications intact to specifically targeted parts of the body through a medium that can control administration by means of a physiological or chemical trigger. To achieve this goal, researchers are turning to advances in the worlds of micro- and nanotechnology. During the past decade, polymeric microspheres, polymer micelles, and hydrogel-type materials have all been shown to be effective in enhancing drug targeting specificity, lowering systemic drug toxicity, improving treatment absorption rates, and providing protection for pharmaceuticals against biochemical degradation.

Several other experimental drug delivery systems show exciting signs of promise, including those composed of biodegradable polymers, dendrimers (so-called star polymers), electro-active polymers, and modified carbon-60 fullerenes (also known as buckyballs).

A team of inorganic chemists led by Lon Wilson at Rice University in Houston, Texas (where the carbon-60, soccer ball–shaped fullerene was discovered) is studying metallofullerene materials (all-carbon fullerene cages enclosing metal ions) for various applications including the treatment of cancer tumors. In this application, the goal is to use selected metallofullerenes to deliver radioactive atoms directly to diseased tissues. The hoped-for results are (1) increased therapeutic potency and (2) decreased adverse effect profiles for radiation treatments. Fullerenes, in fact, are ideally suited for this goal because of their size and biocompatibility. Thus, radioactive atoms may readily be transported within metallofullerenes and any fear of stray radiation damaging otherwise healthy tissue is minimized.

At present, Sitharaman et al. (2008) have successfully shown that modified soluble metallofullerenes will preferentially bind to human bone while slowly being cleared from other tissues. As an interesting side note, the systems may also have application as contrast agents for magnetic resonance imaging because of an unpaired electron that renders magnetic the as-synthesized metallofullerenes.

However, problems exist with existing approaches using carbon-60 fullerenes and metallofullerenes. Conventional fabrication methods for carbon fullerenes produce mixtures of metallic and semiconducting fullerenes that have different functional characteristics, including the presence of unwanted metal ions. In addition, existing surface modification methods for carbon-60 fullerenes are not well established but are essential for preparing interfaces selective for binding a wide range of chemical and biological molecules. Moreover, to fabricate carbon-60 fullerenes, very expensive and sophisticated equipment that operates under high temperature and high vacuum conditions is required.

Choi et al. (2000) proposed an intelligent drug delivery system based on self-assembled nanopore membranes and biocompatible nanodisc structures that can be functionalized to deploy medications intact to specifically targeted parts of the body, control administration via a physiological or chemical trigger, and can be fabricated using electrodeposition and chemical etching processes.

Figure 13.7 shows an intelligent drug delivery system based on magnetic nanodiscs manipulated in nanochannels. The system consists of a small silicon (Si) container lidded by an alumina nanopore membrane in which animal cells and biocompatible magnetic nanodiscs are filled. The animal cells inside the Si container release the right hormones responding to environmental conditions and the hormone molecules re-attach to the magnetic nanodiscs inside the

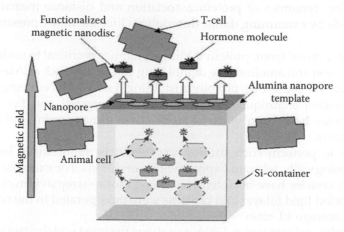

FIGURE 13.7
(See color insert.) Drug delivery system based on magnetic nanodiscs manipulated in nanochannels. The animal cells produce hormones or antibodies responding to environmental changes caused by T cells. The functionalized magnetic nanodiscs attached by hormone molecules are transported out of the container via alumina nanopores that selectively apply a magnetic field.

container. A magnetic field is applied to transport magnetic nanodiscs out of the container. The nanopore sizes are designed so that only comparable-sized magnetic nanodiscs attached by hormone molecules pass through the pores.

The system containing animal cells is implanted in a patient suffering from diabetes. The animal cells produce the insulin hormones responding to environmental change in insufficient sugar concentration caused by T-cells. The generated insulin molecules are attached to the functionalized magnetic nanodiscs and transported out of the Si container by an applied magnetic field. The pore size is designed to allow magnetic nanodiscs but not animal cells to pass through. The transported magnetic nanodiscs attached by hormone molecules interact with T cells and the resulting products are evacuated to another location by applying a different directional magnetic field. Mechanisms of discharging by-products resulting from the reaction of hormones and T cells will be studied while this work is being performed.

13.5.1.2 Biomolecular Analysis

The National Institutes of Health (NIH) identified "imaging biological processes and the effects of disease" as a leading opportunity for nanoscience in biomedical research. This is not surprising considering the variety of applications that have benefited from novel imaging tools.

Multiple color fluorescence imaging spectroscopy using dyes with overlapping spectra has been used to follow pH-sensitive liposomes within cells. Ruan et al. (2002) used two-photon excitation fluorescence resonance energy transfer (FRET) microscopy to image protein interactions in cells and tissues in vivo. The dynamics of protein association and distance measurements can be made by examining donor fluorophore lifetime in the presence of an acceptor.

On the structural front, protein folding dynamics is critical to understanding its function and ameliorating debilitating conditions such as Alzheimer's disease that are caused by protein misfolding. Laser-induced temperature-jump relaxation techniques and time-resolved infrared and fluorescence spectroscopies have been used to probe these dynamics in peptides and small proteins.

In order to perform such studies however, it is important to be able to provide biologically relevant experimental conditions. For example, 100-nm sized lipid vesicles have been tethered using biotin–streptavidin chemistry to a supported lipid bilayer and proteins were incorporated in the vesicles to obtain an average 1:1 ratio.

Of all biological molecules, DNA is perhaps the most widely studied using imaging tools at the single molecule level. Kim et al. demonstrated the use of scanning near-field optical microscopy to detect damaged bases in DNA at a scale of tens of nanometers. In an elegant experiment, single molecules of fluorescently labeled nucleotides were detected following exonuclease digestion. DNA sequencing has also been achieved using electrochemical detection by passing single strands through a nanopore and a solid state nanopore microscope has been described that can probe individual molecules. The application of nanopores in sensors for biomolecular characterization led to enhanced techniques to fabricate robust pores with nanometer precision.

13.5.1.3 Tissue Engineering

Another emerging important area in nanobiotechnology is tissue engineering; the aim is to regenerate new tissue by using biomaterials and medicines. According to the National Science Foundation, tissue engineering is defined as the application of principles and methods of engineering and life sciences toward fundamental understanding of structure–function relationships in normal and pathological mammalian tissues and the development of biological substitutes to recover tissue functions.

Researchers in this field are developing complicated structures of real tissues outside the body that can then mature into complete tissues. In this field, all kinds of human tissues including skin, liver, intestine, pancreas, bones, and muscles have been studies. Researchers can grow the cells into full layers of skin to achieve regeneration.

Tissue engineering is very attractive in terms of cost. Clearly, transplantation is one of the most expensive therapies. If we can have artificial organs and tissues in our cabinets, significant savings for patients will be achieved. Current research focuses on growing cells in three-dimensional structures instead of in beakers and testing methods for growing organ and tissue cells in 3D scaffolds.

Bone substitutes are necessary to enable repair and replace diseased tissues. A Carnegie Mellon University team has been creating engineered bone by using a computer-aided design and computer-aided manufacturing (CAD/CAM) bioreactor system capable of growing large-scale, customized bone substitutes. A CAD model of the desired bone substitute is derived from patient MRI data. The synthetic bone is to be fabricated in vitro by depositing layers of scaffolding material.

Lavik and Langer (2004) use cells combined with special plastics that act as the scaffolding on which living tissue is built. They are also designing organs (liver, kidney, heart, and intestine) that have their own circulation systems. All these attempts in the biological molecular regime are pursued on nanometer scale. In other words, nanotechnology should enable tissue engineering processes.

13.5.1.4 Lab-on-a-Chip Technology

The lab-on-a-chip (LOAC) is based on a micro- and nanofluidic system. The concept originated from microfluidics-related research and now falls into the nanofluidic area because of shrinking sizes of devices and reaction volumes of fluidics in order to increase compatibility and sensitivity.

The LOAC consists of flow channels etched onto a glass or silicon substrate integrated with a flow injection or pumping system that allows fluid transport within the chip and sample processing for a variety of detection techniques. In the area of nanobiotechnology, an LOAC is a complete system that can perform complete biosample processing and analysis on a chip scale.

A biosample with a small amount of fluid is introduced to the chip, then mixed with reagents and buffers that react to form a product. The product is mobilized to a separation unit for analysis. The LOAC will have a significant impact on the diagnostics industry, both for centralized laboratory analysis and point-of-care testing. With the current worldwide market for diagnostics well over U.S. $25 billion, LOACs exhibit immense potential in this area (Laboratory Talk, 2004).

LOACs are also making their mark in high throughput drug screening. Analysis of potential drug candidates is a large-scale automated process that requires greater accuracy and throughput than can be achieved with automated macroscale equipment. Kaigala et al. (2008) are developing a handheld device utilizing a micro- and nanofluidic chip embedded with channels in which fluids thousands of times smaller than a droplet can travel to detect multiple myeloma, follicular lymphoma, and colorectal cancer and also prescreen for susceptibility to breast cancer. Their research has already demonstrated significant practical and commercial potential.

In the near future, patients will walk into a doctor's office, give a few drops of blood, and be diagnosed within minutes. These quick tests will offer significant cost reductions and save valuable time. Pilarski said, "The new device will have the potential to quickly determine the genetic properties of a specific cancer right in the doctor's office. This means that treatment can be tailored to most effectively target specific characteristic of the disease for each patient. It also means that as the cancer changes over time, these changes can be identified and the therapy quickly adapted to target more aggressive cancer cells."

LOAC also has potential for space exploration. Researchers at NASA's Jet Propulsion Laboratory are developing a miniature analytical chemistry instrument that can be used to search for life and characterize possible biological environments during planetary missions. Micro- and nanofluidic technologies are integrated into modular chemical analysis systems at the wafer level. NASA's Marshall Space Flight Center is also using LOAC technology for medical testing and customizing the devices for use in space to detect life forms on other planets. The ability of the chip to detect microbes and contaminants will protect astronauts.

Nanotechnology serves as a solid platform for advancing other sciences. It is the science that fabricates and characterizes materials, devices, and systems that can function at the nanometer scale. Biotechnology and nanotechnology recently intersected to build another important platform. The convergence of two technology fields is opening new avenues for improving the medical aspects of quality of life. Nanobiotechnology is a booming field supported by extraordinary growth in R&D budgets. Consistent efforts of various disciplines should continue to prevent losing momentum.

13.5.2 Energy Applications

The "hype" surrounding the use of batteries and ultracapacitors in a wide range of energy markets seem to be intensifying of late. A variety of recent articles and reports have discussed the potential of nanotechnology for energy production, storage, and distribution.

13.5.2.1 Nanomaterial-Based Batteries

At a fundamental level, a battery consists of three parts: an anode (negative pole), a cathode (positive pole), and an electrolyte. When a connection is made between the positive and negative poles, a chemical reaction creates power. The amount of energy a battery can produce is determined by the masses of the anode and cathode—the more material there is, the more atoms it contains, and thus more electrons are available to provide power—and the chemical reactivities of the materials.

The surface area between the active materials (anode and cathode) and the electrolyte acts as a doorway for the current flow. The larger the surface area, the lower the battery internal resistance, and the greater the current flow. Choi et al. (2003) developed a process of fabricating vertical arrays of 1D manganese oxide materials on silicon chips using anodized alumina templates.

The advantages of nanowire-based active materials are (1) the 1D geometry of nanowires better accommodates the large volume changes that result in improved cycle performance of the cathode materials; and (2) nanowires provide a very large electrode surface area and surface-to-volume ratio for electrolyte contact. The resulting conducting pathways for lithium ions and electrons enables efficient charge transport through the electrodes. These advantages improve process kinetics and thus improve the rate capability.

Electrochemical properties of MnO_2 nanowires fabricated by low-cost electrodeposition can provide superior performance. However, electrochemically deposited MnO_2 (electrolytic manganese dioxide or EMD) contains certain number of water molecules incorporated during electrodeposition. The EMD nanowire array electrodes are heat treated to minimize the water contents of MnO_2 nanowires. The obtained samples were investigated using x-ray diffraction (XRD) to verify crystal quality and chemistry of EMD nanowires to determine microbattery performance as shown in Figure 13.8.

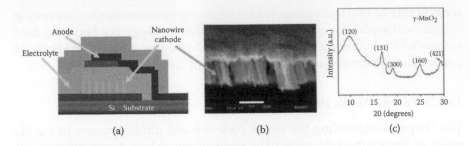

FIGURE 13.8
(a) Cross-sectional view of nanowire-based on-chip microbattery. (b) FESEM image of vertical array of EMD nanowire cathodes. (c) XRD pattern of EMD nanowires prepared by anodic electrodeposition technique.

13.5.2.2 *Nanomaterial-Based Ultracapacitors*

Capacitors (formerly known as condensers) store electric energy charges and release them as needed. Traditional electrolytic capacitors work by utilizing two conducting plates (usually made of metals that are capable of being charged) and a thin film dielectric (insulating material) as a separator in an electrolyte. The amount of capacitance that can be achieved is described in this simple equation:

$$\text{Capacitance} = \text{dielectric constant of medium} \times \text{area of plate}/\text{distance between plates}$$

Leonard et al. (2009) have demonstrated high electrostatic double layer (EDL) capacity with porous carbon supports as well as the potential that is developed on the insulating oxides. The difference here is that most of the charge is stored by the insulating film of nanoparticulate insulating oxides. The charge on the oxide is developed by a potential determining ion such as a proton.

13.5.2.3 *Nanomaterial-Based Solar Cell*

Since the world's oil and other fuel resources are finite and depleting, the focus is on finding alternative energy sources. Among the contenders competing to replace fossil fuels are solar cells that offer many advantages, for example they require little maintenance and are environmentally benign.

Solar cells are already used in a great number of applications and present the potential for many more, particularly in the energy, communications, military, and space fields. The major drawbacks of solar cells to date are cost and performance (power conversion efficiency).

Solar radiation is a plentiful and clean source of power, but because of the high cost of electrical conversion via conventional solar cells, it has been exploited to its full potential when measured on a per-watt basis. We propose a solution to these cost and performance problems through new applications for solar power.

Hybrid organic–inorganic solar cells allowed fabrication at lost cost but did not solve the problem of low power conversion efficiency. Sacriciffici et al. (1993) reported a high fill factor of 0.48 and power conversion efficiency of 0.04% under monochromatic illumination in hybrid solar cells of fullerenes and conjugated polymer. They explained the improved performance as fast electron injection from the photoexcited state of the conjugated polymer into the fullerene that should result in charge separation at the large donor–acceptor interface area.

Another type of hybrid solar cell was proposed by Coakley et al. (2005). They proposed ordered array organic–inorganic bulk heterojunction solar cells combining a light-absorbing conjugated polymer with a nanostructured, large band gap inorganic semiconductor such as TiO2 or ZnO. Their work demonstrated that most excitons are generated close enough to the organic–inorganic interface to be dissociated by electron transfer and all carriers have an uninterrupted pathway to the electrodes. Their devices showed a power conversion efficiency of 0.5%.

One problem associated with conjugated polymers and nanocrystalline TiO2 bilayers is the inefficient filling of the TiO2 pores that limits charge transport. Increasing the size of the TiO2 pore leads to devices with smaller interfaces for charge generation.

Further improvement of hybrid solar cells using nanotechnology was reported by Hyunh et al. (2002). They combined CdSe nanorods with poly-(3-hexylthiophene). Their devices showed an open circuit voltage of 0.7 V, a short circuit current of 5.7 mA, a fill factor of 0.4, and a power conversion efficiency of 1.7%. Their devices exhibited improved optical absorption in the visible range by the CdSe nanorods.

One reason for the low quantum efficiencies is the poor transport of charges resulting from phase separation in the composite films. Alivisatos et al. proposed that as the aspect ratio of the nanorods increases from 1 to 10, the charge transport must improve substantially to yield external quantum efficiency enhancement (EQE) by a factor of approximately three. Therefore, for plastic–nanorod devices to achieve typical power conversion efficiencies of conventional inorganic solar cells, it is necessary to reduce charge recombination, which decreases EQE at solar light intensities. Further enhancement of carrier mobilities can be accomplished by improving the nanocrystal–polymer interface to remove nanorod surface traps, aligning the nanorods perpendicular to the substrate and further increasing their length. However, the colloidal chemistry used for preparing elongated nanocrystals has a limitation of fabricating high aspect ratio nanocrystals. Hybrid organic–inorganic solar cells based on nanowires have several advantages over other types of solar cells:

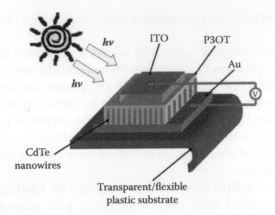

FIGURE 13.9
Nanowire-based hybrid solar cell on flexible plastic substrate. CdTe nanowires are embedded in the matrix of poly(3-octylithiophene).

1. Absorption of light having a wide range of wavelengths
2. Increased carrier mobilities in organic matrices
3. Low cost process compatible with roll-to-roll method

Figure 13.9 illustrates a nanowire-based solar cell on a transparent flexible plastic substrate that can be realized through electrodeposition technology.

13.6 Summary and Future Directions

Novel properties of nanoengineered materials enable us to build devices with useful and specific abilities. These innovative devices require synthesis of nanomaterials and control of their structures at nanometer scale. By improving design, fabrication, processing, characterization, and analysis of nanometer-scale properties of materials, specialized devices and systems for a variety of scientific and engineering applications can be engineered.

We see nanoengineered materials and devices employed for drug delivery, energy storage, and environmental clean-up within 10 years. After that, we expect to see nanomaterials in molecular computers and manufacturing and more advanced engineering applications. Nanomaterials will effect major changes in people's lives, particularly in the areas of electronics, energy, and healthcare. However, a number of difficulties must be overcome, for example, high costs and reliability issues, both of which require more attention and effort.

Acknowledgments

The authors acknowledge the financial support of the University of Idaho's New Faculty Start-up Fund, its Biological Applications of Nanotechnology (BANTech) Fund, and the Korea Science and Engineering Foundation through a pioneer program funded by the Ministry of Education.

References

Alain, C.P. 2002. *Introduction to Sol-Gel Processing*. Amsterdam: Kluwer Academic.

Anderson, M.A., Zeltner, W.A., and Merritt, C.M. 1995. Green technology for the twenty-first century. In Iglesia, E., Lednor, P.W., Nagaki, D.A. et al., Eds., *Proceedings of Materials Research Society Symposium*. Amsterdam: North Holland Publishing, pp. 377–387.

Berber, S., Kwon, Y.K., and Tománek, D. 2000. Unusually high thermal conductivity of carbon nanotubes. *Physical Review Letters*, 84, 4613–4616.

Bethune, D.S., Kiang, C.H., DeVries, M.S. et al. 1993. Cobalt-catalysed growth of carbon nanotubes with single-atomic-layer walls. *Nature*, 363, 605–607.

Bhushan, B. 2007. *Springer Handbook of Nanotechnology*. New York: Springer.

Boon, K., Teo. T., and Sun, X.H. 2007. Classification and representations of low-dimensional nanomaterials: terms and symbols. *Journal of Cluster Science*, 18, 346–357.

Brinker, C.J. and Scherer, G.W. 1990. *Sol-Gel Science: The Physics and Chemistry of Sol–Gel Processing*. San Diego: Academic Press.

Cahay, M., Leburton, J.P., Lockwood, J.G. et al. 2001. *Quantum Confinement IV: Nanostuctured Materials and Devices*. New Jersey: Electrochemical Society.

Cao, G. 2008. Nanostructures and nanomaterials: synthesis, properties and applications. *International Journal of Microstructure and Materials Properties*, 3, 704–706.

Choi, D. et al. 2003. High-Q mechanical resonator arrays based on carbon nanotubes. *IEEE Nano2003 Conference*, Seoul.

Choi, D., Park, M., Wang, K. et al. 2000. Fabrication of nanometer-scale photoresist wire patterns with a silver nanocrystal shadow mask. *Journal of Vacuum Science and Technology*, 174, 1425–1429.

Coakley, K., Liu, Y., Goh, C. et al. 2005. Ordered organic–inorganic bulk heterojunction photovoltaic cells. *MRS Bulletin*, 30, 37–40 .

Costescu, R.M., Cahill, D.G., Fabreguette, F. H. et al. 2004. Ultra-low thermal conductivity in W/Al_2O_3 nanolaminates. *Science*, 303. 989–990.

David, G.C., Wayne, K.F., Kenneth, E.G. et al. 2003. Thermal conductivity of individual silicon nanowires. *Journal of Applied Physics*, 93, 793–818.

Davis, J., Bronikowski, M., Choi, D. et al. 2003. High-Q mechanical resonator arrays based on carbon nanotubes. *IEEE Nano2003 Conference*, San Francisco, pp. 635–638.

Delzeit, L., Nguyen, C., Chen, B. et al. 2002. Growth of carbon nanotubes: a combinatorial method to study the effects of catalysts and underlayers. *Journal of Physical Chemistry B,* 106, 5629–5635.

Deyu, L., Yiying, W., Philip, K. et al. 2003. Thermal conductivity of individual silicon nanowires. *Applied Physics Letters,* 83, 2934–2946.

Dresselhaus, M.S., Dresselhaus, G., and Avouris, P. 2001. *Carbon Nanotubes Synthesis, Structure, Properties, and Applications.* Berlin: Springer.

Gerberich, W.W., Mooka, W.M., Perreya, C.R. et al. 2003. Superhard silicon nanospheres. *Journal of the Mechanics and Physics of Solids,* 51, 979–992.

Hartmut, H. and Stephan, W.K. 1994. *Quantum Theory of the Optical and Electronic Properties of Semiconductor.* New Jersey: World Scientific.

Herring, C. and Galt, J.K. 1952. Elastic and plastic properties of very small metal specimens. *Physical Review,* 85, 1060–1061.

http://media.rice.edu/media/NewsBot.asp?MODE=VIEWandID=9032

http://www.3rd1000.com/bucky/bucky.htm

http://www.bentham.orh/biomeng/samples

http://www.ciam.unibo.it/photochem/MNano06.pdf P20

http://www.nanotech-now.com/nanotube-buckyball-sites.htm

http://www.nd.edu/~gezelter/Main/Gallery1.html

Hussain, M., Nakahira, A., Nishijima, S. et al. 1996. Fracture behavior and fracture toughness of particulate filled epoxy composites. *Material Letters,* 27, 21–25.

Hyunh, W.U., Dittmer, J.J., Alivisatos, A.P. 2002. Hybrid nanorod-polymer solar cells, *Science,* 295, 2425–2427.

Hyunh, W., Dittmer, J., Teclemariam, N. et al. 2003. Charge transport in hybrid nanorod–polymer composite photovoltaic cells. *Physical Reviews B,* 67, 115326–115337.

Iijima, S. and Ichihashi, T. 1993. Single-shell carbon nanotubes of 1-nm diameter. *Nature,* 363, 603–605.

Joon-Gu, L. 2010. U.S. Patent 7,696,687 B2.

Kaigala, G., Hoang, V., Stickel, A. et al. 2008. An inexpensive and portable microchip-based platform for integrated RT-PCR and capillary electrodes. *The Analyst,* 133, 331–338.

Keblinski, P., Eastman, J.A., and Cahill, D.G. 2005. Nanofluids for thermal transport. *Materials Today,* 8, 36–44.

Kim, A., Tarak, G.V., and Choi, D. 2008. Fabrication and electrochemical characterization of a vertical array of MnO_2 nanowires grown on silicon substrates as a cathode material for lithium rechargeable batteries. *Journal of Power Sources,* 183, 366–369.

Kim, J., Khanal, S., Khatri, A. et al. 2008. Electrochemical characterization of vertical arrays of tin nanowires grown on silicon substrates as anode materials for lithium rechargeable microbatteries. *Electrochemistry Communications,* 10, 1688–1690.

Kim, J.M., Ohtani, T., Sugiyama, S. et al. 2001. Simultaneous topographic and fluorescence imaging of single DNA molecules for DNA analysis with a scanning near-field optical/atomic force microscope. *Analytical Chemistry,* 73, 5984–5991.

Krishnan, A., Dujardin, E., Ebbesen, T.W. et al. 1998. Young's modulus of single-walled nanotubes. *Physical Reviews B,* 58, 14013–14019.

Kobayashi, M., Rharbi, Y., Brauge, L. et al. 2002. Effect of silica as fillers on polymer interdiffusion in polybutyl methacrylate latex films. *Macromolecule,* 35, 7387–7399.

Kymakie, E. and Amaratunga, G. 2002. Single-wall carbon nanotube–conjugated polymer photovoltaic devices. *Applied Physics Letters,* 80, 112–115.

Lavik, E. and Langer, R. 2004. Tissue engineering: current state and perspectives. *Applied Microbiology Biotechnology,* 65, 1–8.

Leonard, K., Genthe, J., Sanfilippo, J. et al. 2009. Synthesis and characterization of asymmetric electrochemical capacitive deionization materials using nanoporous silicon dioxide and magnesium doped aluminum oxide. *Electrochimica Acta,* 54, 5286–5291.

Li, D. and Xia, Y. 2004. Electrospinning of nanofibers: reinventing the wheel. *Advanced Materials,* 16, 1151–1170.

Lu, Y.C. and Zhong, J. 2004. *Semiconductor Nanostructures for Optoelectronic Applications.* Norwood, MA: Artech House.

Myung, N., Lim, J., Fleurial, J. et al. 2004. Alumina nanopore template fabrication on silicon substrate. *Nanotechnology,* 15, 1–6.

Penza, M., Rossi, R., and Serra, E. 2010. Metal-modified and vertically aligned carbon nanotube sensors array for landfill gas monitoring applications. *Nanotechnology,* 21, 1–14.

Petrovic, Z.S. and Zhang, W. 2000. Glassy and elastomeric polyurethanes filled with nano-silica particles. *Material Science Forum,* 352, 171-176.

Robertson, D.H., Brenner, D.W. and Mintmire, J.W. 1992. Energetics of nanoscale graphitic tubules. *Physical Review B,* 45, 12592–12595.

Ruan, Q., Chen, Y., Gratton, E. et al. 2002. Cellular characterization of Adenylate Kinase and its isoform: Two-photon excitation fluorescence imaging and fluorescence correlation spectroscopy. *Biophysical Journal,* 83, 3177–3187.

Sacriciffici, N.S., Braun, D., Zang, C. et al. 1993. Semiconducting polymer-buckminsterfullerene heterojunctions: diodes, photodiodes, and photovoltaic cells. *Applied Physics Letters,* 62, 585–587.

Schedin, F., Geim, A.K., Morozov, S.V. et al. 2007. Detection of individual gas molecules adsorbed on grapheme. *Nature Materials,* 6, 652–655.

Simkin, M.V. and Mahan, G.D. 2000. Minimum thermal conductivity of superlattices. *Physical Review Letters,* 84, 927–930.

Sitharaman, B., Shi, X., Walboomers, F., Liao, H., Cuijpers, V., Wilson, L.J. et al. 2008. *In vivo* biocompatibility of ultra-short single-walled carbon nanotube/biodegradable polymer nanocomposites for bone tissue engineering. *Bone,* 43, 2, 362–370.

Sumita, M. 1984. Dynamic mechanical properties of polypropylene composites filled with ultrafine particles. *Journal of Applied Polymer Science,* 29, 1523–1530.

Sumita, M., Tsukumo, Y., Miyasaka, K. et al. 1983. Tensile yield stress of polypropylene composites filled with ultrafine particles. *Journal of Materials Science,* 18, 1758–1764.

Thostenson, E. and Chou, T. 2002. Aligned multi-walled carbon nanotube-reinforced composites: processing and mechanical characterization. *Journal of Physics D,* 35, L77–L80.

Treacy, M.M.J, Ebbesen, T.W., and Gibson, J.M. 1996. Exceptionally high Young's modulus observed for individual carbon nanotubes. *Nature,* 381, 678–680.

Veprek, S. and Argon, A. S. 2001. Mechanical properties of superhard nanocomposites. *Surface and Coatings Technology,* 146/147, 175–182.

Veprek, S., Nesladek, P., Niederhofer, A. et al. 1998. Recent progress in the superhard nanocrystalline composites: towards their industrialization and understanding of the origin of the superhardness. *Surface and Coatings Technology,* 108/109, 138–147.

Veprek, S., Niederhofer, A., Moto, K. et al. 2000. Composition, nanostructure and origin of the ultrahardness in nc-tin/a-si3n4/a- and nc-tisi2 nanocomposites. *Surface and Coatings Technology,* 133/134, 152–159.

Weller, H., Schmidt, H.M., Koch, U. et al. 1986. Q-particles: size quantization effects in colloidal semiconductors. *Journal of Chemical and Physical Letters,* 124, 557–560.

Ziabicki, A. 1976. *Fundamentals of Fibre Formation: The Science of Fibre Spinning and Drawing.* London: John Wiley & Sons.

Index

f indicates figure

0D, *see* Zero dimensional (0D)
    nanoparticles
1D, *see* One-dimensional (1D)
    nanomaterials
2-DEG, *see* Two-dimensional electron
    gas (2-DEG) system

A

Accelerated testing conditions, 7
Acceleration sensitivity, 95
Accelerometers
    fabrication, 130
    MEMS application, 62, 69–71
Acoustic matching and interfacial
    thermal resistance, 338
Active circuitry systems, 236–239
Activity factor, *see* Electromigration
    (EM)
AFM, *see* Atomic force microscopy
    (AFM) techniques
Aging-induced variation, 6
Air damping in resonators, 96
Akheiser effect, 98
ALD, (atomic layer deposition), 101
ALE, 333
Alumina etch masks, 143–144
Aluminum nanofluids, 329
Aluminum nitride (AlN)
    for nanowires, 113
    resonator applications, 85
Alzheimer's disease research using
    nanomaterials, 376
Ambipolar behavior, 292
Ambipolar diffusion, 33
Amino acids in viral template
    selectivity, 162
Amphifullerene, 367–368
Analog circuits
    total dose effects in, 54–57
    transients in, 43–44

Anchor loss in resonators, 97
Angular effects of ion strikes, 41
Anisotropic etching
    control of, 134, 138
    in fabrication, 130
Anisotropic growth mechanisms, 358
Annealing of Y-junctions, 295–296
Anodized aluminum oxide (AAO)
    membranes, 360–361
Antimony trioxide enhances toughness
    in polymeric nanoparticles, 371
ARDE, *see* Aspect ratio-dependent
    etching (ARDE)
"Armchair" SWNTs, 181
Armor applications, 354
Arrays of CMS sensors, 239
ARS, 67–69
Arsenic removal from water, 354
Aspect ratio-dependent etching (ARDE),
    138
    for cryogenic etching, 151–153
    predictive model for, 151–153
    vs. pseudo-Bosch etching, 138
Atomic bonding in nanotubes, 365–366
Atomic diffusivity and interconnect
    material failure, 19
Atomic force microscopy (AFM)
    techniques, 280
    for quantum dot studies, 172–173
Atomic layer deposition (ALD), 73
    of alumina, 144
    for gap reduction, 85
    for resonator filters, 101
Atomic resolution storage (ARS), 67–69
Atomistic-level constraints, 6
Average length efficiency (ALE), 333

B

"Babbage machine" at nanoscale, 273
Back end-of-line (BEOL) reliability, 20
Backflow stress, 19